Metabolic Pathways Regulating Weight Loss

Healthy Nutrition
How the Body Stores and Burns Fat

FIRST EDITION

Christopher D Boulton

Metabolic Pathways Regulating Weight Loss

Healthy Nutrition
How the Body Stores and Burns Fat

First Edition

Christopher D Boulton

About the Author

Chris Boulton

Chris received his Honours Degree in Biology from London University and during a distinguished career, achieved status as a Chartered Scientist, Chartered Environmentalist and Chartered Water and Environmental Manager, with over 40 years' experience in business, environmental management, water supply and demand management and is now happily retired living in Staffordshire England. He has a Diploma in Advanced Nutrition Level 5, certified by the UK Association for Nutrition (AfN) and recognised by the UK National Health Service (NHS), the UK General Medical Council (GMC) and the UK Health and Care Professions Council (HCPC).

Chris also holds a Diploma in Environmental Management and Pollution Control and was appointed a Fellow of the Chartered Institute of Water and Environmental Management. He has given lectures in Europe and the United States and has an extensive knowledge of and passion for nutritional biochemistry, human metabolism, endocrinology and the science of nutrition and healthy eating and constantly maintains an avid interest in scientific papers, research and emerging evidence in this field. He is now actively engaged as founder of Healthy Nutrition Advice, helping and advising others who strive for improved diet, weight goals and overall health. Chris has exceptional communication and writing skills, delivering training and technical workshops to engineers, technicians and all levels of management spanning 4 decades.

A significant record of professional achievement, encompassing managing directorship, running his own successful business and board level reporting with a solid background in the environmental management, Water Industry and Government sectors. Chris spent 15 years in UK Defence as Water Resources and Demand Manager for the Ministry of Defence, closely working with the legislative Regulators, the Environment Agency, Defra, DWI and Ofwat.
A sustained and proven delivery of substantial savings and business improvement and efficiency gains for blue chip companies across industrial, commercial and municipal sectors, including petrochemical, steel, rail, textiles, paper, and port authorities. An experienced data analyst, with sales and business development skills and a natural flair for team training.

Chris has been a keen follower and advocate of Whole Food Plant Based Nutrition for many years and his specialist subjects include; Dietary Macronutrient and Micronutrients, Substrate Energy Metabolism, Sustainable Healthy Weight Loss, Regulation of Metabolic Pathways, Chloroplast and Mitochondria Biochemistry and Animal and Plant Physiology and morphology.

"I am passionate about improving future education in Nutritional Science. I want to help people understand the benefits of healthier nutrition and how changing the way we eat can finally start to reverse the devastating environmental impact humans have caused to other species and the forests, land, sea and fresh water resources, due to our demand for food, especially meat and the supply systems feeding this demand across the globe over the past 100 years. I have a continuous driving enthusiasm and insatiable thirst for emerging and evolving scientific knowledge, as the frontiers of science are pushed further and further for the benefit of our fragile Earth"

Copyright

Acknowledgements

There are many people, scientists, broadcasters, podcasters, research organisations, health schools and programs too numerous to mention, that have inspired and had a profound and most positive influence on the author over the past 4 decades and in particular, those within the wonderful Plant-Based community health and wellness organisations such as the McDougall Program, the Barnard Medical Center and research Centers, including the Physicians Committee for Responsible Medicine, the Cleveland Clinic Wellness Institute, the Center for Nutritional Studies, the Real Truth about Health and Nutrition Facts.Org. Also those marvellous people who work so hard to share the nutritional health message and knowledge throughout the world from podcasts, interviews and conferences and to whom the author is indebted for their faith in real science and for the vast array of knowledge, integrity and grace and above all good old common sense on nutritional health, including; Chuck Carroll, Rich Roll, Simon Hill and Chef AJ. For their research and wisdom and pioneering work in pursuit of helping others; Nathan Pritikin, Dr. John McDougall, Heather McDougall, Dr Peter Rogers, Dr Michael Klaper, Dr Michael Greger, Professor Marc Hellerstein and Dr Neal Barnard. Mention of any specific individuals, companies, authorities or products in this document does not imply endorsement by the author, nor does the mention of individuals, companies or authorities imply that they endorse this document or its author. References to some of the most influential science research papers the author has read are flagged and provided at the end of the document. The author declares no vested interest in any product, nor organisation, apart from his own founded Healthy Nutrition Advice © 2024.

Restrictions on Use

Preface

Weight loss and dieting continues as a hot topic of discussion, impacting on a large proportion of populations over many decades. Levels of obesity are now on the rise in most countries. Globally, one in three people are either overweight or obese. This has a massive impact on individuals health, as it is likely that >75% of ill health and serious diseases can be attributed to diet and lifestyle and this further adds enormous pressure on society, productivity, health services and health care costs estimated to grow to a staggering $1.2 trillion per year by 2025. Health authorities are struggling to cope with this growing burden. Even more of a growing concern, is that the number of children that are overweight or obese has more than doubled since 2000. Clearly, the medical treatment, and pharmaceutical organisations, can-not fix this problem, solely through the use of drugs, supplements or surgery, nor will the vested interest food supply industries. Rather, it is up to each individual to improve their diet with better, wholesome nutrition and overall lifestyle and achieve and sustain a healthier weight, both for their own benefit and for the benefit of the whole planet.

Although nutrition is inextricably linked with health, the prime objective of this book is to advise on weight loss, energy requirements, the energy balance and how metabolism works in the utilisation of food, rather than the health issues arising from poor diets and health issues that are associated with certain foods. However, it is highly likely that hypercaloric high fat, high protein, high free sugars, over nutrition, all common in the western style diet, causes cellular overloading of the liver and other organs, leading to the vast majority of illnesses in these societies. This book does not include

detail for every individual's needs, for example pregnant women, infants and young children etc. It is written principally, for those people who wish to lose weight and understand how the body works and why. The book is provided for informational and reference purposes only and not as a source of medical advice and is not intended to supplement, supplant or be a substitute for medical advice or any treatment from your doctor. This first edition focuses on key weight loss strategies and the science behind weight management and is packed with information, knowledge and data to help explain how this can be achieved. It is intended to provide the reader with accurate nutrition and weight loss advice and information, focussing on the body's requirement for energy and is based on actual science to help cut through the considerable amount of misleading portrayals and myths and misinformation, reported in the press, the internet and social media, especially on miracle diets and magic bullet foods and supplements. The text also explores the reasons why the vast majority of diets fail and the multitude of reasons why people put on weight. The author is not a keen advocate of the so called popular diets such as 'keto', paleo, 'low carb' 'high fat' and 'high protein' diets etc, where carbohydrates generally are demonised and rises in insulin should be suppressed at all times and more recently, increasing use of glucose monitors to monitor insulin levels for healthy people without diabetes or prediabetes.

The author also firmly believes that any diet that restricts and deprives wholesome and healthy carbohydrates and starches, whilst replacing this with excessive protein i.e., meat and dairy, animal and plant fats, is unnecessary, unnatural, unhealthy and unsustainable.

However, the vast majority of people trying to lose weight or build muscle, simply need to follow a wholesome, healthy, nutritious, enjoyable, common sense diet, together with some moderate exercise, all backed up with 100 plus years of robust, good science. This book focuses on just that. The key, fundamental foundations for successful, healthy and sustainable weight loss, including nutrition quality, density, caloric density, meal timing, exercise and sleep are discussed. Our ingenious bodies are, in a sense, old fashioned and have evolved to prepare for and protect us from potential famine, using highly integrated and elaborate hormonal, physiological, psychological and metabolic regulatory processes. This evolutionary bedrock is in complete contrast with an ever developing 24/7, societal food café culture and lifestyle revolution, that probably started around 50 years ago and where plenty is in excess and where food variety, quantity and availability of fast-food, junk-food and processed and ultra-processed food, lacking nutritional quality is ubiquitous. This, coupled with constant day and night eating out, across vast so called 'westernised' areas of the world, has now become the default way of life for so many, the antithesis of how humans prepared and ate food prior to the 1970's.

Yet, alarmingly, this plentiful and excess bounty for many, is set against the fact that in 2023, 1 in 11 people in the world i.e., 753 million people are undernourished and go hungry and 2.4 billion people, or 35% of the global population are severely or moderately food insecure and do not have access to sufficient safe, nutritious food.

Christopher D Boulton
Founder of Healthy Nutrition Advice
February 2025

Food system provision around the world requires complete transformation.

This edition is written in 2 parts. Part 1; Healthy Nutrition and Sustainable Weight Loss, discusses current trends in nutrition and obesity. Chapter 2 looks at the energy balance of intake and expenditure and metabolic rate. Chapters 3 to 7 review Nutrition Quality, exercise and lifestyle. Chapter 8 provides the 20 point Healthy and Sustainable Weight Loss Plan, explaining the step-by-step approach to achieve healthy and sustainable weight loss goals, through SMART, realistically achievable changes, that can be adapted to suit the needs of each individual. Chapter 9 provides the recommended, healthy food Menu Guidelines and which foods to reduce or eliminate from the diet to maintain healthy weight. Chapters 10 to 13 discuss the 3 principal macronutrient food groups and fuels, together with micronutrients and vitamins. Part 2, Nutritional Metabolism and Biochemistry, delves a little deeper into some of the key mechanisms and regulatory processes on how the body uses food, that are pivotal to weight management and weight change and Chapter 15 examines how light energy is turned into chemical energy within the carbohydrate, fat and protein fuel substrates that we eat, in order to release the energy held within the carbon atoms enabling every cell in our body to function. The concluding remarks draw attention to how a predominantly plant-based diet can make so much difference to the global environment.

Pre and Post start food diary forms, together with the weight loss record to monitor progress towards the weight loss goals are provided in appendix I, II and III

CONTENTS

Preface

PART 1
HEALTHY NUTRITION & SUSTAINAIBLE WEIGHT LOSS

Contents/2

Contents/3

Contents/4

Contents/5

PART 2
NUTRITIONAL METABOLISM & BIOCHEMISTRY
CHAPTER 14 Metabolic Processes & Weight Management   131

Contents/6

Abstract

In western societies generally and a growing number of other locations around the globe, a significant proportion of people are eating too much, too often. They eat too much unhealthy food and not enough healthy food. They eat out too much and eat too late each day. Portion sizes are often too large. many people eat too much fat, meat, dairy, processed food and free sugars and eat far too little fibre. Many people also drink too much alcohol, too often.

The reasons for this eating and lifestyle behaviour are multi-fold and some of the most common reasons are reviewed in this first edition. Being overweight and obese is a major risk factor and one of the world's leading causes of ill health, including cardiovascular disease and diabetes. In 2024, 3 billion people were overweight (BMI >25 kg/m^2), including more than 1 billion people, i.e., 880 million adults and 159 million children, living with obesity, This corresponds to 37% of the world's population overweight. This has risen dramatically from 25% in just 34 years since 1990. The statistics are particularly alarming in the US, where around 74% of adults are overweight or obese. US adults, ages 40 to 59, have the highest rate of obesity (43%) of any age group, with adults 60 years and older having a 41% rate of obesity. About 40% of children and adolescents are overweight or obese and the rate of obesity increases throughout childhood and teen years.[1] More worrying is that UNICEF estimates that the prevalence of overweight and obesity among children under the age of 5 years, has risen from 30 million in 2000 to 37 million in 2022. There has been a 14 fold rise in children aged 5-19 years old, living with obesity, at a total of 159 million, against just 11 million in 1975. In 2022/23, a total of 42.7 million people, i.e., 64% or almost two thirds of the total population in the UK, were overweight or obese.

Introduction

This book sets out the most relevant aspects of weight gain, weight loss, the science of nutrition, the bioenergetics of energy metabolism and how changes in nutrition and lifestyle can lead to healthy and sustainable weight loss. The result of any weight loss plan is ultimately controlled by the individual and how adherent the individual sustains the new approach long term. This depends on personal commitment, past experiences, motivation, ability to apply information and many other factors. However, the author does not believe in over complicating diets and strict, rigid regimes, or diets based on physiological extremes, nor 100% avoidance of treats or food enjoyed. This has been repeatedly proven to lead to failure. Instead, the step by step process and lifestyle changes recommended, are designed to be SMART; i.e., Sustainable, Achievable, Relevant and Timely and should enable most individuals to achieve their objectives, with far less likelihood of relapse, as experienced with many other diets.

Individuals on Medication or with Other Health Issues

If you are significantly overweight or obese, have pre-diabetes, diabetes, metabolic syndrome or any other health issues, or you are on prescribed medication, you should always consult your doctor before making any significant lifestyle changes, or embarking on any weight loss program, dietary change or exercise regime. Your doctor is best placed to advise on the suitability of change in respect of your individual circumstances and medical history, to ensure this change will be appropriate for you.

PART 1
HEALTHY NUTRITION
&
SUSTAINABLE WEIGHT LOSS

CHAPTER

Dieting & Weight Loss

1

1.1 The Wonder and Complexity of Life

The leaf of a tree, the petals of a flower, the wing of a beetle, the tentacle of an octopus, the gill of a fish, the contents of a sesame seed and the genome of a bacterium are all vastly complex parts of living organisms. The human body is of course, no exception, a wonderful piece of biological architecture, comprising upwards of 30 trillion cells and host to another 37 trillion friendly gut bacteria of the gut microbiome. A closer examination of each and every one of these tissues and cells has confirmed to scientists, and all who study this, just how staggeringly complex biological life is. Each human cell carries out a vast array of functions to enable the organism to grow and survive, comprising proteins, enzymes, carbohydrates, fats, phytonutrients and a whole host of other molecules and cellular components, all working in harmony, orchestrated by the brain and regulated by other key organs, such as the liver, gut and kidneys. The human heart requires and consumes energy carrier ATP molecules 15 times its own weight each day. Human nutrition and the control of cellular metabolism is a perfect example of this complexity and organisation, demonstrating the sheer wonder of nature, that has evolved on the Earth over the past 3.7 billion years. It is against this phenomenally complex background that scientists, researchers, nutritionists, fitness coaches and all interested and passionate students have discovered and learnt much of how things work at cellular, system and organism level, over the past 200 years and especially in the past 80 years, with the advent of sophisticated instruments and techniques such as mass spectrometry and many other developing technologies. Much is still to be learnt and will most probably always be the case, given the size of the task and knowledge sought.

1.2 Nutritional Science Fact Myth or Misinformation

The author has spent most of his adult life, since the mid 1970's, well over 40 years, studying cellular, molecular biochemistry and the role of metabolism in human nutrition. It is clear that there is a considerable volume of knowledge within the scientific and medical communities, plus government health advice and a plethora of knowledge out there in the public domain (including this book). Most of this knowledge is based on sound science and data and evidenced research plus meta analyses of randomised control trials and the gold standard metabolic ward studies, together with the conclusions from epidemiological evidence over many decades, no better for example than the Framingham Heart Study in the United States, with data across 3 generations of people, still running at 76 years since its outset in 1948. It is always 'healthy' and indeed an integral part of real science, to challenge knowledge and research and this is surely the only way to learn and carry out good science. This should aways be the case. However, there are countless reports, podcasts and comments from so many sources, especially on the ubiquitous media power of the internet and social media, most of all who are genuine well-meaning and some well-informed, some with vested interests, some with their own extreme agendas and some, whatever the motive, who irresponsibly advise many people incorrectly that can lead to health issues. These conclusions and opinions are promulgated and preached to many people across the world, often as magic bullet diets. For example, more recently, quoting molecular cell biology and metabolic pathways to promote carbohydrate privation, all with the objective of weight loss by using protein to synthesise glucose and fats as ketones to engender energy stress and ketosis. The author does not endorse these diets. This fashion and fad diets passion is not new and has grown since the 1970's, including such diets as 'Atkins', to 'raw vegan' then 'keto', 'Mediterranean', 'paleo' and recently, even 'carnivore'. Some of these diets will be discussed further in Section 1.10. Over the past 20 years, the professional fitness fraternity have also established a considerable, internet dominating role (mostly responsible) in this vast topic of healthy nutrition, to provide recommendations and guidance on how to improve fitness levels and what to eat or not to eat to lose weight and gain muscle and endurance and high intensity workouts. It is interesting to note that a high proportion (>85%) of all internet searches on nutrition, health and weight loss now returns advice from fitness coaches. There is absolutely nothing wrong with this and the author is a keen advocate of healthy levels of exercise for all, as part of weight management, as this book will attest but it is yet another avenue that can lead to a different way of looking at the various macronutrients, nutrition, weight loss and muscle growth, from the perspective of physical fitness, sometimes promoting supplementation and high levels of protein. The vast majority of people (especially the 'older folk') are not necessarily wishing to engage in hard endurance exercise, nor do they need to go to extreme lengths, to be healthy and reach sustainable weight goals.

Much of these over simplified messages, sourced from popular culture have, understandably, led to confusion and nutrition fog amongst the public. All the sources quoted above, including scientists (author included), the medical and public health professions, physicians, researchers, academic lecturers, governments, politicians, nutritionists, diet specialists, fitness coaches, weight loss gurus, the food industry, big pharma, big farming, podcasters, influencers, celebrities, bloggers, have their opinions and goals. These perspectives range from medical health advice with the experience of managing patients, right through to selling supplements and conspiracy theorists, whether backed by proven good consensus science or not. [43] It is often left to individuals to make up their own minds what is right for them; to understand science, cut through the jargon, misinformation, the agendas, the plethora of contrasting, conflicting and dissonant advice, and last but by no means least, the vested, commercially interested parties. No wonder the debates continue and no wonder the public is largely confused and a good many people are not necessarily equipped to understand if the reported science or is correct. Another issue is flawed data and biased research, often funded by vested interest and also incorrectly reported dietary data from the subjects partaking in the studies. The science of metabolism is highly complex, even for the scientists, so how can the lay person, who is not familiar with research papers, nor scientifically qualified, be expected to navigate through all the mist. For all these reasons, from a moral and ethical perspective, it is incumbent on health specialists and nutrition advisers to uphold professionalism and their codes of practice and help people cut through this growing swamp of misinformation to help them achieve their weight loss goals in a healthy and sustainable way.

Section 1.3, lists the key cornerstones and bedrock behind the most important fundamentals of human nutrition and cellular and energy metabolism that, in the opinion of the author, underpin healthy nutrition, sustainable weight loss, and weight management.

1.3   The Fundamentals of Nutritional Metabolism

The totality of scientific evidence from all weight loss trials and programs to date indicates that there is no single dietary regime or quick fix diet that is proven to reduce weight and sustain that weight loss for all people, for example, in time, both low carb/high fat and high carb/low fat diets ultimately lead to the same weight loss. Moreover, it is far more effective to manage the quality and quantity of nutrients and energy intake with regard to overall health and lifestyle when embarking on a sustainable weight loss program.

Glucose from carbohydrate is an essential energy source for virtually all forms of life on earth and the body's preferred and primary energy source for every cell in the body, especially the brain and red blood cells. It is the pivotal driving force of central metabolism and energy production required for human health and for the synthesis of many critical molecules throughout the body. The brain requires a minimum of 30% glucose to survive without damage.

Apart from some rare genetic disorders, obesity is caused by over eating and chronic excessive food intake. As little as 10 kcal/d, just 0.4% of daily intake excess each day will lead to significant weight gain of 4.5 kg over 10 years. This caloric excess drives fat storage in adipose tissue and is due to excess fat, protein or carbohydrate, or a combination of all 3 macronutrients and alcohol.

Carbohydrate and glucose are necessary for optimal metabolism to secure energy from terminal fatty acid oxidation, through oxidative phosphorylation and the acetyl-CoA, formed in fatty acid (β)-oxidation only enters the TCA cycle if carbohydrate and fat oxidation are appropriately balanced. Acetyl CoA must combine with oxaloacetate to gain entry to the TCA cycle and the availability of oxaloacetate depends on an adequate supply of carbohydrate. Hence burning fats efficiently relies on sufficient carbohydrate.

Dietary fat is stored at a higher rate in the presence of macronutrients with a higher oxidative priority than fat, including alcohol, protein and carbohydrate. Insulin does not cause weight gain, it is overnutrition of food.

Dietary fat is very easily stored and it is not used for energy. Any fat that is burnt, either through normal metabolic processes, for example, used continuously by the heart, skeletal muscle, kidney cortex, or burnt due to exercise, is burnt as mobilised fatty acids, via lipolysis of the fat from the fuel rich adipose tissue and (β)-oxidation and is derived from endogenous body fat stored in adipose tissue.

Carbohydrate (glucose) and fat (fatty acids) are being burned (oxidised) in the body all the time and the proportion of each burnt reflects the amount of feeding, the inter-prandial periods (timing since the last meal) and the level of activity and exercise. During rest or moderate levels of exercise, glucose and fatty acids are the dual sources primary of energy. Fatty acid (β)-oxidation is regulated by the cell's energy requirements.

Carbohydrate. i.e., glucose, is always preferentially burned to supply energy to the brain and many other parts of the body and therefore delays any significant fat burning until glycogen levels begin to be depleted, for example, during sleep and fasting.

When the body is in the fed (absorptive) state and particularly after a large meal that includes fat, carbohydrate and protein, the body burns the carbohydrate first and holds on to its adipose tissue, supresses lipolysis and also adds further dietary fat to the adipose reserves via lipogenesis, reforming triacylglycerol from the constituent fatty acids and glycerol, broken down during digestion.

To lose fat, the body must be in a negative energy balance, hence;

'carbohydrate spares the fat' [14]

It is a natural evolutionary trait for humans to store and hold onto both dietary and endogenous stored fat, which is the most energy dense food. The body will endeavour and fight to hold on to its energy storage, to increase the chances of survival if food availability is short. The body cells will remember times of abundance for a long time, hence;

'the fat cell is a harsh master with a long memory' [2]

The body has several important metabolic tools to preserve and maintain blood glucose levels hour to hour and day to day. One of these tools is the critical process of gluconeogenesis that serves to preserve levels of blood glucose. When glycogen stores and blood glucose levels become depleted, for example every day during overnight fasting, or deliberate carbohydrate privation diets, or prolonged fasting >2 days and real starvation, the liver provides synthesised glucose, that is derived from other non-dietary carbohydrate substrates, such as glycerol - from adipose tissue triacylglycerols, lactate, from pyruvate and amino acids from proteolytic break down of protein.

With the exception of the liver, normal day to day fatty acid beta (β)-oxidation, in body tissues, such as the heart and skeletal muscle, leads to the production of acetyl-CoA for energy via the TCA cycle and oxidative phosphorylation. However, in the liver, during prolonged fasting conditions when glycogen stores are depleted and levels of glucose and OAA are low, the accumulated acetyl-CoA is used to produce significantly more ketones. Significant ketosis, arising due to this carbohydrate deprivation, is how the body reacts to and protects itself during starvation.

Like fatty acids, normal levels of ketogenesis also fulfils the body's second Starvation Priority in sparing muscles from catabolic breakdown. However, it can be dangerous if very high levels of ketones persist. It is far healthier to avoid this deprivation and allow the body to use the glucose it prefers, without over eating to caloric excess.

When the body has a lower level of glucose, glycogen stores and gluconeogenic precursors, such as lactate, glycerol and amino acids, become scarce and levels of insulin are sufficiently low, for example, between meals or during fasting, the body naturally increases the rate of and substrate proportion of (β)-oxidation of fatty acids to produce energy.

Fatty acid (β)-oxidation also fulfils the body's second Starvation Priority in sparing muscles from catabolic breakdown. This proportional rate of 'fat burning' is increased when exercise levels increase to moderate intensity. Chronic, excessive (β)-oxidation of fatty acids for energy, compared with significantly diminished levels of carbohydrates oxidation for energy, leads to cell damage and Reactive Oxygen Species (ROS).

Excess consumption of free sugars, such as fructose and sugars found in pastries and drinks etc. causes hepatic steatosis (fatty liver disease) due to excess fat build up in the liver and mitochondrial dysfunction.

Diets that deliberately induce metabolic-challenging conditions, depriving sufficient levels (<225 gram/d) of dietary carbohydrates, are not healthy, nor recommended, nor sustainable in the long term and can lead to health issues. Low carbohydrate diets necessarily equate to daily ingestion of high levels of fat or protein or both. This can lead to diet related diseases such as insulin resistance, Type 2 diabetes, atherosclerosis and kidney disease. Excess dietary protein is not stored, unlike dietary fat, which is stored in adipose tissue and dietary carbohydrates, which is stored in liver and muscle glycogen. Instead, excess protein is either broken down in the liver and excreted as urine, or a small amount will be converted to fat. Hence it is not necessary to take supplemental protein powders, or eat excessive quantities of meat and cheese. Continued excess dietary protein significantly increases the risk of kidney stones and other kidney diseases.

Carbohydrates are not converted to fat by de novo lipogenesis (DNL) to any significant level in humans, either from healthy complex carbohydrates, or simple free sugars. This process requires a major bioenergetic cost of energy, compared with storing dietary fat. The rate of DNL only increases to any significant extent, when there either a significant daily caloric excess of chronic food intake over long periods of time, or the individual is very obese or has non-alcoholic fatty liver disease hence;

'de novo lipogenesis is the pathway of last resort' [21]

Neither alcohol, nor carbohydrate significantly alters metabolic rate. However, alcohol consumption reduces fat burning by almost 90% and will contribute to weight gain. Consuming carbohydrate after alcohol consumption, stops fat burning for at least 90 minutes and this combination of alcohol and carbohydrate reduces fat burning by 80% for as much as 4 hours after feeding and if these are combined regularly this will lead to weight gain.

Exercise does play a role in weight management but weight gain is far more significantly influenced by food intake compared with that burned from exercise. However, exercise has many other health benefits and moderate exercise should be encouraged for all. Moderate levels of exercise, for example daily walking, prior to refeeding after a normal overnight fast of >12 hours, will promote an increased rate of fat burning from stored adipose tissue, compared with the rate of burning glucose and glycogen stores.

During intensive exercise, more fat from stored adipose tissue will be burnt, even though a higher proportional rate of carbohydrates is burnt compared with fat per unit time. Intensive exercise will increase the proportion of glucose and glycogen burnt compared with fat burning from stored adipose tissue but the overall number of total calories burnt will be higher than that burnt during moderate exercise, as will the overall amount of fat burnt from stored adipose tissue.

1.4 Food Combinations Not Individual Components

For overall nutritional health, the reader should consider nutrition as a whole and not regard food as individual components to be listed and counted out each day. We most often eat combinations of foods and not solely, individual parts. There is a tendency in the nutrition, weight loss, fitness and athletic spheres, to dwell on individual macronutrients, (carbohydrates, proteins and fats), micronutrients and vitamins, plus supplements, often referred to as the *'reductionist approach' or single nutrient approach*. Our diets should be holistic, not reductionist.
Yes, it is necessary to understand how these dietary ingredients work to keep us healthy and we have tremendous scientific knowledge and plenty evidence on this.

These popular approaches tend to focus on the metabolic effects of consuming individual nutrients but overlook the fact that nutrients in our diet interact to influence our health. The body and its component cells comprise a vastly elaborate system of interdependent processes, with countless mechanisms of regulation and metabolism, overseen and conducted via the brain and central nervous system and orchestrated through a wide range of instruments within the organs, tissues and cells, via genetic, neurological, hormonal and enzymatic controls. For the readers that wish to know and understand more of how the body uses and metabolises food energy substrates, some of the most important metabolic pathways and regulatory mechanisms are explored and explained in some detail in Chapters 14 and 15.

Ultimately it is advisable to concentrate on the broader perspective and common sense of balanced nutrition, rather than dwell too much in the detail on the plate and metabolic weeds. A good example is to think of food groups as a whole rainbow of nutrients, such as vegetables, rather than focus on the precise phytonutrients within a single vegetable or 'super foods'. Notwithstanding this, many facets of nutrition, especially how the body uses and conserves energy from these food groups, underpinned by the biology of the cell and the needs of the organism are highly relevant to weight loss.

1.5 Not Dieting but a New Way of Life

To lose weight successfully and sustainably, it is necessary to achieve long-term negative energy balance. This will be achieved, either by a reduction in appetite and energy intake, or by a sustained increase in energy expenditure, or both. Overall caloric intake is the most important contributor and fundamental and successful approach to managing weight, backed with consistent scientific evidence and data over many decades. The most successful weight loss diet change requires relatively small step by step, sustainable changes in eating habits, listening to our highly evolved, intelligent body and understanding how our body naturally works and has evolved to react to change. The body reacts in order to provide long term sufficiency, sustenance, survival, prolong life and protect us from harm.

These small changes culminate in much healthier lifestyles in the long run. If implemented and followed properly, the changes and weight loss plan recommended in this book will, for most individuals, lead to a more achievable, acceptable and sustainable long term management of weight and will lead to a step by step, different approach in eating and moreover, a fresh approach, which simply becomes an enjoyable new way of life.

This can be achieved without the need to constantly overthink food, hunger, counting calories, high levels of exercise or will power or avoidance. Above all, it is far better to reduce the stress on the body and cells by not deliberately putting the body through stress on starvation type diets.

1.6 The Body's Nutritional Metabolic Priorities

The body's first Starvation Priority, in times of nutrition deprivation and starvation, is to provide and maintain safe and sufficient levels of glucose to the 25 trillion red blood cells, the brain and other glucose dependent tissues (such as, the kidney medulla), that are heavily dependent and can-not survive long periods without glucose.

The body's second Starvation Priority during nutrition deprivation and starvation is to to preserve structural protein, which is accomplished by shifting gluconeogenic pathways to use glycerol and lactate, increasing fatty acid (β)-oxidation and eventually the production of ketones for the brain and away from the gluconeogenic and proteolytic catabolism of amino acids used for synthesis of glucose.

These 2 two key metabolic priorities are fundamental to the body's requirements; firstly, to maintain glucose health in normal overnight fasting, then in the last resort, if little or no food is genuinely available, ensure structural protein is protected in emergency starvation conditions. These priorities are the lynch pin of the body's nutritional metabolism for survival. They are, not evolved to be utilised and manipulated as mechanisms to satisfy fad diets, such as 'low carb', 'high protein', 'high fat', 'keto', 'paleo' or 'carnivore' etc. [43] Unless medically necessary, due to specific requirements of certain patients undergoing medically supervised treatment, it is at best, inappropriate to attempt to manipulate these highly complex metabolic systems and in the opinion of the author, irresponsible to advise others to adopt these approaches. Many of these metabolic processes are still not entirely understood, even by the world's leading scientists and researchers.

1.7 Types of Body Fat

Subcutaneous fat is the fat just under the skin, the kind of fat that can be grabbed or pinched between the fingers, collecting around the belly, thighs and backside but can also be stored in the upper arms, neck, face and hands. Approximately 80% of total body fat is located in subcutaneous fat. There is good evidence that each person has a 'personal fat threshold', beyond which additional fat is not stored as subcutaneous fat but instead, stored as visceral and ectopic fat and explains why some people develop type 2 diabetes with a relatively normal or low BMI and the corollary why others do not develop diabetes with a high BMI. [4] However, both can still develop other diseases such as cardiovascular disease. Visceral fat is belly fat that lies deep within the abdominal cavity, surrounding various organs, including the stomach, liver and intestines and can-not be seen from the outside. Approximately 10% of total body fat is located in visceral fat. Visceral fat is more dangerous to health than subcutaneously stored fat, as it can lead to insulin resistance, where the body is unable to facilitate sufficient transport of blood glucose into cells requiring glucose for energy. This poses an increased risk for Type 2 diabetes and developing cardiovascular disease. Ectopic fat (can also refer to non-alcoholic liver disease NAFLD) is the deposition of triacylglycerol (TAG) within cells of non-adipose tissue, that normally contain a small amount of fat i.e., where fat does not belong. Approximately 10% of total body fat is located in ectopic fat. It interferes with cellular functions, causes inflammation and insulin resistance and accumulates in the liver, heart, pancreas and muscles. This type of fat has also been shown to increase the risk of many diseases and high blood pressure and stroke. Individuals with large waist lines and high Body Mass Index (BMI) measurements are most at risk but even people with flat stomachs and normal BMI can have hidden fat that puts them at risk. The signs and of excess fat are described as follows:

Muscles: Excess Ectopic Fat Insulin Resistance and Post Prandial Hyperglycaemia
Liver: Excess Ectopic Fat NAFLD Incorrect Glucose Levels - Excess Gluconeogenesis
Pancreas: Excess Ectopic Fat NAFP Fatty Pancreas High Blood Lipids Damage beta cells
Spleen: Excess Ectopic Fat Hypersplenism from NAFLD Lose WBCs/RBCs -Anaemia

1.8   What Leads to Weight Gain

Although sadly, many people around the age of 21 are now overweight, or are becoming overweight, this was of course, not always the case historically across populations. Some 50 to 70 years ago, in the 1950's to the 1970's, the preponderance of being overweight at the age of 21 was significantly less than it is now in the 2020's. It was then quite rare to see significantly overweight people and extremely rare to see obese individuals. This age group is a useful marker for those folks who are now a little older and who were slim at the age of 21. Comparing your current weight now and when you were at your healthy leanest, (say at 21) is a reasonable way of assessing how much weight you have gained since then. The change over the years tends to be added fat, i.e., in the vast majority of cases, the weight increase is due to an increase in non-lean body mass, rather than in lean body mass. Non-lean body mass increase, through adipose (fat) deposition, can occur at slow, incremental or relatively quickly, depending on caloric excess, lifestyle and habits. For example, an imbalance of energy intake over energy expenditure of just 10 kcal/day, or only 0.4% of average total daily caloric intake, will lead to excess body weight of 1.3 grams/day, or 1 pound (0.45 kg) a year, resulting in a weight increase of 10 pounds (4.5 kg) per decade. This accumulates to a total increase of 30 pounds (13.6 kg) or 2 stones, 2 pounds, over 30 years. It is likely that this is fairly typical in people, now in their 50s, 60's and 70's. As an illustration, 10 kcal is equivalent to just 15% of one plain biscuit or less than 5% of one plain croissant. Clearly, this illustrates that over eating does not have to extreme, in order to lead to substantial increase in weight over long periods of time. The most likely reason for excess weight is through quantum, i.e., hypercaloric excess; from consuming (eating and drinking) more energy; as fat, carbohydrate, protein or alcohol, compared with the total energy burnt. Energy burnt comprises; metabolism, the thermic effect of food, general activity and additional exercise. In addition to the quantum of excess energy, the quality (how nutritiously rich), of the food eaten and the timing of meals are also influential on weight control and overall health. For the majority, weight gain is due to excess dietary fat, laid down in adipose tissue, as either subcutaneous fat, visceral fat or ectopic fat. Too much food, too much fat plus excess of simple refined carbohydrates, fructose, processed food and even too much protein, i.e., over nutrition, all have a role to play in the nutritional quantum and quality, leading to weight gain.

A further reason is the increasing proportion of food and drink from eating out, or collected or delivered as takeaway or fast food. Figure 1.1 illustrates the correlation of both increased fat consumption and total hypercaloric excess intake and increased obesity since the beginning of the 20th Century from 1910. Apart from a very small proportion of people who are extremely fit and lean body builders, this increase is all down to one thing; excess adipose tissue i.e., fat. Note how carbohydrate levels have dropped around 20 years ago but obesity continues to rise.

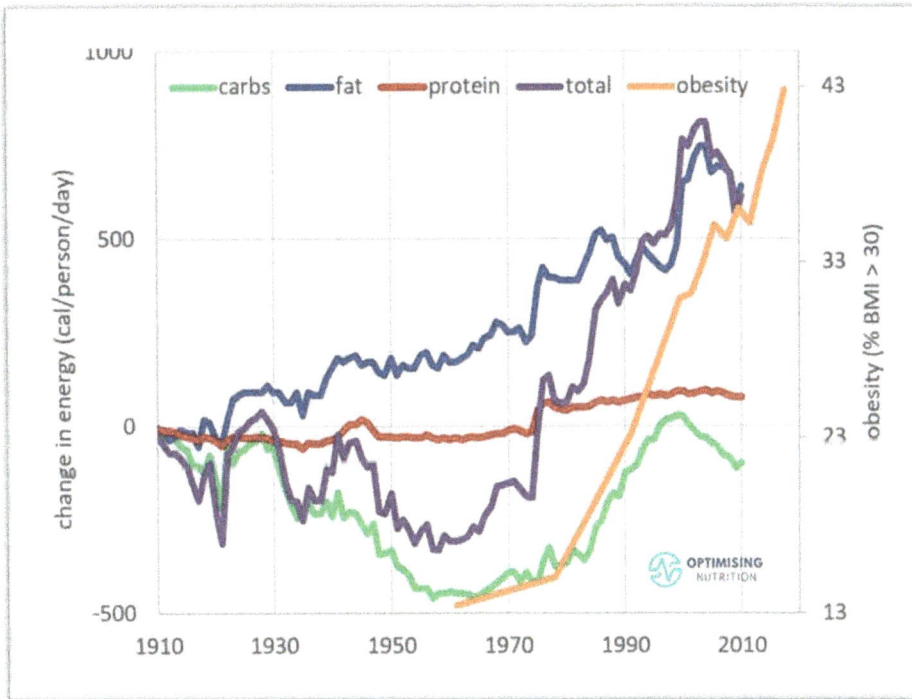

Figure 1.1 Metabolic Macronutrients Energy Production & Obesity: Source Optimising Nutrition, US

As illustrated in Figure 1.2, there is further evidence, aligning the obesity trend with fat and particularly polyunsaturated fat and monounsaturated fat. Note also, the total level of dietary saturated fat intake has not decreased as is often stated. Note also the variability highs and lows in saturated fat since 1970, most probably due to high or low fat fashions but obesity is still rising.

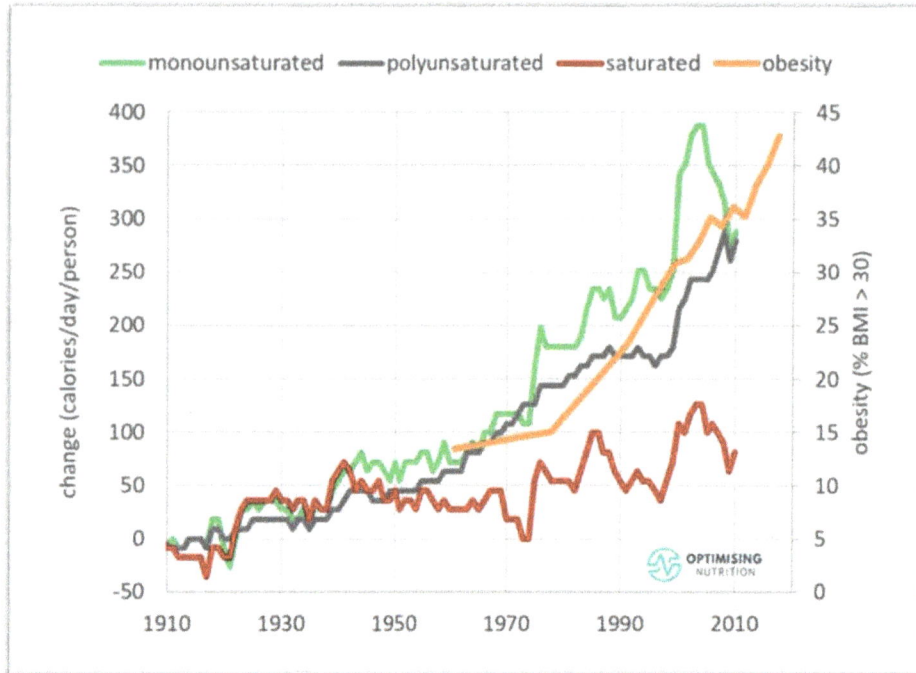

Figure 1.2 Obesity and 3 types of fat: Source Optimising Nutrition, US

This strongly suggests that saturated fat is not the key driver of obesity. Further evidence, illustrated in Figure 1.3, shows an increase in obesity even though consumption of both free sugars dropped from 1998 and a reduction in all carbohydrate from 2000.

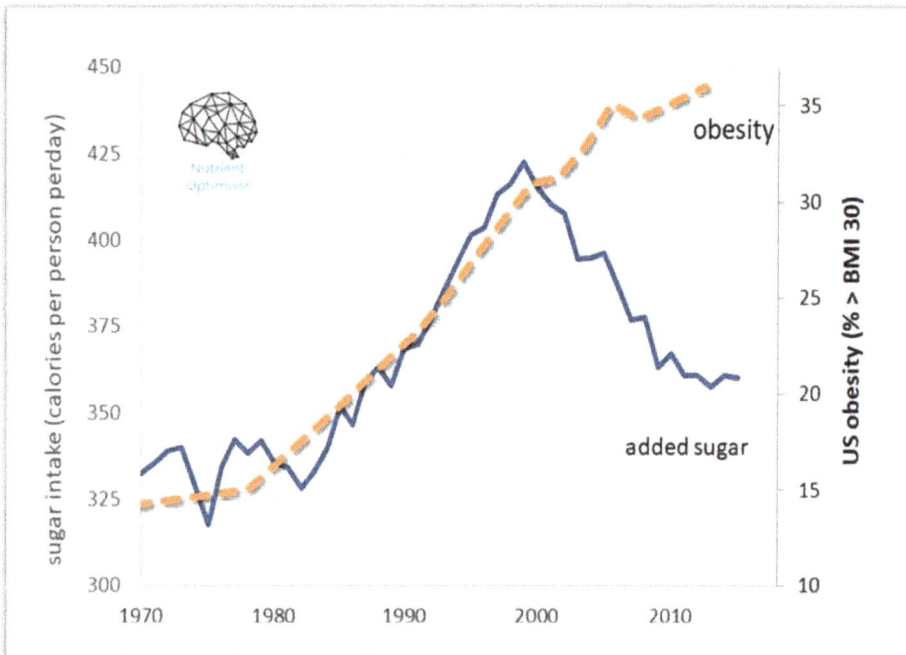

Figure 1.3 Obesity and added sugar: Source Optimising Nutrition, US

Figure 1.4 shows the specific relationship between obesity and total carbohydrate, including both complex and simple added sugars. Note the apparent correlation up until 2000, then carbohydrate levels tailing off significantly, yet since then, obesity rates have continued to rise.

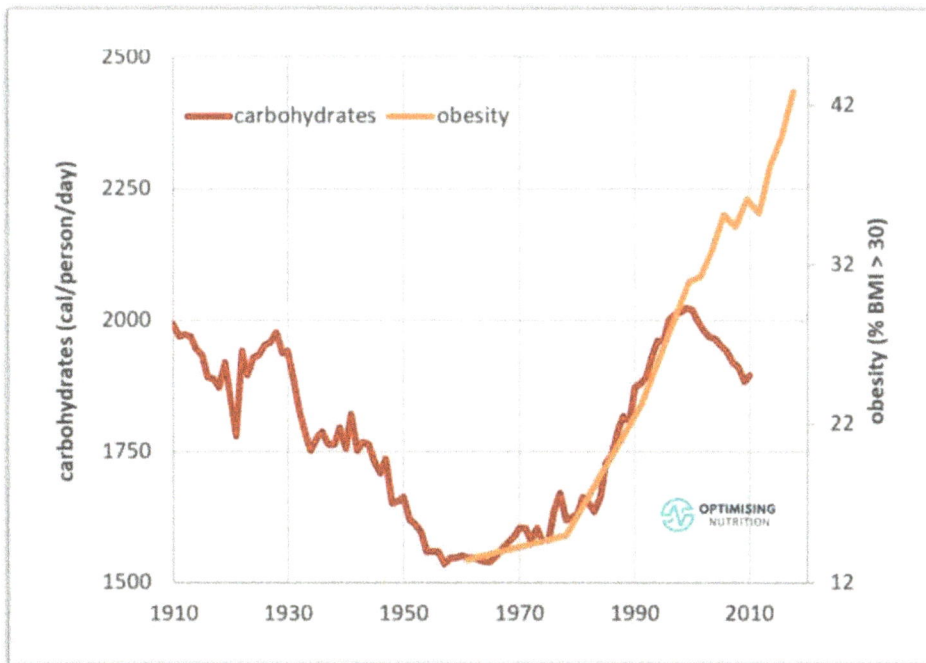

Figure 1.4 Obesity and carbohydrates: Source Optimising Nutrition, US

These plots and a considerable wealth of scientific studies and reports and evidence produced over the past 100 years, clearly show a direct correlation between obesity and hyper caloric feeding, together with a significant increase in dietary polyunsaturated and monounsaturated fat. If saturated fat is not directly responsible for obesity, it does represent a significant risk to cardiovascular health and the development of other diseases.

Humans have a predilection for the taste of fat, probably inherited from ancestors long ago because of its high energy density. Modern days have produced a blitz of high fat, often salty attractive foods which are naturally, very appealing for storage during shortages but the consequence leads to permanent and inexorable weight gain. The relative massive difference between carbohydrate, in the storage of glycogen and blood glucose levels (1 teaspoon full of glucose in the blood), and the substantial adipose storage volume, can also lead to the perception that energy is in short supply and that more energy is needed than it is actually required. All food plays a part. Excess of any macronutrient, leads to weight gain and eventually, obesity, either directly through dietary fats, or indirectly, through reduced fat burning, due to excess carbohydrate and protein, and thus a net growth in adipose tissue with reduced lipolysis and additional storage of fat.

Excess polyunsaturated and monounsaturated fats (Section 12.22) obviously play a significant part in obesity. They have risen exponentially but note monounsaturated fats, such as olive oil show the largest fluctuation and increase around the year 2000. It could be argued that the message to reduce saturated fat and replace this with other fats has been seen as a green light to consume more fat overall. In parallel, the hospitality industry globally has grown exponentially over the past 40 years and currently the majority of people in the UK, USA, Europe and many other parts of the world, eat out more and more at restaurants, cafes, bars and pubs and far more than they eat wholly prepared food at home.

This leaves very little control on actual food ingredients on the plate, with the customer usually unaware just how much unknown oils, fat, free sugars or salt has been used for preparation of the food to enhance taste. The likelihood of the availability of the correct quantity of highly nutritious, whole food choices is very limited at these external hospitality locations. Consequently, there are in fact many reasons for increased weight gain over the past few decades; the principal reasons listed below.

1. Hypercaloric excess; eating too much each day
2. Portion sizes too large
3. Too much fat (cheese, meat, butter, cake, fries, oil, avocado)
4. Too much free sugars and refined carbohydrates
5. Too much processed food, junk food and snacks
6. Too much eating out
7. Too much alcohol, too often
8. Excessively long daily eating window, 17 hours or more
9. Poor sleep patterns
10. Insufficient general activity (not helped with electronic devices)
11. Insufficient moderate exercise
12. Increased stress leading to bad eating habits

In summary, remember that excess dietary fat will be directly stored as fat. Excess dietary carbohydrate increases glucose oxidation, therefore reducing fat oxidation, which will mean that more dietary fat is stored as fat. Excess dietary protein increases amino acid oxidation, also reducing fat oxidation, again leading to more fat being stored. For weight loss, remember these 4 key points.

1. Weight gain is far easier to achieve than weight loss
2. It is easier not to gain weight in the first place than to lose this weight gained later
3. Humans are better equipped to regain weight lost than prevent the gain in the first place
4. Sustaining the weight loss achieved is easier than losing this weight in the first place

1.9   Why Most Weight Loss Diets Fail

Diet-induced weight loss, long term body changes, appetite, gut hormones, insulin resistance, healthy carbohydrate tolerance, hormonal regulation, cell physiology, food quality and quantity choices, food addiction, lifestyle, social pressures, eating with others, or alone, habit and psychology all play a dynamic role in metabolic changes governing weight. [24, 25, 26] Quick-fix diets often fail due to psychological factors like unrealistic expectations and the body's metabolic response to restrictive eating patterns. Metabolic adaptation persists over time and is likely a proportional, but incomplete, response to contemporaneous efforts to reduce body weight. This metabolic adaptation may persist during weight maintenance and predispose to weight regain unless sufficient levels of moderate exercise or caloric restriction are maintained. Metabolic adaptation is related to the degree of energy imbalance and changes in levels of the satiety hormone leptin. During the first phase of dieting and active weight loss, short term reductions in resting metabolic rate are related to extreme calorie restriction. However, the larger, persistent metabolic adaptation that takes place subsequently is more related to the level of sustained increases in physical activity. Maintenance of a reduced or elevated body weight is associated with compensatory changes in energy expenditure, which oppose the maintenance of a body weight that is different from the usual weight. A decrease in weight results in a metabolic resistance leading to a decrease in total energy expenditure at the time of active weight loss. At the time of weight loss, very high levels of exercise do not prevent this dramatic initial slowing of resting metabolism. However, increased use of skeletal muscle and moderate exercise subsequent to the achieved weight loss can help maintain the reduced body weight. This mechanism may be partly due to a reduction in appetite and also follow the theoretical 'constrained total energy expenditure model' [20]. This evidence is provided from the results of the following studies;

Energy compensation and metabolic adaptation: "The Biggest Loser" study reinterpreted 2022 [20]

Comprehensive Assessment of Long Term Effects of Reducing Intake of Energy (CALERIE) 2018 [72]

Changes in energy expenditure resulting from altered body weight 1995 [73]

Metabolic slowing with massive weight loss despite preservation of fat-free mass 2012 [74]

Actions to deliver goals for weight loss and weight management should not be solely restricted to dieting because weight loss is inextricably linked with the wider goal of achieving a healthier overall lifestyle.
The majority of people that lose the weight in the short term, generally follow weight loss diets with the single objective of losing weight and inevitably end up putting the weight back on, sometimes gaining even more weight than lost. Further studies are required to fully clarify the dynamics of weight regulation and how these resetting mechanisms operate.

Forms of restrictive dieting can actually be more harmful to health compared to staying at the heavier weight and often substitute wholesome food for powders, soups or shakes. These so called 'Yo-Yo', quick fix diets can go on repeatedly for years and all tend to fail in the end, sometimes leading to a lowering of self-respect.

These weight loss diets do not address the metabolic reasons why weight has increased over time, nor do they reflect the fundamental importance of the link between healthier nutrition and health in the long term. For example, the metabolic Set Point, applicable to each individual.

Most people maintain a very similar level of weight throughout the majority of their adult life, (>25 years of age), whether overweight or lean. The Set Point theory states that the human body tries to maintain its weight within a preferred or 'set range'. Successful weight loss requires a change in this set point.

Many of the weight loss diets do provide short term results but are unsustainable, restrictive diets, with the emphasis on temporarily reducing dietary intake and calories and restriction to these powders, blended 'smoothies' or soup.

However, losing unwanted, excess weight and then sustaining this reduction in the long term, needs a more enlightened and refined strategy than simply counting calories and needs to be treated as a multi-faceted challenge. One thing is for sure, the evolutionary capacity to store excess fat as storage is very high indeed but naturally leads to considerable problems for the individual in modern times of plenty across the west and increasing elsewhere.

There are, thankfully, physiological mechanisms in the body that limit this process, such that, without these mechanisms, obese people would become even more obese. As a corollary, there is only so much weight loss before the body plateaus to its healthy settling point or new set point i.e., where the body becomes used to the revised level of food intake and activity and therefore responds to adjust metabolism and hunger. In this event the body assumes that this is the current environment that needs to be maintained for the organism's survival. Weight loss plateaus. It is impossible to simply lose a pound a week, every week. Weight loss does not go on for ever, or humans would simply vanish after 2-3 years of dieting. [24] There is a multitude of metabolic and physiological mechanisms controlling the relationship between energy stored and energy used. [20]

As noted in Section 1.8, most weight gain over long periods (decades) occurs incrementally, requiring very small caloric excess day to day over many years, also affected by several surges per year, for example, during holidays and excessive over eating during these 2 or 3 day periods. One of the most important things to understand, is that the body wants to preserve the weight it has gained. This is a natural, evolutionary survival mechanism. Even today, with unlimited food available in shops and food outlets, the body considers this additional weight stored as valuable energy reserves that may be required in order to cope with a potentially lean future, where there could still be a nutrient shortage or famine. It is probable that early humans evolved during regular times of hardship and famine and harsher winters, especially compared to today's 24/7 resources and food availability to many highly obese populations. Genes express themselves according to their environment.

The body has evolved many mechanisms to guard against food shortage during times of 'famine' but the phenomenon of constant excess and over availability of food is a relatively new development and our bodies have not yet adapted efficiently to this time of abundant 'feast', which causes havoc with metabolism and health in so many ways. Scientific knowledge is relatively new in this regard of excess, though fairly well-developed regarding deficiency and starvation.

For example, it is likely that fat (adipocyte) cells are able to register and remember eating patterns earlier in life, via epigenetics, where the expression of the gene is controlled by the environment, in this case which substances were eaten i.e., environmental triggers and signals, without changing the DNA sequence of the genes. This is part of the "*thrifty gene*" hypothesis. The body is more content to keep storing excess hypercaloric dietary, precious energy, for the longer term uncertainty. It is not minded to adjust future intake, to compensate this excess, nor return to the original body weight. Unfortunately, the body does not see any implications in continuing down this anabolic 'growth' path, as it is not aware that there is an ever plentiful and constant supply of energy available 24 hours per day, 7 days per week, 365 days per year, either at the local supermarket or take away, from the large number of 24/7 outlets a couple of blocks away. Some scientists argue against the thrifty gene or fat preservation argument, as not all people in modern western societies are overweight or obese. The answer most probably lies in a whole combination of influencing and metabolic regulatory pathways, serving to protect the individual, together with lifestyle and eating behaviours in earlier years and how the individuals' genes interact with the environment.

Section 14.1 discusses weight regain and metabolism in further detail. Hence, if weight is lost, the body 'remembers' an individual's original weight and inevitably, tends to 'fight back' hard, to restore the original weight and set point. Metabolism and energy expenditure changes also occur. It is against the background of this naturally restorative physiology, that most diets fail. [20]

In addition to metabolic and hormonal triggers, there are a multitude of non-metabolic reasons why people tend to overeat. The need to refeed may be purely physiological, for example, people eat when they are hungry or because they do not feel satiated from an earlier meal. This is usually the case when people start snacking. Their body is telling them to continue eating as it requires more nutrients or energy or it is trying to restore status quo if weight loss is the goal. People also overeat for sensory reasons, such as appearance, odour, taste, colour, texture and other positive emotions associated with the food or the location. This plays a large part of modern day stimulation and food manufacturers invest considerable sums of money to ensure the right colour, odour and taste appeals to as many people as possible to buy as much as possible.

Psychological influences and general mood can be a very complex subject for example, boredom, anxiety, depression, being bullied, feelings of rejection or unhappiness, or for all of life's trials. These symptoms often affect food choices and lead to comfort eating and a preference for so called 'palatable comfort foods', high in fat, sugar and salt, as a way to alleviate their negative mood. These moods, which can unfortunately have a major part in someone's life, also often lead to a change in energy metabolism and lower levels of exercise. There are several studies that suggest there is a close interaction between food, mood and stress.

Stress, whatever the reason, can significantly affect eating behaviour, either resulting in increased or decreased food intake, which is dependent on the types of external or psychological stressors. Intake of more desirable, comfort foods often reduce signs of stress and anxiety, however short lived. This tends to be repeated leading to chronic weight issues. Stress has a major metabolic and hormonal impact also, worsening eating habits as the body is preparing for hardship. Some people eat for habitual, rather than for reasons of physical hunger or anxiety or unhappiness, for example, due to boredom and some may be unconsciously aware of their intake, which often results in eating beyond real hunger. The daily regime becomes habitual and a difficult habit to break, in a similar way to the addictive patterns of other more addictive drugs and alcohol, however the stronger drugs are more chemically addictive whereas the daily eating pattern is more likely to be brought on by boredom. Social norms play a big part in motivations for eating, and can be due to socialising many times per week, including low quality food and alcohol, or religious festivals, holiday breaks or just simply peer pressure, even though, the individual is aware that weight is on the increase. Without such pressure, the individual would, at times, prefer not to engage- at least not so much every day or every week.

Financial concerns can also play a part leading to weight gain or loss. Though more preponderant in many parts of Africa and Asia, it is less common in the Western countries not to eat due to financial constraints. However, it is more common to find those on a low income, purchasing cheap, very highly subsidised, processed foods, including fast, relatively unhealthy eat out or takeaway food (not helped by fast food outlets selling inexpensive burger/fries/drink meal deals etc). Conversely, those people on higher levels of income may also excessively spend on food. Lower cost nutritious foods are discussed further in Chapter 9.

1.10   Popular Fad and Extreme Diets

Given the growing obesity and increasing health problems, described earlier and the changing societal influences, together with the failure of most diets to deliver successful and sustainable weight loss, there has been an increasing trend and indeed populist fashion over the past 2 decades, to adopt magic bullet, fad diets. These diets promise to lead to quick fix solutions delivering weight loss and improvement of other health issues. [43] Fad and extreme crash diets do not work in the long term and there are no special fad or extreme diets that successfully lead to successful and sustained weight loss in the long term.

Fad diets often focus on rapid weight loss and although some of these diets do result in significant weight loss over the first few weeks or months, sited by those that have lost this weight, principally due to an overall reduction in energy intake, loss of water in glycogen and loss of lean muscle mass, it is clear that this rapid weight loss has consequences, including weight regain and issues relating to health due to restriction of nutrients, minerals and fibre. For example, the Paleo diet excludes foods rich in vitamin D and calcium which can lead to osteoporosis and high animal meat and protein consumption increases the risk of many diseases, including cardiovascular diseases and cancers. These diets also often lead to low mood due to low glucose levels, requiring the person to avoid many energy giving foods long term, that the body craves and are therefore unsustainable. Low carbohydrate diets can potentially lead to several health issues including nutrient deficiency, lack of B vitamins, headaches, cramps, constipation, kidney problems, hypoglycemia, dizziness, poor gut health.

The ketogenic and low carbohydrate diets, are predominantly high fat diets, high in LDL cholesterol, significantly restricting carbohydrate to 5% of energy intake, effectively replacing this with high levels of fat at 75% of energy intake and eating very few fruits, vegetables, legumes and grains are often unsustainable. Although the ketogenic diet has recently received a considerable amount of attention because it is purported to help treat obesity and type 2 diabetes, evidence that it actually offers such benefits is limited. Long-term randomized trials have failed to show a clinically significant benefit over comparator diets.[86] Furthermore, the ketogenic diet is not without consequence as it may produce adverse effect such as hyperlipidemia, vitamin and mineral deficiencies, and fatigue. [87]

The carnivore diet, excludes plants and is very high in fat, meat and sodium and in addition to significant increased risk of cardiovascular diseases and cancers, increases the risk of poor gut health, gout, liver and kidney diseases and kidney stones. There have been no controlled studies on the carnivore diet. Most people find these diets impossible to maintain beyond 3 months and often regain weight and even more weight than before staring the new diet.

Most of these diets do not take account of metabolic adaptive responses and hormonal and appetite changes that occur day to day and especially during restrictive unbalanced diets and hence ultimately fail. Some of these diets attempt to mimic stress and starvation conditions over long periods, in order to manipulate the body's biochemistry and metabolic pathways away from natural sources of energy and nutrients, that the body requires and prefers for optimal health. These diets have many things in common and some of these are:

- Short term and rapid weight loss
- Require cellular stress and raised stress hormonal changes
- Important food groups missing or reduced, notably healthy carbohydrates and fruit
- Vilification of all carbohydrates and promotion of high fat and excessive protein
- Difficult to maintain adherence and have long term sustainability issues
- Lack of consistent scientific research, scientific consensus and robust evidence
- Nutritionally inadequate and lacking fibre
- Can lead to potential health risks without medical supervision
- That exercise is not necessary
- Presented on social media using science that may not align with evidence-based practice promoted by those with vested interests in products and supplementation

In contrast, this book and weight loss plan recognises, supports and supplements (pun intended) how the body works and is designed and above all, focuses on, nutritional quality and healthy and sustainable weight management, through eating a wide variety of vegetables, wholegrains, legumes, fruit, seeds and nuts, to provide sufficient and nutritious essential carbohydrate, protein and fat, plus fibre, minerals and vitamins [49].

Free sugars and empty caloric, processed carbohydrates, such as cakes and pastries etc. should be removed or minimised in the diet. This should be the case with all healthy diets. [58] Extreme diets, however, are not endorsed in this book, as they are mostly predicated on 'insufficient healthy calories' and or complex carbohydrate reduction. These diets are generally high in saturated fat and protein, underpinned by the metabolic science of the body's fasting status, both during sleep and starvation, with the prime aim of forcing the body to produce high levels of ketones from fats and glucose from non-carbohydrate gluconeogenic precursors, rather than eating natural, healthy, dietary carbohydrates, which the body requires and prefers for well-being.

For those people who are not sure what to believe, the clue is in the whole body's ethos of Priority 1 in maintaining glucose from carbohydrate and we will see how the human body has evolved to fulfil this.

It does not feel intuitive nor appropriate to add stress to the body and deprive the body of the very substance it needs and then replace this with high quantities of meat, dairy and high levels of fat, all of which have been scientifically proven to increase the risk of poor health during an individuals' lifetime. [56]

Poor Western type diets are most probably responsible for >75% of illnesses across Western populations. Yet, the fundamental reason why most people gain weight is over eating unhealthy foods and not due to undernutrition. Rather than following extreme, fad regimes, dieting and weight management should address these points first. Many of these fad diets enable the individual to continue enjoying aspects of a poor and unhealthy diet. People do not wish to hear bad news about the things they enjoy and have done for a considerable period of their lives, for example, a bacon sandwich, or butter and cheese laden baked potato. This is of course a natural thing but one of the main reasons many people are overweight. We shall explore some of these reasons and behavior in later Chapters.

Much is made of the fact that dietary carbohydrate and glucose can be mostly supplanted by other fuels to supply energy for many cells throughout the body and that glucose can also synthesised by other fuels during carbohydrate reduction and calorie restriction, or by increased burning of fat, for example via the gluconeogenesis break down of proteins, plus lactate, glycerol and in ketogenesis producing ketones from fats. [26] However, this rather extreme, unsustainable status also often leads to brain fog, blood acidity, reactive oxygen species (ROS), high levels of cortisol stress and a whole host of other health issues, both short and long term.

Calorie restriction type, alternative metabolic pathways to lose weight, or even grow muscle, motivate many of the proponents of the low carbohydrate' or so called 'low carb' diets. If people prefer low carbohydrates and it works for them, that is fine but there should be no need for prolonged fasting or reducing carbohydrates to such an extent to invoke unhealthy levels of ketosis, or the use of amino acids from protein to provide significant proportion of energy. Many of these diets fail to recognize how the body really works naturally and are, for the most part, unsustainable in the long term and they are certainly not necessary to achieve and sustain weight loss. In time, both low carb/high fat and high carb/low fat diets ultimately lead to the same weight loss.

Continued avoidance of healthy, complex carbohydrates, for example, starchy vegetables, potatoes, legumes, fruit, nuts, seeds, wholegrains and fibre, for years, is not only completely unnecessary but very difficult to achieve. It also prolongs carbohydrate resistance, meaning that return to eating any carbohydrates, including healthy carbohydrates, alongside the habitual consumption of fat and excess protein will often add weight and may lead to other issues.

The theory of the carbohydrate-insulin model of obesity is that carbohydrates cause insulin spikes which prevent the body from burning fat and that this promotes the passage of sugar and fat from the blood into adipose cells and away from the more active muscle cells.

The theory assumes that this causes a slowing of the metabolism and the feeling of hunger. However, the reality is that fatty acid levels remain in the blood even with high insulin. There is no strong evidence that insulin causes an increase in hunger. But the research also shows that insulin does not decrease but in fact increases fatty acid uptake into muscle cells. If the low carbohydrate model does work and promotes the metabolic advantage over high carbohydrate diets, then people on the low carbohydrate diet should lose more body fat.

However, more research has confirmed no real difference either way and either low carbohydrate ('low carb') or low fat diets can work differently in producing early weight loss for different people. Ultimately, it is most likely that low carbohydrate diets cause people to eat fewer overall calories for various reasons.

For example, less food choices lead to reduced food appeal. Although insulin influences the rate of fat burning, insulin is not the final key driver in weight management. This is in fact driven by caloric intake in excess of caloric expenditure. If the insulin sensitivity is upset due to insulin resistance, caused by an accumulation of fat in many tissues, then insulin will begin to have a diminished effect, leading to elevated levels of insulin required and excessive glucose in the blood stream. This elevated level of insulin inhibits lipolysis, the break-down of stored fat.

The main purpose of Insulin is to react to and take advantage of the fed state, store nutrients and energy, by promoting glucose uptake and lipogenesis, (fat storage) both to ensure the body can use the energy ingested and slow the rate of fat burning. It is bioenergetically more efficient to use ingested nutrients for energy rather than to burn stored energy. The key is that insulin does not act in seclusion, as there are several other hormones acting together, regulating the balance between nutrient use and storage. Protein also stimulates insulin release. (Section 14.3.).

Whilst insulin acts to stimulate lipogenesis (fat storage) and inhibit lipolysis (fat breakdown) these other hormones, including glucagon, cortisol, leptin, epinephrine, norepinephrine, vasopressin and growth hormone, act antagonistically to stimulate lipolysis (fat breakdown).

In addition, leptin, growth hormone and cortisol act to inhibit lipogenesis (fat re-esterification and storage). Cortisol and growth hormone act more slowly than the much speedier effect of glucagon and epinephrine. Another recent discovery is fibroblast growth factor-21 (FGF-21) which is considered to regulate whole body metabolism and energy use and storage by decreasing appetite, reducing carbohydrate oxidation and increasing fatty acid oxidation. Research is ongoing. The synopsis of this interplay is that even with high levels of insulin, fat burning still proceeds but at a slower pace.

Carbohydrates are always burnt preferentially for energy if these foods and glycogen and glucose molecules are available. On top of all this, cholecystokinin (CCK), amylin and glucagon-like-peptide also play an important part in appetite regulation. Hence, the carbohydrate-insulin model is not sufficient to use in order to lose weight and maintain weight loss and is therefore likely to be too simplistic because there are many other metabolic and hormonal mechanisms that influence metabolism, appetite and psychology. [25, 26]

While the carbohydrate-insulin model does offer insights into metabolic processes, current evidence suggests that factors such as metabolic adaptation, appetite, combined hormonal change, genetic predisposition and lifestyle habits all play a larger role in obesity than solely insulin regulation.

Carbohydrate is an integral component of DNA and energy metabolism. There is no convincing evidence that elimination or long-term reduction of dietary carbohydrate is necessary in humans. Primates and humans throughout the world have thrived on carbohydrate for millennia and for humans, evidence confirms at least 800,000 years long before humans developed agriculture [19]. Observe the evolution. The anatomy, structure and morphology of the human being. The gut, the hands, the jaw, the teeth, let alone the enzymes.

Did humans evolve to eat meat and fat and a tiny shred of carbohydrate? Nutrition theories ebb and flow, it is only 80 years ago that research showed that vitamins and minerals are essential. If some people still do not accept the need for dietary carbohydrates and need scientific proof to demonstrate this beyond what is currently known, then one day science may also prove this irrefutably. After all, it took decades of research, thousands of studies and scientific papers for the majority to finally accept that smoking severely damages health.

All 'low carb' advocates and bloggers and even some distinguished scientists, still claim that dietary carbohydrate is 'not necessary, nor 'essential', as glucose can be made from proteins via pyruvate, oxaloacetate and gluconeogenesis and we can gain the energy we need from ketogenesis'. Substituting nutrient dense carbohydrate with high fat and or high protein and putting the body through stressful, metabolism, is neither natural, healthy, nor sustainable and can lead to short and long term health issues. There is a fundamental difference between starving and thriving. In the short term, very low carbohydrate diets may result in more weight loss than higher carbohydrate diets, probably due to the significant reduction in stored glycogen, which is 77% water, plus prolonged caloric restriction but over time, there are no significant advantages, as most people adopting the low carbohydrate diet, find it less satisfying, troublesome and far more difficult to maintain as described in Sections 1.9 and 1.10. The long term problem is often low adherence, as most people eventually find it impossible not to eat their once favourite complex carbohydrates and overall, the weight returns.

The latest research into low carb diets is that these may offer a slightly higher metabolic advantage, presumably due to the higher thermic effect of more protein from higher meat, cheese and egg consumption but this is countered with the fact that people eat fewer total calories on a normal carbohydrate plant-based diet because they feel more satisfied more easily, probably due to the larger amounts of wholegrains, vegetables and fruit and fibre. Very long term (>40 years) chronic effects of a low carbohydrate diet, high in saturated fat and meat and dairy protein are still not yet known, however, given the 2015 World Health Organisation guidelines on meat, these diets are certainly unlikely to benefit the cardiovascular system or kidneys and are however, proven likely to lead to other diseases. Substituting nutrient dense carbohydrate with high fat and or high protein and putting the body through stressful, metabolism, is neither natural, healthy, nor sustainable and can lead to short and long term health issues.

A two-month metabolic ward study was carried out by Dr. Hall and his team in 2015, with 17 male subjects, comparing a low carbohydrate ketogenic diet with a high carbohydrate diet, where both diets reduced total caloric intake by 300 kcal/d and both with the same level of protein and exercise. The study yielded the following results: The low carbohydrate diet produced 22% less insulin, increased calories burnt (energy expenditure) by 57 kcal/d and on average lost 4 pounds body mass, including 1.16 pounds of fat. [25]

The high carbohydrate diet produced no change in insulin, no change in calories burnt or energy expenditure and on average lost 3 pounds body mass including 1.29 pounds of fat. In conclusion, people lost roughly the same amount of weight and fat. The change in insulin did not directly nor indirectly result in any change in weight. There was a slight change (2%) in calories burnt on the low carbohydrate diet, which confirms that this diet does marginally increase metabolic rate.

A further metabolic ward study was carried out by Dr. Hall and his team in 2020, with 20 overweight subjects (11 men and 9 women), this time comparing an animal-based, low carbohydrate ketogenic diet; at 75% fat, 10% carbohydrate and 15% protein; with a plant-based, low fat, high carbohydrate diet; at 11% fat, 75% carbohydrate and 14% protein and both diets with minimal processed food. [26] Each diet provided double the number of calories required by each for energy balance; i.e., EI = 2 TEE. Both groups had the opportunity to eat as much as they wanted to eat 'ad libitum'. The results indicated that for the second week, 544 fewer daily calories were consumed on the plant-based, low fat, high carbohydrate diet compared with the animal-based, low carbohydrate, ketogenic diet. Adding the results of the first week increased this difference overall to 689 fewer daily calories. Energy expenditure was 166 higher (3%) on the animal-based, low carbohydrate, ketogenic diet. Blood glucose and insulin levels were much lower on the animal-based, low carbohydrate ketogenic diet. Both groups lost weight without intentionally reducing food eaten. Those subjects on the animal-based, low carbohydrate, ketogenic diet lost a total of 3.9 pounds, whilst those subjects on the plant-based low fat, high carbohydrate diet lost a total of 2.4 pounds.

The plant-based low fat, high carbohydrate diet yielded a significant reduction of 1.3 pounds of non-lean body fat mass i.e., 54% of total weight loss as fat and 46% as fat free mass including water, whereas the animal-based, low carbohydrate, ketogenic diet yielded a significant reduction of 3.5 pounds in fat free mass from muscle bones and other organs, probably due to a significant reduction in glycogen and water i.e., 10% of total weight loss as fat and 90% as fat free mass including water. [2]

The main conclusion drawn from this more recent metabolic ward study is that although the animal-based, low carbohydrate, ketogenic diet increased metabolism by approximately 2% (consistent with the 2015 study [24]), people on the plant-based, low fat, high carbohydrate diet ate considerably less calories each day. Also, none of the 'ad libitum' diets caused weight gain.

Overall, this study does add weight (pun intended again) to the importance and relevance of the caloric and nutrient density of different foods and the importance of complex carbohydrate starches and grains, fibre and water content. Also, the importance of a more varied content and therefore a higher level of satiation, of a plant-based diet, compared with the animal-based, higher fat, high protein and lower carbohydrate diet.

In high fat, low carbohydrate diets, the individual is burning less carbohydrates and burning more fat but eating much more fat. Therefore, even if a balance or weight loss on such a diet is achieved, the serious, long term healthy implications of excessive dietary fat, are simply not worth substituting fat for carbohydrates in the first place. Far better to let the carbohydrates i.e., glucose drives the body's central metabolism so that all substrates can be burnt and without the metabolic stress. While low-carb diets can certainly yield short-term weight loss, long-term health outcomes often and sustained weight loss indicate a need for a more balanced diet and a more sustainable approach.

1.11   The 5 Pillars of Healthy and Sustainable Weight Loss

In line with the objectives of the Nutrition and Metabolic Guidelines for Weight Loss listed in Section 8.22, there are essentially 5 pillars of nutrition and lifestyle listed below that are required to lose weight in a healthy, sustainable and nutritious way, leading to an overall healthier lifestyle. These are discussed in Chapters 3 to 7.

- Nutrition Quality   Chapter 3
- Nutrition Quantity   Chapter 4
- Nutrition Timing   Chapter 5
- The Benefits of Exercise   Chapter 6
- The Importance of Sleep   Chapter 7

The Healthy and Sustainable Weight Loss Plan in Chapter 8 comprises these 5 fundamental pillars of weight loss, summarised as a step by step plan including nutritious (food quality), marginal caloric deficit of energy intake against energy expenditure (food quantity), adjusted eating period, regular, moderate exercise and improved sleep patterns.

This will provide the most likely success in target weight loss and just as importantly, sustaining the weight loss achieved in the long term. Chapter 2 that follows, examines energy intake and energy expenditure. This is important to understand, in order to set and build the weight loss target for each individual.

CHAPTER

Energy Balance: Intake & Expenditure 2

The food we eat provides energy for all of our body's metabolism and activities. Whether sedentary or moving, writing, thinking, sleeping, eating, exercising, or running, the body is constantly using energy. When we are asleep, our heart is continuously pumping blood to all the tissues requiring oxygen and nutrients. Our digestive system is working, producing enzymes and our cells are releasing energy from the food molecules. Proteins are carrying out maintenance and repair. Our immune system constantly guards against infection and cellular stress. All this uses up around 2,000 kcal of energy per day. On the flip side, the food and drink we consume provides this energy but if we regularly eat more than we use (also referred to as 'burn'), we gain weight. The energy balance is the relationship between energy input, i.e., energy intake from food and drink and energy output, i.e., energy expenditure, the energy used by our body. This Chapter looks at this energy balance and how this balance can be used to help reduce weight.

2.1 Daily Energy Intake (EI)

Daily energy intake (EI) is the total amount of energy consumed by an individual, taken in from food and drink; typically measured in calories, or more correctly, kilocalories (kcal). Healthy weight maintenance is essentially governed by the maintenance of energy intake, (EI) and energy expenditure, (TEE). Daily energy intake (EI), is the energy content of food and drink, consumed from the macronutrient's carbohydrates, fats and proteins and from alcohol each day.

2.2 Daily Total Energy Expenditure (TEE)

Daily total energy expenditure (TEE) is the total amount of energy utilised to maintain all core physiological functions and locomotion. TEE varies on a daily basis and comprises 4 principal components; basal metabolic rate (BMR), the thermic effect of food (TEF), exercise related activity thermogenesis (EAT) and non-exercise activity thermogenesis (NEAT). The average TEE for an adult female is in the order of 2,000 kcal/d and 2,500 kcal/d for an adult male. TEE is sometimes referred to as total daily energy expenditure or (TDEE). For an energy balance:

$$TEE = (BMR + EAT + NEAT + TEF)$$

A eucaloric (or isocaloric) diet is when a person's energy consumption matches energy expenditure i.e., EI equals TEE. Over time, a eucaloric diet is expected to produce no change in body weight and fluctuations in weight arising will be due to daily metabolic processes or possibly some illnesses.

EI = TEE	= Weight Maintenance
EI > TEE	= Weight Gain
EI < TEE	= Weight Loss

2.3 Basal Metabolic Rate (BMR)

BMR is the minimum amount of energy expended or used up by an individual to maintain all the body's vital processes and vital organs (basic organ and cellular function, breathing and respiration), whilst at rest (the basal state), in a temperate environment, when the digestive system is inactive, i.e., the fasted state, also referred to as the post prandial state or post absorptive state.

BMR ranges from 60% to 80% of total energy expenditure, TEE and probably around 70% for the majority of people. BMR is principally a function of lean body mass (LBM). LBM accounts for 80% of BMR and non-lean body mass (fat tissue) accounts for 20% of BMR. BMR is not necessarily the lowest level of expenditure during the day since, for example, during sleep or undernutrition, metabolic functions may be lower than observed under basal resting conditions during the waking hours.

BMR accounts for the main proportion of TEE at approximately 60-70% of TEE for individuals with a normal reasonably sedentary occupation. However, for individuals with very minimal physical activity, it is proportionally larger, up to approximately 85-90% of TEE for example, hospital patients.

The variability of BMR between individuals of similar age and equal gender ranges around 7-9%. The level of BMR depends on many factors including gender, height, weight and age.

BMR is calculated using various formulas. An average lean 70kg male has a BMR in the order of 1,700 kcal/d or around 68% of TEE.

2.4 The Thermic Effect of Food (TEF)

The thermic effect of food (TEF) is the increment of energy expenditure above BMR, following meal ingestion, that reflects the bioenergetic cost burned by the body, to process and use fuel during food digestion, absorption and storage.

TEF typically ranges between 8-15% of TEE but can vary significantly depending on the type of food consumed, for example, dietary fat has a TEF of 0-3%, carbohydrate is 5-10% and protein is 20-30% as protein is more difficult to process and requiring more energy. Hence, protein is a relatively highly thermogenic food, a fact used by some to eat excessive amounts of protein in order to lose fat mass and gain muscle (Section 14.2). Note how easy, (very low energy required) it is to store fat. BMR plus TEF are relatively fixed in amount and together and account for approximately three quarters (75%) of daily TEE variance. Another term commonly used to describe the thermic effect of food is diet induced thermogenesis (DIT).

2.5 Exercise Activity Thermogenesis (EAT)

Exercise activity thermogenesis (EAT), also referred to as exercise related activity thermogenesis, is defined as planned, structured and repetitive physical activity that has the objective of improving health. For example, all high activity sport, visiting the gymnasium, swimming, ball sports, running and also includes golf and brisk daily walks from >30 minutes per day. EAT varies considerably, ranging from 15-30% of TEE, for those individuals who habitually participate in purposeful, physical training, down to only 1-2% of TEE, for those who only exercise for one or two hours per week. On average, this 1-2% of TEE accounts for only 100 kilocalories (kcal) per day.

However, for reasons such as ill health, medication, significant obesity or hip replacements, not all individuals can engage in vigorous or regular EAT. Such is the very limited level of EAT generally measured across whole populations, the overall average contribution of EAT is therefore, usually assessed at just 1-2% of TEE at a population level.

2.6 Non-Exercise Activity Thermogenesis (NEAT)

Non-exercise activity thermogenesis (NEAT) is defined as that portion of daily energy expended for non-sleeping, non-eating, or non-sports like exercise, resulting from spontaneous physical activity, incorporating body posture, ambulation (moving) and all other spontaneous movements such as fidgeting, that is not specifically the result of voluntary exercise.

NEAT also corresponds to all the energy expended with work occupation, general leisure time (not fitness) activity, house and garden maintenance, changing the bed linen, sitting, moving, standing, climbing the stairs, cleaning, washing the car inside and out, clearing out the loft, cleaning the kitchen and hall floor, decorating, walking up the stairs to the office or at the shopping centre, vacuuming, general housework, toe tapping, singing, playing musical instruments, engaging in fun play with the children and grandchildren, etc.

NEAT usually occurs at a low energy workload for minutes and sometimes up to hours. In contrast to EAT, NEAT comprises the largest share of daily activity-related thermogenesis. This is the case overall, even for most individuals that engage in regular physical training.

NEAT accounts for 6-10% of TEE in individuals with an inactive lifestyle and surprisingly accounts for as much as 50% or more in highly active people. NEAT can account for a considerable range anywhere between 100-800 kcal burned per day. This shows just how much moderate levels of daily exercise can play a part in healthy lifestyle and energy balance. It also demonstrates that the gym or hard (EAT) exercise is not essential for good fitness levels. As EAT is believed to be negligible on a population level and BMR and TEF are relatively static overall, NEAT consequently represents the most variable component of TEE, both within populations and across individuals. Hence, NEAT is the factor that can influence general, overall fitness levels for most people most of the time and should be seriously considered as a factor with a potentially major impact on and strategy for, realistic health improvement for most individuals across all societies. As a summary of TEE therefore, the rough order estimated proportions for the average population are:

BMR	=	70% TEE
TEF	=	10% TEE
NEAT	=	8% TEE
EAT	=	2% TEE

2.7 Realistic Weight Loss Targets

For individuals weighing in excess of i.e., >36 kg over ideal, lean body mass, consultation with the doctor is firstly recommended, to discuss and agree the most appropriate options for weight loss targets best suited for each individual. For others currently less than i.e., <36 kg over ideal weight, wishing to lose weight sustainably, this long term lifestyle weight loss strategy, plus incorporating exercise and non-exercise related thermogenesis (EAT) and (NEAT) is recommended. Different approaches to weight loss will apply to different individuals and each individual will find what works best and feels right for them. This exercise choice selection mantra is the same for nutrition. It is what works best and is most comfortable and enjoyable for each individual that counts and which will ultimately be most likely to succeed.

Generally, a gradual, non-lean body weight loss rate in the range of 0.5 - 1.0 kg/week is recommended.

Anything in excess of this may be too difficult and will be likely to invoke a stronger (fight back) reaction from the body's metabolism. [3] Fashionable, popular diets often cause significant weight loss in the first few weeks, principally caused by physiological, semi-starvation status, glycogen depletion and glycogen water loss (glycogen is approximately 77% water by weight), then slowing and plateauing, leading to frustration, reversion to other unsuccessful diets and finally despair and often resulting in weight regain. Many individuals have gained this excess adipose tissue weight over many years and sometimes decades. A gradual weight loss strategy to reverse this weight gain is therefore more likely to be successful in the longer term (Chapter 1).

Six weight loss scenario weight loss plans are shown below using the 'fatcalc' reference calculations model https://www.fatcalc.com. Note that TEE metabolism slows in reaction to a reduction in EI food energy intake. After the 3 or 6 month periods of significant change, the long term sustained change for each is around 5% to 9% reduction in TEE and EI and this can be reduced further by approximately 2% if exercise is moderately increased. These are average values, calculated for most individuals but the actual impacts may vary depending on each individuals' set point. Note, Plan 1a below is shown as the baseline represented by an adult female, aged 40 with body mass 68 kg and a height of 162 cm who normally engages in 'very light' NEAT and little EAT exercise. The current TEE is 1,868 kcal/d and BMR is 1,335 kcal/d. She aims to lose 9.64 kg (1.5 stones or 21 pounds) in non-lean body mass, towards a goal weight of 58.46 kg, over a period of 3 months (92 days).

Note, Plan 1a requires an overall caloric deficit change (delta) of 861 kcal/d, each day, for the 92 days i.e., around 46% reduction in TEE and therefore the same reduction in EI caloric intake, compared with baseline, to match this drop. As the body adjusts to the new weight, the EI caloric intake and TEE, for the new long term lifestyle after the 3 months and thereafter, would need to be maintained at the lower level of 1,705 kcal/d i.e., a reduction delta of 163 kcal/d, compared with the current 1,868 kcal/d, or a new sustained TEE reduction of 9%. It can be seen that this target objective, requiring an 861 kcal/d and 46% reduction in TEE caloric intake for 3 months, would be difficult for most (though possibly achievable for some) to achieve and such a large deficit is not therefore, recommended, given the potential for some health issues, plus the long term adherence likelihood. If individuals can achieve and sustain larger weight loss, at quicker pace and this does not impact adversely on their health, then that is fine also. To demonstrate the relative inequity between the relative contribution of food intake and exercise towards fat reduction and weight loss, note Plan 1b below compared to Plan 1a. Both produce a weight loss of 9.64 kg over 92 days. Plan 1b assumes an increase to 'Moderate' NEAT and EAT and this needs a 7% lower change (46% to 39%) over the 3 months but an additional 2% (9% to 11%) long term. Note therefore, if the same individual significantly increases TEE (increasing NEAT and EAT from very light to moderate exercise) this requires more fuel, thus raising TEE by 533 kcal/d to 2401 kcal/d and would now need to reduce caloric intake by 938 kcal/d i.e., 39% of the new TEE, to reach this goal. Therefore, neither scenario Plan 1a or Plan 1b is recommended for most, nor probably sustainable for many.[3]

Plan 1a: -9.64 kg loss Very Light NEAT and EAT Over 92 Days, 3 Months		Plan1b: - 9.64 kg loss Moderate NEAT and EAT Over 92 Days, 3 Months	
Current Weight:	68.00 kg	Current Weight:	68.00 kg
Target Weight:	58.46 kg	Target Weight:	58.46 kg
Target Reduction:	-9.64 kg	Target Reduction:	-9.64 kg
Duration:	92 days (3 Months)	Duration:	92 days (3 Months)
TEE at Start:	1,868 kcal/d	TEE at Start:	2,401 kcal/d
TEE at End:	1,007 kcal/d	TEE at End:	1,463 kcal/d
TEE delta:	-861 kcal/d	TEE delta:	-938 kcal/d
TEE delta:	-46%	TEE delta:	-39%
TEE to Maintain:	1,705 kcal/d	TEE to Maintain:	2,146 kcal/d
TEE delta:	-163 kcal/d	TEE delta:	-255 kcal/d
TEE delta:	- 9%	TEE delta:	-11%

It is therefore recommended to aim for more realistic long term weight loss objectives, either less weight loss, or over a longer term, or a combination of both tools. For this lower weight loss, more realistically achievable goal, note Plan 2a. With the same exercise as baseline Plan 1a, this time a weight loss of 6.00 kg over the 92 days (13 pounds), would result in a TEE reduction of 548 kcal/d i.e., a drop of 29%, rather than 46% for Plan 1a and a new sustained TEE reduction of 5%. Then, for Plan 2b, adding a change in exercise to moderate NEAT and EAT, this time a 25% drop in caloric EI, rather than 39% required in Plan 1b and a new sustained TEE reduction of 7% overall and probably, more achievable.

Plan 2a: -6.00 kg loss Very Light NEAT and EAT Over 92 Days, 3 Months		Plan 2b: - 6.00 kg loss Moderate NEAT and EAT Over 92 Days, 3 Months	
Current Weight:	68.00 kg	Current Weight:	68.00 kg
Target Weight:	62.00 kg	Target Weight:	62.00 kg
Target Reduction:	-6.00 kg	Target Reduction:	-6.00 kg
Duration:	92 days (3 Months)	Duration:	92 days (3 Months)
TEE at Start:	1,868 kcal/d	TEE at Start:	2,401 kcal/d
TEE at End:	1,320 kcal/d	TEE at End:	1,806 kcal/d
TEE delta:	-548 kcal/d	TEE delta:	-595 kcal/d
TEE delta:	-29%	TEE delta:	-25%
TEE to Maintain:	1,766 kcal/d	TEE to Maintain:	2,242 kcal/d
TEE delta:	-102 kcal/d	TEE delta:	-159 kcal/d
TEE delta:	- 5%	TEE delta:	- 7%

Either of the target tools, weight loss or time, can be varied to suit the individual. Plan 3a and Plan 3b below, show the same weight loss of 6 kg as for Plans 2a and 2b but this time, over a significantly longer period of 184 days (6 months) rather than 3 months. In Plan 3, the required overall daily caloric reduction is less (18% or 15%) and may be more manageable for some. Set what you feel is right for you. For example, you could try either lower loss over shorter period, or a higher loss over a longer period. If it works and is sustainable for you it will be successful in the long run. These tools will enable you to see what fits best.

Plan 3a: -6.00 kg loss Very Light NEAT and EAT Over 184 Days, 6 Months		Plan 3b: - 6.00 kg loss Moderate NEAT and EAT Over 184 Days, 6 Months	
Current Weight:	68.00 kg	Current Weight:	68.00 kg
Target Weight:	62.00 kg	Target Weight:	62.00 kg
Target Reduction:	6.00 kg	Target Reduction:	6.00 kg
Duration:	184 days (6 Months)	Duration:	184 days (6 Months)
TEE at Start:	1,868 kcal/d	TEE at Start:	2,401 kcal/d
TEE at End:	1,539 kcal/d	TEE at End:	2,035 kcal/d
TEE delta:	-329 kcal/d	TEE delta:	-366 kcal/d
TEE delta:	-18%	TEE delta:	-15%
TEE to Maintain:	1,761 kcal/d	TEE to Maintain:	2,238 kcal/d
TEE delta:	-107 kcal/d	TEE delta:	-163 kcal/d
TEE delta:	- 6%	TEE delta:	- 7%

To maintain nutrient sufficiency, women should not eat less than 1,200 kcal/d and men should not eat less than 1,500 kcal/d and hence, illustrating the difficulty succeeding with Plan 1a. Once the long term goal is achieved, additional targets could be set to suit the individual if required.

These scenario examples are provided as illustrations of the different impacts of the 2 variables i.e., target amount of weight lost and the time taken to lose this weight. They are not necessarily intended for actual use because calorie counting is not the overall goal and difficult if not impossible to follow without medical or expert assistance but those individuals wanting to utilise or review and look at this can utilise on line calculation models and enter in and work through other scenarios can refer to calculation models, for example, the fat calc model https://www.fatcalc.com used here. There are other weight loss calculation models available. In following the weight loss plan in Chapter 8, you should see the impact of lower overall energy intake through healthier eating and there should be no need to count the calories each day or week.

You will become familiar with your change in total intake on the new regime and providing the total intake is a little less than your current (depending on the required deficit for your goals) intake plus exercise more and you follow the plan, you will start to see the results you wish to see. Exercise is always recommended for those that are reasonably fit and able. Exercise coupled with healthy carbohydrate stokes up the fire of metabolism and keeps the cogs of substrate (carbohydrates and fat) burning in tune and well-oiled and hence fires and maintains good health.

'Fat burns in the flame of carbohydrate....'

2.8   The Bioenergetic Cost of Burning Adipose Fat

There is a common belief that enough exercise will control weight gain and that regular gym membership, for example, will manage weight, whilst continuing 'normal' eating and drinking habits. It is true that a sufficient amount of very hard exercise and strength training can burn significant amounts of energy, including a little fat and help with the caloric balance.

However, the bioenergetic cost of this mantra is very high indeed. Table 1.1 shows the amount of energy required from different exercise regimes, to counter the body's fat storage default pathway of consuming various food types of low nutritious quality.

Note, 1 hour 40 minutes swimming and 182 lengths just to burn off one cheese burger and fries. Walking 2 and a half hours and swimming for half an hour, just to burn off one chocolate bar. It is clear that for the vast majority of people, using exercise as the tool to reduce or maintain weight, especially with high calorie, low nutritious food and snacks, is not bioenergetic cost effective and gives very low return for the amount of work invested or work done.

Regular consumption of high caloric food close to or above normal TEE, can-not be sustained without gaining weight, unless the individual is a professional athlete, or trains very intensively every day, or in one whose work demands significant levels of exercise, for example, military (>5,000 kcal/d) personnel.

Far better to eat nutritious food, adding nutrient value to the body whilst ensuring calories are not excessive and hence a far better 'bang for your buck'.

Table 1.1 The bioenergetic cost in Hours of Burning Food (40 year old male at 70kg) kcal burned/h

Exercise Type		Kcal/h	Bag of Crisps [a]	Brownie [b]	Cheeseburger [c]	Pizza [d]
Walking	- Moderate (4.8 km/h)	188	1 hr 13 min	2hr 28 min	5 hr 37 min	2 hr 49m'
Running	- Jogging (7.25 km/h)	560	25 min	50 min	1 hr 48 min	57 min
Swimming	- F/Crawl (2.75 km/h)[e]	610	23 min	46 min	1 hr 40 min	52 min
Cycling	- Road (20 km/h)	700	20 min	40 min	1 hr 27 min	46 min

[a] Bag Cheese and Onion Crisps 32.5g= 231 kcal (42 lengths swimming)
[b] Chocolate Brownie = 466 Kcal (84 lengths swimming)
[c] Cheese Burger and Regular Fries 23 g = 1009 Kcal (No Drink) (182 lengths swimming)
[d] Pizza based on 3 slices of Cheese and Tomato Pizza = 531kcal (96 lengths swimming)
[e] Swimming Freestyle 2.75 km/h covering 110 lengths in 1 hour at 1.83 lengths/min, or 5.54 kcal/length, or 64 lengths in 35 minutes

Sufficient, moderate levels of exercise and some gentle resistance training is however, strongly recommended for all (on approval from the doctor based on individual health status) and has many benefits (Chapter 8) but should not be used as a mechanism to burn off food, especially if eating low nutrient value food.

It should not be used as a rebalance between food intake and exercise. Some people eat more than necessary because they feel they 'deserve it' after a hard workout and then find that they need to do more exercise to rebalance this, which is a very common reason why people struggle to maintain weight even after high amounts of exercise over long periods.

2.9   Significant Caloric Depletion and Health Impacts

Eating insufficient calories and or low quality nutrients mostly leads to a fall in metabolism that can make weight loss slow, as the body senses that food is in short supply. This can occur even for those overweight individuals who have high levels of excess adipose tissue stored as fat. Different people respond differently to very low-caloric restricted diets.

For some, the metabolic rate is slightly reduced, whereas for others it is a much greater fall, hence variability in fat loss rates between individuals.

Recent studies at Newcastle University have shown that each individual has their own *Personal Fat Threshold* (PFT), probably due to genes and the impact of the individuals' eating patterns during earlier life.

This determines how much fat they can put on before the problems begin to develop, such as the accumulation of ectopic fat around the liver and pancreas, which can then lead to insulin resistance and then Type 2 diabetes. [4]

This strongly suggests, along with considerable other evidence, proven by many researchers over the past 60 years, that it is possible to slow, stop and even reverse the development of Type 2 diabetes, reset blood glucose levels and reverse and reduce many cardiovascular and other health issues and by shedding weight.

However, significantly depleting the normal level of consumed energy calories of around 2,000 for women and 2,500 calories for men by >600 kcal change, in the absence of proper medical supervision, coupled with failure to eat nutritious whole foods, such as those listed in Chapter 4, may also lead to increased risk of the following health issues.

This extreme change without medical supervision should always be avoided.

Constantly Feeling Cold
This is the most common symptom of caloric restriction.
A slower metabolism can decrease the body core temperature, causing the individual to feel cold regularly and diversion of blood from extremities, especially away from the hands and feet to the vital organs to protect internal core metabolic function.

Nutrition Fatigue and Muscle Loss
A long term reduced metabolism can also have harmful effects such as muscle loss i.e., lean body mass reduction. As muscle has a higher metabolic rate than adipose tissue (Table 14.1), this causes a further reduction in metabolism.

Hair Loss
B vitamin deficiencies have been associated with hair loss. Furthermore, sudden fluctuations in weight loss can also lead to telogen effluvium, a form of temporary hair loss.

Reduced Fertility

Severe caloric restriction can lead to a disruption in the menstrual cycle for women and to a lower sperm count and motility plus reduced testosterone levels in in men, making it more difficult for couples to conceive.

Weakened Bones

Extreme caloric restriction may reduce oestrogen and testosterone and also calcium levels which can reduce bone density, mass and strength.

Lowered Immunity

Sufficient nutrients are essential to support the immune system that may lead to decreased infection resistance and longer recovery times from illness.

Insomnia

Insufficiency of magnesium can contribute to sleep problems. Magnesium helps regulate melatonin, a hormone regulating sleep and wake cycles and it also regulates neurotransmitters responsible for reducing anxiety and improving sleep quality.

Constipation

Most current diets, lower caloric diets and some extreme diets all lack sufficient dietary fibre, leading to digestion issues and constipation. A very low caloric diet slows the digestive process down significantly, also reducing bowel movements, exacerbating constipation.

Nutrient Deficiency

Significant poor nutrition caloric deficit may also lead to deficiency in macro and micronutrients and vitamins causing increased fatigue, brain fog and low blood sugar (hypoglycaemia).

By following and observing the Nutrition and Metabolism Weight Loss Guidelines (Section 8.22) and properly adopting and sustaining the Healthy and Sustainable Weight Loss Plan (Chapter 8), most individuals will steadily retrain and influence their physiology to burn more and lose excess adipose tissue (excess fat storage), whilst maintaining healthy level of moderate exercise and food intake, substrate metabolism, correct hormonal regulation and lower cellular inflammation.

Continuing this new 'way of life' lifestyle, rather than reverting to or adopting temporary dieting, will provide the best dietary influence for sustainable non-lean body mass weight loss and a healthier life.

CHAPTER

Nutrition Quality

3

Nutrition quality refers to the content of the food and how densely packed the food is regarding both nutrients and calories, i.e., nutrient density and calorie density. There are foods and beverages which are both calorically dense (containing a significant number of calories) and nutrient dense (packed with healthy nutrients, including macronutrients and minerals and vitamins), for example, avocado, salmon, nuts and seeds. There are also foods and beverages that are calorically dense but nutrient light, for example, alcohol and high fat foods such as fries and pastries. There are also foods that calorically light and nutrient dense, for example, potatoes (baked) oats, beans lentils, fish, soups, egg whites, water melon, chia seeds. Hence, a calorie is not simply a calorie, it depends on the content. [24]

3.1 Calorie Density

Calorie density is a measure of how many calories are in a given weight of food, usually expressed as calories per gram. Table 3.1 lists foods high in caloric density. A food high in calorie density has a large number of calories in a small mass of food, whereas a food low in calorie density has much fewer calories in the same weight of food but can have a much higher nutrient density. Table 3.2 lists 120 foods that are low medium and high in caloric density, per pound weight and illustrates the large range of density. Compare olive oil, broccoli, Plain Crackers Cheddar Cheese and an apple.

3.2 Nutrient Density

A food with a high nutrient density is the opposite and has high nutritional quality with a low number of calories in a large mass of food. (Table 3.1) Therefore, a person can consume a larger portion of a low-calorie dense food than a high-calorie dense food, for the same number of calories. People generally eat a similar amount of food, by weight each day. Therefore, choosing foods with a lower calorie density and higher nutrient density, allows and promotes the consumption of the normal amount of food or even more, whilst providing the opportunity to reduce the total caloric intake and thus promoting heathier weight.

Table 3.1 Calorie Dense & Nutrient Dense Foods. Note one tablespoon of olive oil or just 14 grams is 119 calories

Nutrient Dense Foods	Caloric Dense Foods	Empty Calory (Low Quality) Foods
Kale	Lamb	Soft drinks/Energy drinks
Potatoes	Chicken Leg	Ice Cream
Salmon	Butter	Pies
Sweet Potatoes	Chocolate	White Bread
Brussel Sprouts	Granola Cereal	Cakes/Pastries/doughnuts
Pumpkin	Dried Fruit	Junk Food
Quinoa	Fish and Chips	Candy/Sweets
Berries	Avocado	Crackers/Cereal bars
Legumes	Sausages	Crisps/Potato Crisps
Nuts	Cheese	Alcohol

Note Table 3.2. Eating more low caloric density, high nutrient dense foods, will be healthier, more satiating and will lead to less weight gain.

Oils are by far the highest caloric density and vegetables and fruit, the lowest.

The key is to select foods high in nutrient density and not too high in caloric density, listed in the first column and avoid the high caloric, low nutrient dense foods, highlighted in red font listed in the second and third columns.

Note some food should be eaten in moderation shown in blue font, due to other content of the food; saturated fat, unsaturated fat, total fat, cholesterol, salt or additives.
Table 9.1 lists and prioritises these foods in order of nutrition and health.

Table 3.2 The Caloric Density of Food: Low Medium and High Calorically Dense Food

Low Density Foods		Medium Density Foods		High Density Foods	
Food	kcal/lb	Food	kcal/lb	Food	kcal/lb
Watercress	50	Lentils	516	Salami	1532
Cucumber	62	Chicken Curry and White Rice	521	Wheat Cereals (Whole grain)	1592
Courgette (Zucchini)	71	Black Beans	599	Wholegrain Rice	1592
Celery	77	Yoghurt (Oat Milk)	645	Pastries	1593
Lettuce	82	Tuna (Canned)	721	Whole Grain Pasta	1596
Bamboo Shoots	86	Avocado	725	Brie Cheese	1601
Cabbage	91	Chicken (Roasted Breast)	739	Cornflakes	1619
Tomato (Raw)	91	Eggs (Fried/Scrambled/Omlette)	744	White Flour	1650
Turnip	93	Chick Peas	753	Chocolate Éclair	1659
Rocket (Arugula)	112	Soy Beans	765	Oats (Steel Cut)	1664
Cauliflower	113	Houmous	787	Quinoa	1669
Broccoli	118	Salmon	825	White Pasta	1682
Mushrooms	118	Lamb (Leg Roasted Lean Only)	871	Table Sugar Sucrose	1746
Radish	127	Fried Fish (Deep Fried Cod)	907	Cake Chocolate	1748
Spinach	132	Ice Cream	939	Chinese (Sweet/Sour Chicken)	1765
Kale	141	Beef Steak (Sirloin Broiled Lean)	943	Doughnut	1778
Beetroot	145	Potato (Deep Fried Chips/Fries)	977	Crunchy Nut Cornflakes	1805
Peppers (Green/Red Grilled)	158	Tortillas	984	Cheddar Cheese	1828
Onion	159	Potato (hash Brown)	989	Cake Victoria Sponge	1864
Carrots	204	Pork Chop (Loin Broiled Lean)	1039	Croissant (Butter)	1869
Orange	213	Beef (Roast Lean and Fat)	1093	Granola Cereal	1871
Blackberries	236	White Bread	1105	Chocolate Bar (Milk)	1968
Apple	268	Wholemeal Bread	1134	Chia Seeds	1977
Pear	272	Pizza Margareta	1136	Crackers (Plain)	1991
Aubergine (Egg Plant)	289	Hamburger in Bun	1161	Croissant (Almond)	2033
Corn (Tinned)	303	Chicken (Fried/Batter/Breast)	1184	Black Coloured Biscuits	2118
Garden Peas	354	Wholegrain Bread (Rye)	1188	Double Cream	2118
Corn on the Cob	362	Dried Fruit	1192	Brownies (Chocolate)	2155
Blueberries	363	Cheese (Feta)	1202	Flax Seeds	2267
Potatoes (Boiled)	386	Sour Dough Bread	1234	Plain Crisps (Any Flavour)	2305
Baked Beans	390	Pitta Bread	1247	Peanut Butter	2598
Kidney Beans	408	Dates	1256	Sesame Seeds	2599
Banana	417	Bacon	1269	Sunflower Seeds	2648
Strawberries	435	Cheese (Mozzarella)	1279	Peanuts	2789
Yoghurt Greek (Cow's Milk)	449	Maple Syrup	1315	Walnuts	2880
Sweet Potato (Baked)	458	Camembert Cheese	1356	Butter	3239
Raspberries	463	Cheese Cake	1379	Coconut Oil	4000
Potatoes (Baked)	494	Pork Sausage	1396	Olive Oil	4000
White Rice	499	Honey	1455	Avocado Oil	4000
Potato (Mashed)	508	Beef Curry & White Rice	1478	Sunflower Oil	4000

3.3 Empty Calories

The third type of food groups are those described as having empty calories. These are foods or fluids that have little, or no nutrient content, they are the antithesis of nutrient dense foods and are virtually or totally lacking in any healthy nutrients whatsoever, i.e., devoid of complex carbohydrate, fibre, protein, healthy fats, minerals and vitamins. Table 3.1 lists some of these foods.

Current Western type diets contain energy i.e., calorie-dense foods that are generally low in fibre <(10 g/d) and high in sugars and fats. There are numerous examples in today's world, such as, butter, oils, alcohol, crackers (a tremendous increase in availability and variety available since 2022) but also cakes and doughnuts, junk food, fast food, sweets, chocolate bars, granola, energy bars, ice cream, ketchup, energy drinks, soda, tonic water, sports drinks and lemonade etc.

Many of these foods and fluids are laden with calories i.e., calory dense and not only that but also packed with large amounts of unhealthy fats, sugar, salt and additives. They are highly calorific and often addictive, leading to a very significant proportion of total daily energy intake (EI). A dietary survey in the United States in 2012 [9] showed that adult men consume an astonishing average of 923 empty calories per day, and women, 624 empty calories per day.

This adds up to a staggering 37% and 31% of total daily energy intake (EI) for men and women respectively. It is effectively weight gain without any nutrients and therefore well worth getting to grips with on your weight loss journey. This, often together with a reduction in oils, is the key for some in reaching their weight target.

3.4 Macronutrient Caloric Value

Each of the 3 food macronutrient types, contains a different amount of energy calories per gram.

•	Carbohydrates	4 kcal/gram [10]
•	Protein at	4 kcal/gram [10]
•	Fat	9 kcal/gram [10]
•	Alcohol	7 kcal/gram [10]

The caloric value of fat is > double the other two macronutrients. (Glucose is 3.76 kcal/g)

Alcohol is 7kcal/g, though not considered a nutrient, nor adding any nutrients to the diet, is still relevant to weight management, as it does feature significantly on some peoples overall diets every week, for example, in the order of 20% or more of total energy dietary intake EI, and this increasing trend, to include alcohol in every-day life, shows little or no sign of diminishing.

High caloric dense, high caloric foods are often addictive, due to the so called 'sweet spot' (purposefully designed by the food manufacturers) of fat, sugar and salt, hence more attractive and addictive to the body's sensory reward centres.

Often a sugar fix in the morning with a pastry or cereal followed by a fat and salt fix later in the day with fast food, finished off with another sugar fix of chocolate in the evening.

However, these foods are overall less satisfying and satiating and often lead to additional cravings for other food soon after each meal.

On the other hand, low calorie, less calorie dense, high nutrient dense foods, such as vegetables and wholegrains, keep you feeling fuller for much longer and will promote better satiation, much better weight maintenance and if required, weight loss (Table 3.1).

High calorie dense foods such as processed foods or take away foods, are often far smaller in volume than the equivalent number of calories derived from a more nutritious, less dense food such as broccoli.

Compare 300 kcal of broccoli with 300 kcal of Indian take-away or Chinese take-away food! The processed, calorie dense, low nutritious dense food, will be much lower in volume and lead to much lower satiety.

Understanding calorie density is the key to managing weight whether you need to lose weight, maintain weight or add lean mass. You do not need to practise portion control when you eat foods that are high in fibre, high in nutrients, low in fat and calories, far healthier and naturally satiating.

CHAPTER

4

Nutrition Quantity

Each person should eat as much as they feel they need for satiety. Satiety level can change gradually depending on the nutritional quality of the food.

For example, healthy complex starch carbohydrates, such as baked potatoes (minus the butter and cheese), sweet potatoes or wholegrain rice, with vegetables and legumes, or oatmeal (steel cut porridge oats) and fruit is an excellent way to supply wholesome, fibre rich satiety, whilst fuelling energy requirements that feed the body with a multitude of nutrients, rather than a meal disproportionately and heavily laden with fat, oil and protein.

4.1 Portion Sizes

It is not mandatory, nor necessary to empty the plate each time, or all the contents on the table. Portion sizes of unhealthy foods have increased very significantly over the past 20 years. Standard portion size in foods and drinks has increased very markedly in the past two decades. Table 4.1 lists food portion sizes from 2002 to 2022.

Table 4.1   Food Increase in Portion Sizes from 2002 and to 2022

Food Type/meal	Kcal/meal 2002	Kcal/meal 2022	Difference Kcal	Increase %
Cheeseburger	333	590	257	77%
Pizza	500	850	350	70%
Meatballs and Spaghetti	500	1025	525	105%
Brownie Cake/Muffin	210	500	290	138%
Biscuit	55	275	220	400%
Coffee	45	350	305	678%
Bagel	140	350	210	150%
Popcorn	270	630	360	133%
Cola Drink	85	250	165	194%
Chips/French Fries	210	610	400	190%
Ham salad sandwich	300	1000	700	233%
Chocolate Bar (Normal to King Size	250	365	115	46%
Total	2898	6795	3897	
Average % Increase				134%

This average increase of more than double i.e. 134% increase in the quantity of calories is a further clear example demonstrating why increased obesity and weight gain has occurred. A coffee is not just a coffee anymore, it often comes with extra sugar, milk, syrups etc and of course the inevitable muffin. Sandwiches have increased significantly with much more processed bread and meat disguised with a little green salad.

Chocolate bars have gone large as have the meal deals etc. People now have second or third servings at carvery restaurants where the plate is more than double the size in previous decades. Cinema trips are now accompanied by gigantic volumes of popcorn and sugar drinks.
In the United States this change has reached a staggering 5 fold increase compared with 1970.

The emphasis on the now common terms such as; 'super-size; 'go large' 'meal deal' 'value for money pack', 'free carvery second helpings' and 'all you can eat' has contributed significantly.

People need to eat smaller portions. Aim to fill half or three quarters of your plate with vegetables. Leave some space, your entire digestive system will and waist line appreciate the reduced onslaught!

4.2 Meal Times

Meals with family at the table are advised where possible, as this encourages slower eating, leading to improved digestion.

Each person should pause and think as they are eating, rather than continuously loading and reloading the fork whilst, for example, watching television.

If it is practical and suits the individual, then a change to eating twice a day on a few days per week, for example, a large late breakfast and dinner at 18:00, or lunch and dinner only, or breakfast and lunch only with a snack at dinner, can be tried.

Your body is driven by your mind and convention. Eat when you are nicely hungry and forget convention where practical to do so.

4.3 Food Sufficiency and Satiety

It is imperative to ensure people:-

a) Consume sufficient, healthy, nutritious food (Table 3.1 and Table 9.1), taking time to eat
b) Are engaged in sufficient activity (NEAT) plus any additional exercise (EAT) for all that can
c) Set an achievable, steady, sensible weight loss target
d) Aim to achieve that the sum of (a + b + c) are less than TEE

For example, as discussed in Chapter 2, if TEE (BMR + TEF +NEAT +EAT) is 2,000 kcal/d, excess body fat will be reduced by eating nutritiously and maintaining total energy intake less that 1,800 kcal/d. (assumes a 10% value for TEF).

Therefore, the best strategy to lose weight is to achieve and maintain a daily metabolism close to but less than TEE, then gradually decrease the gap between intake EI and expenditure TEE. For example, during weight loss intake, EI may be at 70-85% of TEE and then following achievement of the weight loss target, maintain intake at 95-100% TEE. In any weight loss plan, it is essential to combine maintain healthy satiating, metabolic driving nutrition, whilst reducing intake below TEE.

The most important aspect of satiation is bulk. Therefore, adding good wholesome quantities of salads and greens, which have relatively high quantities of water and fibre, vitamins minerals with the added benefit low caloric density, to grains and legumes will help considerably with optimal satiation weight management and overall health. In this world of plenty - at least in the consumer driven west - this strategy could essentially be seen as a way to fool the body into giving up some of its energy stores (oxidising fat), by stimulating the metabolic engine (with healthy food and exercise) to drive substrate (β)-oxidation of the adipose (fat) tissue.

4.4   Over Eating Leads to More Fat Storage

It takes the human stomach around 20 minutes to determine that it no longer requires food. However, if meal consumption takes 15 minutes, this can lead to over eating. Taking time when eating meals helps this process. Foods higher in vegetable content increases satiety. The body has evolved naturally through time to store excess dietary high quantum energy as fat (9 kcal/g) for the rainy days, around the corner, when food is in short supply or no longer available. Stored adipose tissue fat, direct from dietary fat, is physiologically laid down very easily in the body during lipogenesis, costing only c2% to 3% of ingested fat energy to metabolise and store. Carbohydrate however, is not generally, to any significant extent, turned into fat via De Novo Lipogenesis (DNL) (Section 14.40), as this costs a considerable amount of energy at around 25-30% of ingested carbohydrate energy to metabolise in the liver but instead, carbohydrate is used to stoke the metabolic flame, and maintain energy levels for every day processes including burning fat. [36, 38, 39]

Even high overloading on carbohydrates (>+500 g/d) only marginally increases DNL and it is thought that the function of DNL may be to preserve some necessary levels of fat in case of low dietary fat availability, though clearly not a situation that arises often in the west in the 21st Century. The vast majority, (>96%) of carbohydrates will either be stored as glycogen, oxidised for cellular energy or burnt as heat. However, caloric excess >TEE overloading on all 3 (actually 4 including alcohol at 7 kcal/gram) macronutrients, including carbohydrates or fats or protein, means that any fat burning to reduce existing stores adipose tissue, will be considerably diminished, whilst dietary fat continues to be laid down as adipose tissue. [37] Energy sufficiency in the fed state will signal to the body to reduce fat burning.

Whilst there are sufficient glycogen stores in the liver and muscles and sufficient blood glucose levels for the tissues relying on glucose (3.88-5.82 mmol/L), the body will not need to burn additional levels of stored fat for fuel than it normally does in its day to day role to supply energy to the heart, skeletal muscle and other tissues, for example, in order to preserve glucose for the brain and other tissues. By reducing the total intake EI to <TEE, especially dietary fat, whilst maintaining sufficient and sufficient protein of 0.8 g/kg/d to 1.2 g/kg/d (Section 11.2 and 11.3) and carbohydrates, for the cellular metabolic mitochondria engines, additional fat burning of stored adipose tissue, will be promoted as an efficient dual fuel burning system with carbohydrate. Mitochondria are the vital cell organelles located within every cell (apart from red blood cells) that operate as power houses and enable animals and plants to obtain energy from food. There are approximately 100,000 trillion mitochondria in the human body and 2.4 billion are synthsised each second.

CHAPTER

5

Nutrition Timing

Eat when you feel hungry, not because it is time to eat. Convention across many societies promotes eating 3 meals a day at set times each day, for example, breakfast (07:00-09:00), Lunch (12:00-14:00) and dinner (17:00-19:00). For many, this leads to an anticipation or expectation habit, 'It's one o'clock and time for lunch' and 'I must be hungry as I have not eaten since breakfast' etc. Hence, many people eat because they think they should eat, due to time or social norms and habit and not because they are actually hungry.

5.1 Time Restricted Eating (TRE)

There have been numerous studies conducted over the past 10 years (mostly on rodents, a few with humans) that indicate several health and even weight loss benefits in following various types of fasting. A great deal of this world leading research has been led by Dr. Prashant Regmi, 'Time-Restricted Eating: Benefits, Mechanisms, and Challenges in Translation by Prashant Regmi, Leonie K. Heilbronn. [11] The TRE time frames include; (a) long term fasting; 3 - 5 or even up to 14 days fasting; (b) day restricted fasting, i.e., 5 days eating normally, then 2 days eating less; (c) short term intermittent fasting; up to 2 days or 48 hours fasting and finally, (d) Time Restricted Eating (TRE) i.e., different maximum periods of eating during each 24 hour day, with variations in the gap between the last meal on day one and the first meal on day two (Figure 5.1). Time-restricted eating (TRE) is an additional dietary approach that consolidates all calorie intake, EI, to a total daily eating window, from 6 to 10 hour periods, during the active phase of the day, without necessarily altering diet quality and or quantity.

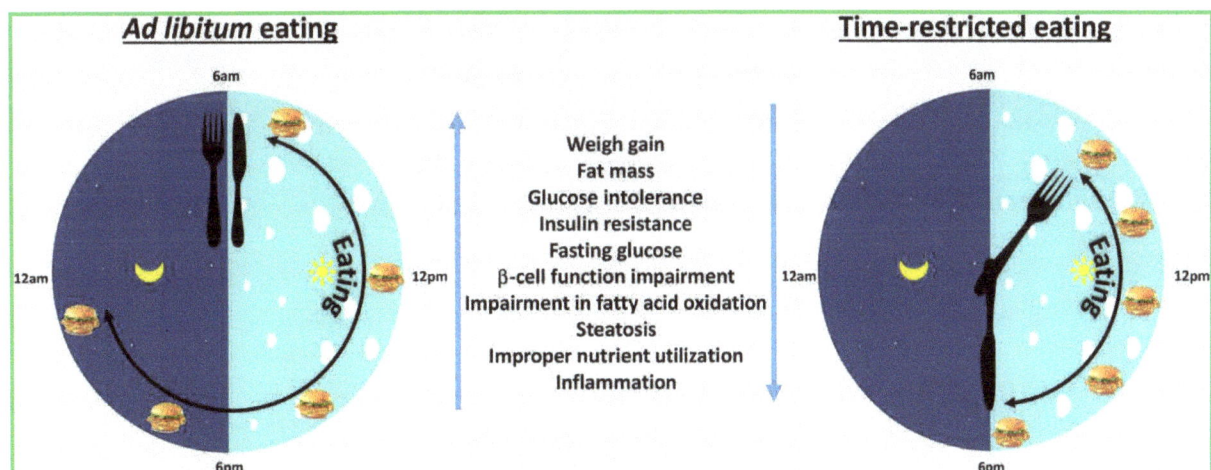

Figure 5.1 Regmi P, Heilbronn LK. Time-Restricted Eating: Benefits, Mechanisms, and Challenges in Translation. iScience. 2020 Jun 26;23(6):101161. doi: 10.1016/j.isci.2020.101161. Epub 2020 May 15. PMID: 32480126; PMCID: PMC7262456. [11]

Fasting has been practised for centuries across many world cultures, for religious and other reasons. Natural fasting, during sleep and the period between dinner and breakfast (break the fast, TRE), has been shown to deliver several metabolic health benefits providing the privation has not been too long. TRE also has the advantage of better adherence and being easier to follow. For example, some find the short term fasting diet, not eating for 48 hours very difficult, then having to be careful to correctly gradually restore eating, for example on certain light broths.

TRE however, can be more practical and flexible, adapted to suit each individuals' lifestyle and as stated above, TRE is largely comprised of the natural break in time of sleep every overnight.

Some consider that any weight loss achieved by fasting or TRE is simply because the individuals adopting this pattern of eating are more conscious of weight management and that these individuals are therefore more likely to be mindful of overall caloric deficit management and may well eat less i.e., a lower EI. There is certainly some truth in that logic and assumption.

Others think that fasting and TRE weight management and other health benefits are due to the longer break between eating that enables the body to 'heal' physiologically through cellular autophagy. It is also thought that some of these benefits of TRE may simply be due to overall mediated calorie restriction (EI<TEE) compared with normal habits and therefore yielding weight loss.

However, the recent studies on rodents fed a high fat-high sugar diet showed that TRE improved glucose tolerance, without any weight loss and this study was supported by a study on humans that showed a benefit on metabolism independent of change in body weight. No doubt further evidence either way will arise as further research, with larger human subject numbers, is conducted. Dr Jason Fung, the renowned nephrologist has also written extensively on fasting. The author considers that a daily healing, maintenance period is healthy for the body.

5.2 Combining TRE Improved Sleep and Exercise

In addition to the benefit of reducing EI compared with TEE, weight loss will be enhanced to utilise the combination of:

(a) The health promoting benefits of TRE, extending the period of the post absorptive (post prandial) state and

(b) Introducing sufficient moderate and regular EAT exercise, (Section 6.5) and

Improved sleep patterns, enabling the body to recover further and maintain processes more efficiently and is likely to promote further stored fat mobilisation and lipolysis (fat mobilisation and break down) for energy, along with glucose and this may be summarised and represented as:

$$(EI \ <TEE + TRE + >EAT + Improved\ Sleep) \ = \ Increased\ Fat\ Burning\ \&\ Weight\ Loss$$

5.3 Time Restricted Eating Options

There are several options or forms of TRE listed as follows usually 7 days per week if practicable:

18:6	18 hour period not eating :	6 hour period during which eating occurs
16:8	16 hour period not eating :	8 hour period during which eating occurs
14:10	14 hour period not eating :	10 hour period during which eating occurs
12:12	12 hour period not eating :	12 hour period during which eating occurs

As an example, the 14:10 TRE option above would mean finishing dinner on day 1 at around 19:00 then starting the first meal of the next day, i.e., day 2, at around 09:00. This gives a 14 hour fasting window and a 10 hour eating window.

Water and tea and coffee consumed between 19:00 and 09:00 are considered not to break the fast, providing there is no milk or sugar or any other additives in the beverages.

Aim to consume lighter meals in the evenings and not to drink beverages too close (no later than 4 hours before bed) to bed time.

Each individual will be comfortable with different fasting and eating windows.

Some people will prefer an earlier breakfast, as this is what they enjoy, or fits in best within the family regime.

Others may prefer to go for a later (homemade) brunch plus a salad, in the early afternoon and then a light dinner in early evening. For example, the author has found that the 16:8 TRE has been suitable for him personally and has now fully adopted this for the past 18 months, together with moderate daily walking (EAT) prior to breakfast (usually steel cut porridge oats and fruit). This gives a 16 hour fasting window and an 8 hour eating window.

The daily walk can provide a metabolism boost, promote increased fatty acid oxidation and lead to a much anticipated and most enjoyable breakfast. Once TRE is successfully introduced, each individual may aim for increasing the fasting period from 12:12, eventually up to 16:8, but again stressing to reach what is comfortable, practical within each person's lifestyle and social network.

This TRE fasting option is the fasting option that will be most likely to achieve sustained adherence for each person as for example, compared with multi day 100% fasting. Those looking to reduce weight and change lifestyle may therefore, wish to introduce TRE step by step, to establish how this can work for them. If a person is not comfortable with adopting TRE and for example, eating at different times than they are comfortable with, or delaying breakfast, then do not change. Everyone is different and individuals should only adopt change if it is suitable, practical and enjoyable and sustainable for them.

An 18:6 TRE is not recommended because it is likely to be practically difficult, though certainly not impossible, to meet sufficient quantitative, caloric TEE requirement and moreover provide sufficient, healthy, wholly nutritious macronutrient and micro nutrient quality. It is imperative that TRE, or any other form of fasting, is not simply used to reduce calories i.e., EI total daily intake. Some people have adopted an OMAD diet, One Meal a Day. This may be very difficult to achieve, requiring most daily calories in one sitting (assuming additional fruit snacks etc) and maintained and above all how would the individual ensure all requisite and sufficient nutrients are obtained in just one sitting. This can possibly lead to potential under nutrition, which is not recommended for anyone unless specifically necessary and under medical supervision (Section 2.9). The reader should still aim to eat a nutritious, healthy and caloric sufficient, wholesome diet packed with all the benefits this brings and with a marginal reduction in EI compared with TEE, towards the weight loss target as discussed in Section 2.7. Perhaps a 2MAD (2 meals a day) is more appropriate.

TRE can be used as one of the components of the overall weight loss goal.

5.4 Physiological and Cellular Benefits of TRE

These benefits include cellular autophagy (cell repair and maintenance) and mitophagy (mitochondrial repair and maintenance) i.e., the body maintaining and renewing structures during the post prandial break from eating to fasting. The majority of TRE studies in people of both normal weight and overweight, report modest reductions in body weight and fat mass and waist circumference. TRE also reduces plasma triacylglycerols (fats) and inflammatory markers.

Correctly implemented TRE is likely to reduce bodyweight, improves glucose tolerance, protects from hepatosteatosis (fatty liver disease), increases metabolic flexibility, reduces atherogenic (plaque formation in arteries) lipids and blood pressure and improves gut function and cardiometabolic health.

At the cellular and molecular level, it is postulated [11] that TRE increases the amplitude of expression of 5 adenosine mono phosphate activated protein kinase (AMPK) and mechanistic target of rapamycin (mTOR) [5] and nicotinamide phosphoribosyl transferase (NAMPT) in the liver. TRE also increases the amplitude of ribosomal protein phosphor-S6 in skeletal muscle during the active phase (eating phase).

TRE reduces the amplitude of pyruvate carboxylase and glucose 6-phosphotase and increases the glucokinase during the active phase, potentially underpinning reductions in hepatic (liver) glucose production and increases glucose utilisation.

TRE also reduces the amplitude of genes fatty acid synthase, stearoyl CoA desaturase and fatty acid elongase during the active phase and increases the amplitude of hepatic triglyceride lipase during the inactive phase (fasting phase), which is associated with reduced lipid (fat) storage and increased triacylglycerol hydrolysis and (β)-oxidation, or fat burning.

CHAPTER

The Benefits of Exercise

6

As discussed in Section 1.14, besides adding relatively minor contribution to weight loss (kg lost per kilometre and the sheer effort) for the vast majority of people, there are many more benefits derived from moderate exercise aside from weight loss, that are key to a healthy lifestyle as listed in 8.1 below. Exercise i.e., exercise activity thermogenesis (EAT), does contribute a little towards weight loss but only so much because the vast number of calories burnt are due to the basal metabolic rate (BMR), at 70% of total Energy Burnt (TEE), which is very similar to the total intake (EI), from food and general exercise, through non exercise activity thermogenesis (NEAT). Therefore, for the majority, EAT as an avenue for calorie burning, is relatively small by comparison. For those people wishing to introduce new or additional exercise into their lifestyle, or for that matter, anyone increasing EAT exercise activity levels, it is always wise to discuss what you wish to do with your doctor and check any concerns you, or the doctor may have, before setting out on the new regime. This will provide you with further confidence on what is appropriate for you and what level of exercise will suit you best.

6.1 Benefits of Regular Exercise and Resistance Training

1. Improved cardiovascular circulation
2. Increased life expectancy
3. Reduction of diabetes Type 2 risk
4. Improved mood
5. Strengthens muscles and muscle endurance
6. Joint support
7. Increases bone density
8. Improved balance
9. Improved sleep
10. Stimulates hormone regeneration
11. Controls stress hormones
12. Reduces hypertension
13. Increased coordination
14. Stabilises blood sugar
15. Decreased Risk of developing Alzheimer's
16. Increased memory retention
17. Reduces risk of falls
18. Improved cholesterol levels
19. Improved breathing
20. Increased fitness and stamina
21. Can help maintain weight loss
22. Improved Cellular & Metabolic Health

6.2   How to Introduce Moderate Exercise

In order to function healthily, it is not necessary to take up hard, extended or strenuous exercise, for example, in the gym doing heavy weights, or running regularly, or running marathons, or swimming hundreds of lengths in the pool each week, or 'iron man' or '3 peaks or countless kilometres on the bike.

If the individual already enjoys a high level of fitness, or regular trips to the gym then this should of course continue for so many fitness, health and lifestyle benefits.

If, however, the individual is not currently, or has not previously engaged in much (EAT) exercise but wishes to introduce some exercise into their lifestyle, or as part of the weight loss target plan and if appropriate and agreed with the doctor, then the good news is that this can be done without excessive work, discomfort, pain or excessive perspiration!

As well as increasing general activity around the house and garden and at work (NEAT), the introduction of a new daily regime of moderate walking exercise is strongly recommended. If you wish to run eventually, but are currently overweight, it may be best for you and perfectly fine for you to walk for a good few weeks or months before you run.

For those people not used to exercise, this should be introduced gradually, at gentle walking pace, say 15 mins per day initially, then steadily increasing to 40 minutes each day, perhaps adding a gentle hill or two.

Once fitness improves and the individual is able to cope, this can be increased to say, an hour a couple of times per week, or an even longer walk once or twice per week, perhaps combined with a friend or social groups whichever suits. Increasing to 2 walks per day sometimes can be considered for some.

Smart phones or smart watches can be used to record steps/distance/kcal burnt etc, these devices are useful to give feedback and encourage regular EAT and NEAT whilst doing house and garden chores. It is quite surprising and heartening, just how many NEAT steps (kcal) can be burnt doing every day chores.

Safely walking to the shops, a mile or more from home can be introduced, perhaps taking a light back pack to do a little shopping. This moderate form of EAT will work wonders for lifestyle, self-esteem and fitness.

If breakfast is delayed until after the first daily walk (Chapter 7), both carbohydrates, from glycogen stores (energy burning substrate proportion of 30% - 50%) and also adipose tissue fat burning (70% - 50%) will be promoted, which will support the weight loss goals, encouraging the body to burn more fat than carbohydrate. There are several studies that show how periodical increased rate of walking pace (one minute) followed by decreased walking pace (for the next 5 minutes) and repeat, improves fitness levels and also, may therefore increase fat burning due to increased heart rate.

6.3 Exercise and Fat Burning

The bioenergetic cost of burning fat is discussed in Section 2.8. We have seen that using exercise for the purpose of losing weight and burning fat is a challenging way to deliver weight loss. In the post absorptive or fasted state, the higher percentage and proportion of energy burned as fat generally occurs during rest, low intensity and moderate exercise periods because the body relies more on burning stored fat compared with burning stored glycogen (carbohydrate) as its primary fuel source when working at lower intensity, compared with using more carbohydrate at higher intensity. The body naturally preserves precious glycogen i.e., stored glucose for other more important tissues, whereas fats are utilised by other organs such as the heart, skeletal muscle, the liver, adipose tissue and kidney cortex.

Although fat is oxidised more, as a proportion, compared with glucose at rest and low intensity activity, this does not mean that more overall fat stores, or a higher total quantity of fat is burned during low intensity activity compared with higher intensity or endurance training. Table 6.1 shows the relationship between fat burning at moderate and higher intensity cardiovascular exercise for a female weighing 58.96 kg exercising for 30 minutes. If the body changes from moderate intensity exercise and then is suddenly required to 'step on the gas', due to a fright or flight scenario (sprint), the body will switch to burning glucose (CHO), released from glycogen stores, as the main fuel substrate source. This is because glucose is mobilized far more efficiently and rapidly than fatty acids (from fat adipose tissue). At lower intensity exercise, a higher proportion of the total energy calories is burnt at fat, matching CHO at 50% each. However, at higher intensity exercise, even though the total energy proportion of fat burnt drops to around 40% of total energy burnt, with CHO at 60%, the total amount of fat burned overall is 12% higher at 82 kcal, against 73 kcal. Still, the fact that 73 kcal of fat was burned overall and that this is almost 90% compared with higher intensity exercise, shows the value and importance of lower intensity exercise that can be introduced and is probably within reach of and can be followed and sustained by more individuals overall.

Table 6.1 Calories burned by a 59 Kg female during exercise at different MHR (maximum heart rates)

Kcal Burnt	Low Intensity (60% - 65% MHR)	High Intensity (80% - 85% MHR)
Total kcal burnt per minute	4.86	6.86
Fat kcal burnt per minute	2.43	2.7
Total kcal burnt in 30 minutes	146	206
Total fat kcal burnt in 30 minutes	73	82
Total CHO kcal burnt in 30 minutes	73	124
Percentage fat of total kcal burnt	50.00%	39.85%
Percentage CHO of total kcal burnt	50.00%	60.15%

Thus, although a significantly higher proportion of fat is burned at lower intensity exercise, more total kilocalories and more fat kilocalories are actually burned when working out a higher intensity. The extra exercise intensity burns 41% more total calories at 206 kcal compared with 146 kcal burned at lower intensity. Low intensity is fine for most people and it is also likely that going slower may means the person will be able to exercise longer, so burning more kilocalories and fat in that way. Hence, when safe and appropriate for the individual, longer duration walks are strongly recommended.

6.4   VO₂ max and Maximum Heart Rate (MHR)

VO_2 max is the maximal rate of oxygen uptake and represents the highest volume of oxygen consumed, during peak physical exertion, per minute per kilogram of body weight. Additional fat burning will be achieved if the individual, (safely, as agreed with the doctor), gradually, manages to increase walking or relatively slow running speed, to reach anywhere from 40% to 50% VO_2 max for 15 to 20 minutes. Then gradually extending this to longer periods and, for the fitter people, that are more used to exercise, up to 72% of VO_2 max for 20-30 minutes. Even 40% VO_2 is likely to improve health, if not strictly necessarily improving aerobic fitness. Of course, most people are not able to measure VO_2 max but are able to assess heart rate. A smart device may help in determining a person's VO_2 max but if this is not available, it is ok to measure resting heart rate (RHR) i.e., pulse rate over 60 seconds at rest, then measure again at the beginning of the walk or run. Figure 6.1 shows the crossover concept relationship between carbohydrate and fatty acid burning at varying activity and exercise intensity

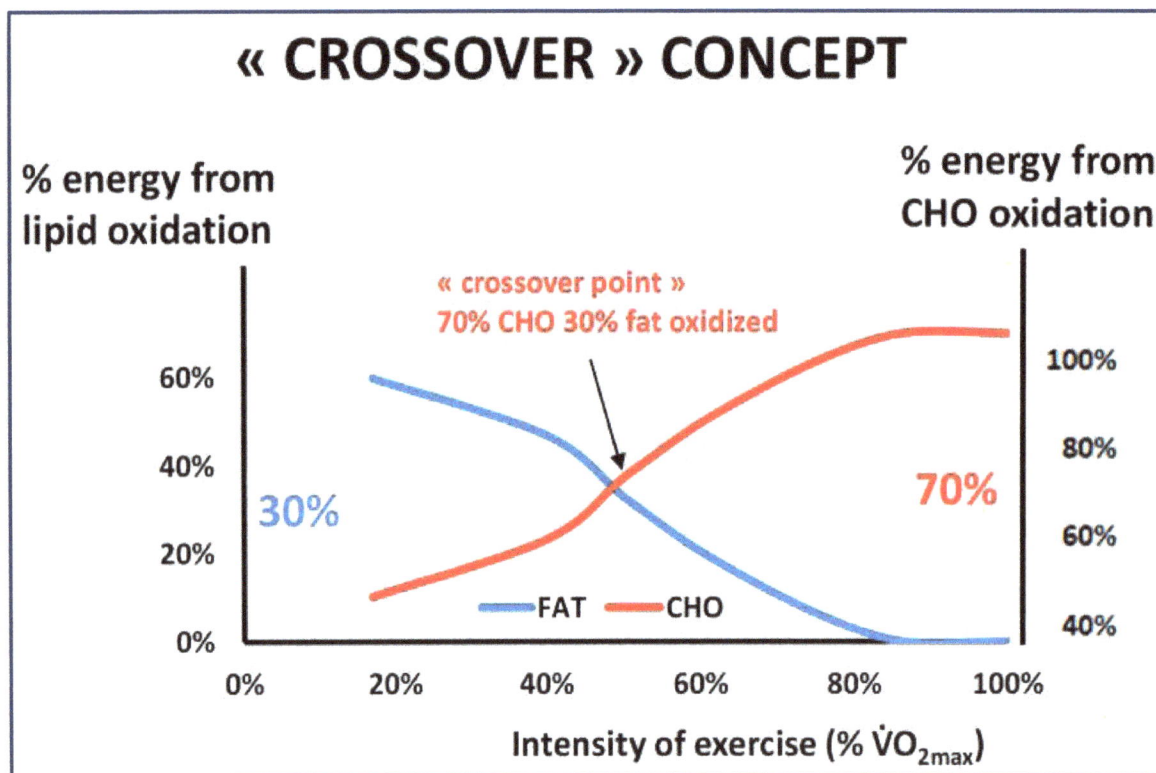

Figure 6.1 Brooks and Mercer's Crossover concept' redrawn from the personal database of more than 5000 exercise calorimetries (see text). This hypothesis assumes that there is a major shift in the balance of substrates used for oxidation during exercise grossly around 50% of the maximal aerobic capacity, when carbohydrates represent more than 70% of the sources of energy for the exercising body. Oxidative use of fat culminates below this level, and close to it or slightly above it, blood lactate increases and the ventilatory threshold occurs. Note that the coordinates for % of fat oxidation and % of CHO oxidation are not symmetric, in order to better visualize the crossover. In a series of more than 5000 exercise calorimetry's, find that the crossover point is, on average, at 55.1% of VO2 max but exhibits a wide variability among individuals. Open access. Brun JF, Myzia J, Varlet-Marie E, Raynaud de Mauverger E, Mercier J. Beyond the Calorie Paradigm: Taking into Account in Practice the Balance of Fat and Carbohydrate Oxidation during Exercise? Nutrients. 2022 Apr 12;14(8):1605. doi: 10.3390/nu14081605. PMID: 35458167; PMCID: PMC9027421.

However, VO_2 max differs from maximum heart rate (HR max) due to differences in heart anatomy, age, gender, exercise type and training level.

There is a linear relationship between heart rate and VO_2 max.

From work by Professor David Swain (Old Dominion University) and the US based research team 1994, the actual relationship between VO_2 max and HR max is shown in Table 6.2 and can be summarised as follows [12] :

Regression Equation: % HR max   = (0.64 x % VO_2 max + 37)

(1) For Some fat loss and overall health benefits is 40% VO_2 max, then:

% HR max   = (0.64 x 40) + 37
% HR max   **= 63%**

(2) For Maximum fat burning during higher intensity exercise at around 70% VO_2 max, then:

% HR max   = (0.64 x 70) + 37
% HR max   = 82%

This most probably applies to very fit individuals who are less likely to be overweight. However, for most people who do not wish to achieve maximum fat burning or higher fitness levels, an increase to 50% VO_2 max will still pay dividends and most probably be more achievable and sustainable. The third example is therefore provided.

(3) For More fat burning and general overall health benefits is 50% VO_2 max:

% HR max   = (0.64 x 50) + 37
% HR max   **= 70%**

Table 6.2   Relationship between VO_2 max % HR max and speed and type of exercise

% VO_2 max	% HR max	Exercise Speed and Type
40%	63%	Walking
50%	70%	Very Slow Running
60%	75%	Slow Running
70%	82%	Steady Running
80%	88%	Half Marathon Pace
90%	95%	10k Speed
95%	98%	5k Speed
100%	100%	3k Speed
110%	100%	1500 m Speed

Average, normal resting heart rate (RHR) is usually in the order of 60 to 80 beats per minute (BPM). To determine an individual's HR max, use the Miller formula of [13] :

$$VO_2 \max = 217 - (0.85 \times age)$$

As shown in the following example for someone who is 40 years old.

$$HR \max = 217 - (0.85 \times 40)$$
$$HR \max = \mathbf{183\ BPM}$$

Then, to establish the relationship between VO_2 max and HR max for this person and individuals aged 50 and 60, substituting in the Swain equation above:

(1) For Some fat loss and overall health benefits at 40% VO_2 max

$$= HR \max \times 63\%$$
$$= 183 \qquad \times 63\%$$

Heart rate target Aged 40:	= **115 BPM** (for 20-30 minutes)
Aged 50:	= **110 BPM** (for 20-30 minutes)
Aged 60:	= **105 BPM** (for 20-30 minutes)

(2) For More fat burning and general overall health benefits at 50% VO_2 max

$$= HR \max \times 70\%$$
$$= 183 \qquad \times 70\%$$

Heart rate target Aged 40:	= **128 BPM** (for 20-30 minutes)
Aged 50:	= **122 BPM** (for 20-30 minutes)
Aged 60:	= **116 BPM** (for 20-30 minutes)

Thus (providing this is safe to do so and agreed with the doctor) this 40 year old person could aim to reach a heart rate of 115 to 128 BPM for a few minutes during the daily walk. And for the 50 and 60 year old people, 110 BPM and 105 BMP then up to 122 BPM and 116 BPM respectively. This is a useful guide and used more by athletes and those involved with fitness training. It is, however, advised that most people should not become obsessed with these measurements, nor attempt to reach elevated heart rates for the general purpose of fat burning, especially if the individual is not used to exercise or is just starting out engaging in a little more exercise.

One knows if one's heart rate is increased. It is more important that the person just enjoys daily EAT exercise, rather than obsessing about fat burning. Despite all the benefits of exercise, including some level of fat burning, it is essential to keep in mind that the quantity and quality of nutrients and beverages that you ingest is by far the most influential factor on the quantum of your weight, fat stores, management of your weight and weight loss results.

6.5   Resistance and Strength Training

Incorporating bodyweight-strength resistance training exercises and Yoga, stretching, Aerobics or Pilates into the daily routine at home is a really excellent good way to build and maintain muscle.

In addition to cardiovascular exercise, for example, from your daily walk, strength training actually boosts the body's metabolism and energy levels, as well as improving mood and sleep quality. This boost helps contribute to fat burning and weight loss.

There is no requirement to invest in equipment and it takes just 15 or 20 minutes per day for 5 days per week. The exercises can be done on the carpet or a yoga mat anywhere in your house or garage.

The following 15 exercises are recommended and details of each can be found in literature or the internet (U-tube is an excellent source of a wide variety of these).

Several of these (your favourites) can be selected and different ones can be used each day for working different parts of the body. Many can be done without any equipment and in 15 minute time slots.

1.	Walking squat
2.	Squat with knee lift
3.	Side lunge with a twist
4.	Half lunge walking lunge
5.	Standing leg lift to the back
6.	Push-ups on the knees
7.	Triceps dips
8.	Plank to down dog
9.	Lat pulldowns
10.	Arm circles
11.	Crunch with knees up
12.	Criss cross and reach with knees up
13.	Single leg lower and lift
14.	Single leg straight toe touch
15.	Modified side plank elbow to knee

CHAPTER

The Importance of Sleep

7

Good quality sleep is strongly associated with more successful weight loss and weight maintenance.

7.1 Sleep and Weight Loss

Sleep is an important component of weight loss, influencing hunger and satiety hormones ghrelin and leptin respectively, plus influencing metabolism, energy production and also motivation for physical activity (NEAT and EAT).

Good quality sleep, around 7-8 hours uninterrupted sleep per night, brings so much physiological repair, muscle repair, cell maintenance (especially the energy burning mitochondria body cell power houses) and restores and improves so much more, including emotional and mental health.

Deep sleep is associated with activation of the immune system.

Short sleep duration increases ghrelin levels and decreases leptin levels, increasing hunger throughout the day and leaving individuals less likely to feel satiated and hence can lead to weight increase.

Furthermore, poor sleep patterns tend to lead to increases in snacking on low quality, high processed fat, sugar based foods and sweets, as the body and mind crave a reward and something to look forward to, again leading to weight gain.

Studies have shown that those sleeping less than 7 hours per night are more likely to be overweight, leading to additional fat deposits around the torso, possibly due to the effects of the fat storage promoting hormone ghrelin, interfering with the body's normal fat release and storage mechanisms.

Low quality sleep also negatively impacts on long term weight loss maintenance. In a fatigued state, the natural cyclical circadian rhythm can be unbalanced. This can then unduly influence hormonal regulation of energy processing, particularly on the regulation of glucose and cholesterol levels.

Each hour of sleep deprivation is associated with a 14% higher prevalence of Type 2 diabetes and those with less than five to six hours sleep per night have higher blood glucose and 62% prevalence of Type 2 diabetes.

Section 7.2 steps 1) to 16) provide recommendations to help improve sleep quality and quantity.

7.2 How to Improve Sleep Patterns

1. Establish regular times for going to bed, going to sleep and getting up

2. Be active throughout the day at work and home more NEAT and EAT

3. Aim to perform exercise (EAT) earlier in the day if practical

4. Try not to exercise too late in the day

5. Eat more, earlier in the day and less, later in the day

6. Establish regular eating patterns

7. Go outside each day to reset the brain and body day and night clocks

8. Avoid eating and drinking late at night or during the night

9. Avoid caffeine, tea and coffee in the evening

10. Eat and drink the last mouthful at least 4 hours before going to bed

11. Don't drink water after 3 hours before bed, especially those who regularly wake to urinate

12. Avoid gadgets (phone, tablet, iPad, television etc) in bed, read a book instead

13. Adjust the bedroom so it is dark and quiet with a temperature around 20 deg. C if possible

14. For those with much on their mind, make a to-do list before bed to return to the next day

15. Aim to generate the same time of retiring, sleeping, waking and rising pattern each day

16. Those who work shifts, should try to maintain alternative patterns and gaps between eating

CHAPTER

The Healthy & Sustainable Weight Loss Plan 8

8.1 Planning and Beginning Your Journey

This Weight Loss Plan is now summarised into 20 key steps (Sections 8.2 to 8.21) using the knowledge, information and tools discussed in this book and observing the Nutrition and Metabolism for Weight Loss Guidelines (Section 8.22). Every individual has their own unique set of circumstances, diet and eating patterns, lifestyle and different goals they wish to achieve.

You will soon know what fits best for you and what makes you feel comfortable in adopting new approaches to diet and lifestyle going forwards in the long term. It is far more likely to yield benefits and success towards your goal in the longer term, if you feel it suits you and you are enjoying your food.

Once you have made the decision to change and as necessary and depending on your overall health, consulted and agreed the changes with your doctor, then it is recommended that adjustments towards your long term goal are introduced as short term goals, step by step, steadily and gradually, as the new lifestyle and eating pattern, SMART goals become the natural way of life, where the individual feels comfortable and enjoying the benefits.

It is important to remember that if caloric reduction is too severe, it will cause a decrease in muscle mass, slow basal metabolism and will more likely to lead to weight increase down the line.

Be comfortable in your efforts so that you will be able to sustain and enjoy your food. If you make very few, inadequate changes, you may not see the benefits you seek.

Further changes can be made if this suits until the long term goal is reached. If any elements of the new eating plan or exercise regime do not suit, for whatever reason, then use other means, or do not adopt these specific elements. If you are eating a well-balanced, wholesome variety of healthy food, you should not feel too hungry, or need to go without food unnecessarily, nor feel too tired.

Weight loss and increased moderate exercise should be accompanied by improved energy self-esteem and bring many other health and lifestyle benefits, so it is worth endeavouring and continuing on this journey, but do it in a relaxed, enjoyable way and do not become obsessive.

Do not try to drastically change things, all or nothing, all at once. Make step by step, small changes and improvements as the weeks progress towards your short term weight loss objectives and each time make the changes to a level that suits you. Just ensure you get enough nutritious food to eat; remember drastically reducing overall caloric intake over a short period of weeks is not a good idea and may well be likely to fail. Successful, healthy and sustained weight loss is a balance between change and enjoyment of the new regime. It should be done gradually and the information and guidance in this book should guide you how best to achieve this.

Do not be concerned about occasional lapses or set-backs and get right back on track. This is not a 100% all or nothing change, it is a gradual and overall improvement change that will suit you for many years to come.

Remember you may have been eating this current way for many years or decades, so be realistic in the time you allow yourself to achieve these goals. Be kind and true to yourself, nourish, listen to and respect your body and your body will respond positively.

8.2   Assess and Record Your Pre-Change Diet

Before setting out on this journey of change, it is important that you firstly review what your current food and drink intake is.

This is what has led you to your current weight in recent years and also how the rest of your lifestyle fits around and has possibly influenced your current diet.
For the first week, make no changes and record everything you currently eat and drink and what time you eat and drink each day for 7 days. Include everything, including snacks and drinks including alcohol. Do not change anything this week. Carry on as normal but record everything.

Regarding the amounts of food and drink to record, if possible, make an assessment of the approximate quantity, for example a sandwich of x and y with z spread on the bread, with 2 slices of a certain type of bread, or a banana, or a plate full of x with y and z vegetables, or a half a litre of beer, or one chocolate bar, or 2 coffees with milk and a doughnut, or one cup of black tea with 2 sugars etc.

Record where the food was eaten, whether it was eaten outside the home in a café, restaurant or bar or at friends or family. Record how the food or beverage was prepared, for example, prepared at home from fresh whole ingredients or ready meal cooked in the microwave or oven, or purchased from or delivered from a fast food outlet, or prepared at the restaurant etc.

Remember, be 100% honest with yourself. This is for your benefit, so not including everything is only misleading yourself. By recording everything truthfully, you will significantly improve your chances of real change and a sustained success in the future.

Appendix I provides a diet proforma for this record; *7 day Pre-Change Diet Record.*

Assess how much exercise related activity (EAT) you did (if any) each day for the 7 days; that is anything outside the house (include any activity on your own gym/weights/exercise equipment if installed in the home should be included as EAT). Just keep this knowledge for your own assessment and there is no need to record unless you wish to make a separate log or note for your own use.

Over these 7 days, assess roughly what time you went to bed, what time you went to sleep and what time you got up to start the day. Assess the approximate number of hours sleep you had each night. Again, no need to record this unless you wish. Once you have completed your diet and lifestyle assessment, re-read Chapters 1 to 8 in this book.

Then assess what changes you think you can make and should make towards your ultimate diet, target weight, exercise and sleep goal. These changes you introduce (where you feel is necessary) should include;

- Nutrient quality
- Nutrition quantity
- Who you eat with
- Where you eat
- Where you source your food
- Timing when you eat the first and last meal/snack/drink of the day per 24 hour period
- Exercise and sleep patterns as you feel appropriate towards your weight loss goal

8.3   Record Your Baseline Weight

Record your current pre-commencement weight in kilograms as your baseline or start weight. Thereafter, it is recommended that you maintain your record over the first 6 months to gain a true trend of change, then perhaps continued monthly thereafter.

It is recommended that weighing on scales is done on a weekly basis, roughly every 7 days at around the same time of day and wearing the same minimal amount of clothes to record your weight. Daily weighing is not recommended as short term daily fluctuations can be misleading, for example, initial weight loss over the first week or two is principally due to loss of water as glycogen may be depleted and there will be other fluctuations in metabolism.

Appendix II provides a proforma to record this; *Weight Loss Record*

Add any comments regarding events or change or your feelings over the period.

8.4   Set Your Realistic Long Term Weight Loss Goal

Review and set your SMART objectives and goal.

I want to lose x kg and weigh y kg by z date. Refer to Section 2.7 and consider the likelihood of achieving the quantum of the change (total amount of weight to be lost) and the timing (over what period) to achieve this. For example, and depending on your current weight, you may feel that a sustained 20% reduction in current total daily current caloric intake (EI), over 6 months, is more realistic, rather than 40% reduction of EI over 3 months. Throughout your weight loss and dietary change plan, continue to record your weight.

8.5   Short Term Weight Loss Goals

Set each short term weight loss goal and change in nutrition and exercise as a component and part of the long term goal, for example, this week, a weight loss of *x* kg/week, 20 mins/d exercise and reduction of less healthy food types; a maximum of 2 portions of meat or a maximum of 2 portions or cheese, or only one meal of fried food and fries, or only eating out a maximum of one meal this week, whilst allowing yourself say, 2 'treats' per week. Once you are ready to start and if necessary, have discussed changes with your doctor, your new weight loss journey can begin. Start your new regime on the day when it suits you and you are ready to change.

8.6   Reassess and Record Your Post-Change Diet

After you have been on your new regime for around one month, make a second 7 day record to log what you are now eating and drinking, including amounts and timing etc as before. Again be 100% truthful. Appendix III provides a diet proforma for this record; *7 Day Post-Change Diet Record.* A comparison of the pre and post change diet will provide a very valuable assessment of the change you have made for future review. There will be no need to continue the diet record after day 30. By then, your new diet and regime should be well established and set the scene for the coming months and hopefully years, if you feel comfortable and you are enjoying the new way of life. By all means continue your log if you prefer, as it will always a useful record. As before, there is no need to record sleep or exercise in your new regime, as you are best placed to assess this change over time as part of your new lifestyle. This knowledge will, however, provide a very useful record for you, or your friends and family and your doctor.

8.7   Manage Nutrition Quantity

Whilst ensuring you eat sufficient amounts of wholesome and nutritious quality food (Chapter 9; Healthy Food Menu and Chapter 3; - Nutrition Quality), aim to reduce the portion sizes, enough to attain satiety but not so much as to completely overfill your stomach. Aim to try filling your plate to just half full, rather than three quarters or a full plate or dish. Eat until you feel satisfied. If you feel full do not be tempted to finish the plate just to avoid waste in this instance. These nutritious foods will improve your satiety level compared to any previous type of low quality, low nutritious, highly calorific foods you ate and should also discourage the need to overeat. The new more fibre rich foods will leave you feeling better and fuller for longer. Depending on which time restricted eating (TRE) regime you select (Chapter 5), stick to that for the first meal and see what comes for the rest of the day. Ultimately, you may find that breakfast and dinner or lunch and dinner i.e., 2 meals per day is sufficient. If you do require or prefer 3 meals per day, ensure you eat healthfully. If you do wish to snack between these 2 or 3 meals, then ensure you only eat healthy snacks (Section 9.2) and do not snack after the 'evening' dinner i.e., the last meal of the day.

If you feel you wish to have the odd treat, such as a chocolate bar, potato chips, crisps, or a biscuit or slice of cake, make this once or twice a week, as something to look forward to, rather than the old days of having treats on most (if not all) days. Again, do not have any snacks after the last meal of the day. Remember only snack if you really need to, rather than revert to old habits too regularly.

8.8 Aim for Nutrition Quality

Make the following changes to your diet and food choices:

1. Change to a predominantly plant-based diet as detailed in Section 9.1. Eventually you may like to try a fully *Whole Food Plant Based* way of life discussed in Sections 9.4.

2. Plant-based eating is based on eating a wide variety of more vegetables, more starch based foods, more delicious salads, more wholegrains, more fruit, more fibre, more legumes (beans) and a small quantity of seeds and nuts.

3. If you need to eat meat, eat less meat (of any kind), to a maximum of 2 portions per week.

4. Remove sausages, bacon, ham and salami from your diet, even on pizza.

5. If you like fish, eat a portion of oily fish (not fried) a maximum of twice per week.

6. If you need to continue eating cheese, reduce this to a maximum of 2 portions per week.

7. If you eat eggs, reduce the number of eggs you eat to a maximum of 4 per week.

8. Where possible and practical, avoid all calorie dense, low nutrient dense, processed food and ultra-processed food purchased (or delivered) from supermarkets, shops or any other retail outlet. These foods include any prepacked foods such as 'ready meals' pastries and pies.

9. Minimise processed food snacks (maximum of each item once per week) such as crackers, crisps, chocolate bars, sweets, so called energy bars, all orange juice, all cola based drinks, all carbonated sugar free or low sugar drinks, all cakes, croissants, pastries, scones etc. Instead, and if you wish to snack between meals, avoid crackers, crisps, 'energy bars or chocolate etc or keep these to occasional special treats and instead, choose healthy snacks.

10. Avoid sugar or any artificial or any other sweeteners in tea or coffee or any other drink.

11. Eliminate entirely or use as little butter in cooking as much as possible. It is assumed you no longer eat any form of margarine or other spreads. If so, eliminate these from your diet entirely. Butter and oil will add weight.

12. Use houmous instead of butter to spread on bread or for dressing salads.

13. Also use houmous and warmed chopped tomatoes or other delicious sauces, also for adding salsa type sauces to baked potatoes and other vegetables, legumes and wholegrain rice.

14. If you like to eat avocado and also wish to lose weight, eat sparingly and limit to one avocado per week. As well as some healthy attributes it is also a highly calorific fruit (320 kcal/avocado) and high fat content will contribute to weight gain if not burnt off.

15. Minimise the use of all oils, including olive oil or other cooking oils. Apply less oil in dressing salads. Do not fry food. Bake more foods and steam more foods. Aim to reduce oil as much as possible over the long term if you still feel you need to add oil to food.

16. Minimise eating chips/fries (oil fried, air fried or oven baked). If you wish to continue eating, some chips/fries, eat a portion at most, once per week, then once in every 2 weeks, then once per month etc.

17. Avoid eating food from take away outlets; there are many of these including (but by no means limited to); burgers, fried chicken, pizza, kebabs, curry houses fish and chips etc. If you do need to eat some of this type of food, early in your new regime, limit this to 'a treat' of a maximum of once per month.

18. Avoid white bread, or minimise as far as possible, plus any other bread, described as 'brown' or 'wholemeal' or 'wheatgerm' or 'seed sensations' or 'high seed' etc, whether sliced, unsliced, baguettes or rolls etc as these bread types may be marginally healthier than white bread but nowhere near as healthy as true wholegrain breads. Instead, if you wish to continue eating bread and a healthy, delicious salad sandwich is certainly recommended, then eat wholegrain bread only, such as seeded, high fibre sourdough, or true wholegrain rye bread, or Ezekiel bread. It can be a little tricky to find healthier breads as they are not common in the UK supermarkets, for example, but a very healthy, wholesome variety can be found in one main chain store (crafted sourdough and seeds made over 35 hours) and there may be other outlets of local bakers that will bake wholegrain bread. Enjoy your lunch time salad sandwich or some home-made soup bread. The truth is that 90% of bread in shops is essentially free sugar.

19. Make your own sandwiches from home and avoid purchasing sandwiches from shops, supermarkets, fuel stations or other outlets.

20. Do not drink smoothies or juiced fruit made in electric blenders from blended fruit as this is the same as relatively quickly drinking simple free sugars, without the digestive and glucose level tempering benefits of the fibre in the fruit. The healthy minerals will still be there but this practice will thwart weight loss and lead to weight gain as the sugars will reduce fat burned.

21. Making vegetable smoothies is less unhealthy and still nutritious and certainly to be encouraged as long as this is consumed slowly during a morning or afternoon, if it helps you increase your intake if vegetables but this is still not as healthy as eating the whole foods.

22. After cooking the carbohydrate foods; potato, sweet potato, rice and corn then refrigerate overnight before eating the next day. The amount of resistant starch in these foods will be then be significantly increased which will slow absorption and digestion and reduce insulin spikes.

23. By the same principle, freeze wholegrain bread first before then defrosting and eating thus substantially increasing the amount of resistant starch in the bread. The actions of steps 22. will decrease the rise in blood sugar levels by around 40%.

24. Add lemon or lime juice or vinegar to cooked starches to reduce the rise in blood sugar slow digestion. This works by the (weak) acid supressing or neutralising the digestive salivary amylose enzymes and the acid slows stomach emptying, again slowing blood sugar rises

8.9 Introduce Nutrition Timing and TRE

Refer to Chapter 5, Nutrition Timing and introduce your Time Restricted Eating (TRE) pattern on a daily basis, at least 5 days per week if that suits you.

Try and lengthen the water and black tea and coffee only fasting phase up to a maximum of 16 hours water only fasting per day (including sleep time and eating window of 8 hours per day, (16:8) if this is appropriate for you.

To introduce this into your lifestyle for the first time, gradually work from a TRE of 12 hours eating window (12:12) through to a 14:10, then finally to a TRE Fasting/Eating pattern of (16:8). Endeavour to stop eating at least 3 hours before going to bed (preferably more if possible).

Avoid tea or coffee or alcohol at least 4 hours before going to bed. If you are a shift worker, try and fit a TRE pattern around your work and sleep times. Centre this fasting period around the hours you sleep.

8.10 Increase Activity (NEAT)

Refer to Section 2.6. Consider how to safely increase activity (NEAT) generally around the home and garden.

8.11 Increase Exercise (EAT)

Refer to Chapter 6, The Benefits of Exercise. If you are not already doing so, safely introduce a daily moderate exercise (EAT) routine, at least 5 days per week. This can all be done starting from home, or around your neighbourhood.

If you are feeling energetic and it suits you, join a local walking group or some gentle exercise or yoga/aerobics/Pilates classes.

Also consider introducing some gentle swimming once or twice per week if practical. Introduce some gentle resistance training (Section 6.5), for example a few minutes each day and also some yoga at home several times per week.

8.12   Improve Your Sleep Quality

Refer to Chapter 7 on Sleep Quality. Aim for 7-8 hours uninterrupted sleep per night and if you are not achieving this regularly, try the recommended steps listed from step 1 to step 16. Try and standardise bed time and getting up time to a similar time each day.

8.13   Lifestyle Changes: Reduce Eating Out

Significantly reduce the number of times you eat out when these events are just normal meal times for you, your family or part of your social life. Make this a treat at a maximum of once per month unless, for example, there are 2 family birthdays, or friends or family's weddings in that month. If you do go out to eat, or have a take away meal on occasions, have a maximum of one eating out meal per month and one take away meal per month as your first goal. If you do eat out, only have a maximum of 2 courses and try to eat healthfully with wholesome ingredients. Do you really need that desert? Are you really still hungry?

8.14   Prepare More Food at Home

Increase the number of meals prepared at home and using whole ingredients rather than packaged. Aim to prepare meals at home as much as possible, as this will also encourage you to minimise eating out or having processed calorie dense, low nutrient dense food delivered to your door. Prepare food from fresh ingredients such as potatoes, whole grain pasta, beans (legumes such as lentils and kidney beans) and rice, vegetables and large salads with just a little oil.

8.15   Plan Food Purchases and Cooking

Certain fruits and some vegetables can be expensive, especially out of season, but overall, very healthy food can be both nutritious and relatively inexpensive and always more economical than eating out most days and nights of the week. Purchasing individual ingredients in season and not packaged is the way forward and no better than adding highly nutritious (dried) lentils (cooked at home), canned legumes, potatoes, carrots, oranges, apples, pears, brown rice, bananas, watercress, kale, broccoli to your shopping and the diet. Planning meals in this way will enable improved nutrition on a relatively low budget.

Make a couple of long cook casseroles in the oven each week. Prepare delicious soups, using vegetables such as kale, celery, mushrooms, potatoes, sweet potatoes, canned kidney beans, black beans lentils, carrots, parsnips with low salt vegetable stock etc.

A tip is to add a very thin film of oil (ensure suitable for oven baking at 200 degrees C, for example avocado oil) around a half cut baking potato, giving a very nutritious and delicious harder skin, adding herbs and spices, then after cooking, dressing this with chopped tomatoes or a little pesto or houmous. Very heathy and very satiating. Your brain will love the glucose derived from the healthy potato. Have a healthy stir fry, using a little water only (there is absolutely no need for oils for frying) and add mushrooms and onions to add taste nutrients and moisture).

8.16 Use Leftovers for The Rest of the Week

Once you are preparing more food at home, get into the habit of storing leftovers to use later in the week. For example, legumes such as lentils, kidney beans, black beans chickpeas, pasta, rice, salads, roast vegetables etc. This means that you do not need to prepare every meal from scratch and this is an excellent way to add highly nutrient dense food to every meal. Store in Tupperware containers to keep safe and fresh and use as lunches at work or picnics.

8.17 Avoid Pastries when Going Out For a Coffee

If you go for a coffee with a friend, avoid sugar or sweeteners in your coffee or tea. Minimise consumption (maximum of each item once per month) of any pastries, cake, wraps, heated sandwiches in coffee outlets and cafés. If you like milk in your coffee, use plant-based milks as a substitute for dairy milk. Request that your drink is made with oat or soya or almond milk.

8.18 Ensure Adequate Water and Hydration

Refer to Section 13.6 and ensure you are sufficiently hydrated each day. Do not consume or include carbonated drinks as part of your hydration, including those labelled as sugar free or low sugar. Check the colour of your urine regularly and maintain the light straw colour. Do not over hydrate. There is a good quantity of water in vegetables, fruit and salads.

8.19 Lifestyle Changes: How You Eat

Eat more slowly, chew every mouthful fully and slowly before swallowing. Try to minimise eating in front of the television and try and eat with the family or friends at home, when possible, a few times per week, to enjoy a leisurely social meal eaten over a longer time. Above all, slow down. Aim for a minimum of 20 minutes, preferably 30 minutes to eat your main meal and even longer if eating with the family and or friends. This will train you to eat sufficiently but not overeat.

8.20 Lifestyle Changes: Minimise Alcohol

Refer to Section 13.7 on alcohol. Avoid alcohol consumption, if possible. It is a toxin and poison to the body. If you wish to have a drink of alcohol, reduce the number of days on which you drink (to a maximum of 3 days per week) and reduce the number of drinks to a maximum of 2 drinks or 4 units of alcohol per day. Alcohol, especially regular consumption, will contribute to weight gain directly and indirectly. Avoid snacks such as crisps and nuts whilst drinking alcohol. Even after just 2 days of eliminating alcohol, you will feel much fresher overall.

8.21 Sustaining Your Achieved Weight Goal

Once you have reached your long term weight goal, firstly congratulate yourself and have what you now call a treat. You may well find that low nutrient dense food such as cakes taste very sweet, as your taste buds and perception will change over time. This is a sign of good progress, improved diet and health. Even the relatively healthy type cereals (14% sugars) can taste very sweet once you have re-tuned your taste buds.

You may find that you begin to really love the satiating taste of delicious baked potatoes or steel cut porridge oats or whole grain rice or kale on a piece of wholegrain sour dough bread with houmous and diced carrot.

Continue to enjoy this wide variety of healthy nutrient dense, calorie light foods to match your energy needs and continue your new lifestyle regime and you will maintain your new weight.

You will probably find your total energy balance has gradually shifted to burning more healthy complex carbohydrates that stokes the metabolic fire to promote healthy burning of fats coupled with a higher total NEAT and EAT, meaning that your body has been re-tuned to work more efficiently and more healthfully, now with lower, still more than adequate fat stores. Remember, you still have at least 11-15 kg of fat, plenty indeed!

This is your new way of life, not a diet and there should be no need to revert to dieting in the future. If you do find times when you may lapse and have a few too many treats (yes it can happen to anyone), don't be concerned, just stop the treats and eat natural, healthy foods most of the time. You will begin to be more tuned in to your body.

You will begin to sense when your body has not enjoyed any unhealthy food eaten and your body will communicate this to you. A good long term watchword is to aim for a realistic (SMART), long term lifestyle and build in odd treats if you still wish to indulge in low nutritious food from time to time.

A realistic aim is to understand the following ratio based on adherence, volume and type of treats and the frequency i.e., number of days per week or month that you consume those treats.

```
A) 100% Healthy Nutrient Dense foods &  0% Processed/Ultra Processed foods & Zero Treats
B)  95% Healthy Nutrient Dense foods &  5% Processed/Ultra Processed foods & Treats 1 day/month
C)  90% Healthy Nutrient Dense foods & 10% Processed/Ultra Processed foods & Treats 3 days/month
D)  80% Healthy Nutrient Dense foods & 20% Processed/Ultra Processed foods & Treats 5 days/month
```

All four scenarios will probably be much healthier than the current diet and maintain a reasonably healthy body weight overall.

Scenario A is unrealistic for 99.9% of people and can only be followed with true 100% adherence by a dedicated (very, very) few Saints! The people that are able to continue this and maintain sanity and friends are truly incredible! You may find them on some nutrition u tube podcasts. But this level of adherence is not necessary to lose weight healthfully.

Scenarios B and C may be a more realistic goal for most people, including birthday treats with families etc. Scenario D will be the easiest to manage but may delay the time to reach your goals and may not produce all the long term dividends you aspire to. Perhaps a range between B, C and D will be the eventual reality and allow you to monitor and adjust your progress.

It is up to each individual to assess how much each wants to achieve their goal against the desire to have treats. Ultimately, you may well find that the natural, wholesome, healthy, highly nutritious, low calorie density food, provides both weight loss and a feast of delicious culinary delights you enjoy long term in your new, healthier and fitter way of life.

8.22   Nutrition and Metabolism Guidelines for Weight Loss

The 10 nutrition and metabolic guidelines that should be followed for healthy and sustained weight loss are:-

1. A gradual, step by step, reset of the body to healthier, less calorically dense, higher nutrition quality of food and eating out less.

2. A marginal reduction in total calories consumed depending on total weight loss targets.

3. A simultaneous increase in energy expenditure greater than the new caloric intake by increasing activity, both within the home and outside, via walking or other exercise.

4. Turning the cogs of healthy cell metabolism, through glucose led total fuels full oxidation of carbohydrate and fat.

5. Retraining the appetite to be more easily satiated, due to the improved nutrient and satiating quality of the new eating regime.

6. Reducing the length of the daily eating window, taking advantage of the natural fasting period during sleep and introducing regular time restricted eating.

7. Reaching a new and healthy metabolic set point and ultimately convincing the body that the body is safe, healthy and thriving.

8. Once the revised, healthier regime is fully adopted, returning to an isocaloric (or equicaloric) diet where the revised energy consumption matches the revised energy expenditure but this time, eating healthier food.

9. Improve sleep quality and regularity.

10. Enjoying food and the new way of life and lifestyle.

With the right nutrition and improved lifestyle, the body will detect the improvement and is likely be content and oblige. This healthier approach has the effect of burning the metabolic fires (full fuel oxidation) in a healthy body in a more efficient way, both at the cellular and whole organism level, so that the body is content with its new metabolic position. This may also help reduce overall stress, reinforcing adherence. This book explores some of these mechanisms.

8.23 Summary of the Weight Loss Plan

1. Set SMART realistically achievable weight loss goals over sufficient time
2. Do not reduce daily caloric quantity by too much to ensure adequate nourishment
3. Eat nutrient dense lower calorie whole foods Vegetables/Grains/Fruit/Legumes
4. Minimise caloric dense, low nutrient, fast/take out/processed food
5. Reduce all animal & plant-based fats including meet, eggs, cheese oils butter & avocado
6. Minimise alcohol and unhealthy empty calorie snacks crackers crisps etc
7. Make treats just that - infrequent treats on occasions - not on regular basis
8. Eat within a time window not exceeding 12 hours per day leading to 16:8 TRE
9. Increase NEAT & moderate EAT exercise in the morning before first meal
10. Aim for regular bed time, sleep time and wake time

CHAPTER

9

Healthy Food Menu Guidelines

The food ideas and suggestions provided in this Chapter are selected from a range of foods that meet the criteria of this weight loss plan. Remember Nutrition Quality in Chapter 3 and caloric density listed in Table 3.1. Aim for the foods highlighted blue. All these foods are nutrient rich and packed with energy, phytonutrients, antioxidants, vitamins, minerals and fibre to nourish your body with sufficient, good quality calories, to help you reach and sustain your new weight goal. Sections 9.3 list foods to minimise or avoid altogether and will assist you in reducing weight, whilst maintaining good health way of life. Section 9.4 provides some basic information on the Whole Food Plant-Based (WFPB) ethos.

This may not necessarily suit everyone, particularly those that still want to eat meat, cheese and eggs, dairy, butter and fish, however, the author has completely changed his life since eliminating these foods, fully adopting the culinary delights and delicious wide variety of WFPB foods. By all means have a look and give it a try, after all it is completely nourishing and a fabulous sustainable way to manage your weight. Have a look at some of the more established plant-based u-tube podcast regular host sites and you will learn so much more about this alternative way of eating and looking after your body. [50] The fact that the Standard American Diet consists of 63% processed foods, 25% derived from animal-based products and just 6% of total daily calories (EI) from healing, whole plant sources, is the reason for the obesity epidemic. Less than 50% of children are eating fruit.

On balance, the evidence on optimal health and weight management over the past 100 years suggests the optimal diet should be based mainly on plants with a low consumption of fish, animal meat, dairy, eggs and cheese. Total daily dietary intake of macronutrients should be in the order of 60-75% carbohydrate, 10-15% protein and 10-30% fat (where total fat should be preferably nearer 10-15%). Fats should preferably be omega 3 unsaturated fats and saturated fat should be minimised and not exceed 10% of total dietary intake.

9.1 Recommended Foods for Your Diet

The overall diet should be largely focussed on plants. These contain all the macronutrients, micronutrients, minerals and fibre needed for good health. Incorporating vegetables, whole grains, legumes, fruits and nuts and seeds into your diet supports long-term health and weight management. Limit processed foods and empty calories to enhance nutrient intake.

The significant benefits of eating a predominantly plant-based diet provides considerable variety and of natural phytonutrients and magnesium in chlorophyll contained in plants. These valuable nutrients provide antioxidants and many other properties that improve and maintain good health, for example, polyphenols, alkaloids, phenolics, genistein, curcumin, resveratrol, sulforaphane, isothiocyanates, silymarin, diallyl sulphide and lycopene.

If you do wish to include animal based food in your diet for example, meat, fish, cheese, eggs and dairy, ensure these are consumed in moderation, for example, oily fish such as salmon twice per week and lean meat twice per week with limited eggs and cheese. Minimise saturated fat and oils.

Be sure to incorporate all green leafy vegetables such as watercress, kale, spinach, rocket (arugula), chicory, leaf lettuce, iceberg lettuce. Eat plenty of starch vegetables , an ideal source of healthy slow burn carbohydrates, for example, potato, sweet potato, corn on the cob, corn, green peas, pumpkin, butternut squash, acorn squash, winter squash, yams. Avoid adding butter or cheese to baked potatoes, brush a little avocado oil to the skins 30 minutes before (total 90 minute) baking is complete. Dress with a can of tomatoes and garden peas rather than saturated fat from cheese and butter. Your diet should also include non-starch vegetables such as Bean sprouts, green beans, Italian beans, string beans, wax beans. If you use olive oil to dress salads use in moderation as all oils are highly caloric and will add weight is consumed in excess (<1 table spoon per day).

For more rich sources of protein and minerals valuable protein introduce plenty of legumes, for example, lentils, kidney beans, pinto beans, lima beans, navy beans, black beans, cannellini beans, chickpeas, soybeans (and edamame). If you are not used to eating legumes, introduce these gradually starting with one portion per week then increase gradually to several times per week and then daily.

Adding in plenty of healthy whole grains will significantly enhance your new diet, providing excellent sources of long term energy and sustenance, for example, oats (steel cut best for porridge), wholegrain rice, brown rice, red rice, black rice, wild rice, rye, barley, whole grain pasta, couscous (couscous is a pasta high in fibre and selenium), millet, quinoa, whole grain wheat based cereals (including bite size), bulgur wheat, buck wheat (cracked wheat), spelt, amaranth.

Fruit should always be a central part of your diet and aim to eat fruit on a daily basis. Fruits provides excellent sustenance, antioxidants and minerals and vitamins at any time of day and is a great snack between meals. All fruit is healthy in its whole form as the natural sugars are slowly absorbed due to high fibre content. Add plenty of citrus fruits, berries, and a date each day, as well as the humble (but exceptionally healthy) apple and melon, bananas and pears. Whole fresh fruit is better than juiced or dried or frozen fruit if possible but any fruit, whatever its form, is always preferable than free sugar snacks such as pastries. Remember that drinking daily quantities of smoothies may contribute to weight gain, as the natural free sugars may be consumed too quickly due to the significantly diminished quantity of fibre compared to eating whole fruits.

Add some nuts and seeds into your daily diet, for example walnuts, almonds, cashews, hazel nuts, and ground flax seeds, hemp seeds, chia seeds, sunflower and sesame seeds, as these promote heart health, lower diabetes risk and provide an excellent source of fibre, protein, healthy unsaturated fats, (omega 3) phytochemicals, vitamin E, B1, B2, B3, B6, niacin folate and minerals such as healthy plant based iron, calcium, zinc, magnesium, copper and selenium, potassium and phosphorous. Approximately 30 grams/day (a handful) is a good, healthy daily measure. Do not have too much as they are calorically dense at over 2,000 kcal/pound(table 3.2) and can thwart your weight loss program if consumed in excess. Do not however, be concerned regarding weight impact if you use these convenient power houses wisely. Eat one Brazil nut per day for a reliable source of selenium, improvement of gut health, thyroid hormone metabolism, and anti-inflammatory properties but do not overdo these (>3/day) as too much selenium may lead to brittle nails, hair loss and other issues.

9.2 Healthy Meal Plans

The author has put together some ideas on choices for the main meals of the day. Select from any of the choices to provide wholesome meals that will help you with your weight loss target and sustained new way of life. You may well discover wholesome foods not listed and you will also find recipes in predominantly plant or wholefood plant based cook books and, on the internet, (example, *Forks Over Knives*).

For breakfast, select any from; Oatmeal (steel cut porridge oats), whole grain wheat based cereals (including bite size). fruit, such as blueberries, strawberries, blackberries, raspberries, red apple, banana, pear, pineapple, kiwi, dates, plus a little chia or flax or hemp or sunflower or pumpkin or sesame seeds. Tofu scrambled is nutritious, fried with chick pea juice, turmeric on sourdough or wholegrain toast with a little houmous and mushrooms fried gently in a little water. If you can-not source steel cut oats, any porridge is fine and together with whole grain wheat based cereals (the best ones are all virtually devoid of free sugars (check the label - around 1% sugars), far healthier than the vast array of other cereals out there.

Then for lunch, try a large salad sandwich on wholegrain or sourdough bread with houmous, including celery, beetroot, radish, rocket (arugula), water cress, spinach. Add balsamic vinegar or fresh lemon juice to the green leaves to help release and absorb the nitric oxide for improved endothelial cells in the walls of your blood vessels. Mashed chickpeas, red onion and houmous sandwich. Chickpeas and salad, half an avocado (twice per week) any legumes on toast with houmous. Quinoa or couscous with salad. Carrot, kale and cucumber sandwich with dill.

And for dinner, there is so much delicious and nutritious food available, including green beans and whole grain pasta with cashews and pesto. Wholegrain rice, quinoa, noodle soup.
Add red cabbage and sweet corn tinned or on the cob to any meal. Baked sweet potato or baked potato with chopped tomatoes are superbly satiating. Make mushroom sauces for any meal.

All soups containing any vegetables, carrots, potatoes, squashes, lentils pumpkin, chickpeas provide excellent nutrition. Add some chili red beans, bell peppers, mushrooms. If you like fish incorporate salmon or tuna (twice per week). Look up the fat-free healthy version of *'Creamy Golden Gravy'* (source Dr McDougall health and medical centre).

If you need to snack between meals, much healthier weight managing options are; wholegrain crackers with cucumber sticks or carrot sticks and houmous, sugar snaps on crackers with grated carrots, half a stick of celery with ground black pepper and paprika. Or some fruit anytime, such as an apple, some melon or banana or orange. Plant based yoghurt is nice with fruit. Add some roasted chickpeas to any snack. Plant-based yoghurt and fruit is a great filler.

9.3   Foods to Avoid, or Reduce or Minimize

The following guidelines, though by no means exhaustive, provides some ideas for people to consider for both the support of weight loss and improved health. Refer to Table 3.1 and caloric density. Some individuals may wish to make small adjustments, and this is encouraged, as it will certainly help improve health and weight loss, no matter how small the adjustment. Some may wish to make further reductions and more adjustments, either because of health concerns due to increased knowledge regarding animal food products, or for those wishing to gradually transfer, or even fully transfer to a predominantly plant-based, i.e., WFPB way of life.

Significantly reduce or remove Microwave meals, ready meals, all processed meats such as sausages, bacon, ham and salami, hot dogs. [57]. Avoid or reduce butter, margarine, ultra-processed food, white flour, refined white pasta. carbonated drinks, cola drinks, orange juice, fruit juice, sports drinks 'sugar free' drinks. Avoid protein powders, diet powders, doughnuts. white bread, bread baps, coconut oil, energy bars.

Avoid all breakfast cereals apart from oats and whole grain wheat based cereals. Reduce all free sugars including added sugar or any sweeteners in coffee or tea. [58] Avoid all foods high in sugar (>22g.100g). Reduce or avoid low fat flavoured yoghurt and all foods high in fat. (>17g/100g). Avoid all foods high in saturated fat (>5g/100g). [58] Avoid foods high in salt (>1.5g/100g) or expressed sodium as (>0.6g/100g).

Reduce your consumption of 'takeaway food' and limit eating out to special infrequent occasions such as family birthdays. Reduce all meat including red meat, chicken, venison, game, poultry lamb and pork. Reduce the number of pastries, bagels, wraps, takeaway shop bought sandwiches, all cakes, alcohol, ice cream, tinned soup, unless very low salt (see above) cheese, eggs, sweets, fried foods, burgers, fries, pizza, baked beans, fried chicken take away food, desserts, crisps, crackers, so called health bars and snacks. Cow's or goats' milk in coffee or tea or to drink or on cereal. Minimise chocolate bars or chocolates for a special bi-weekly treat if you like chocolate.

Regarding healthy fats, by all means enjoy an avocado once per week or a little olive oil on salads, but moderate this for your weight loss target as these are highly caloric, low volume foods and will to some extent certainly thwart a proportion of your weight loss effort.

9.4   Whole Food Plant Based

It is recommended that everyone should consider trying a Whole Food Plant Based Diet (WFPB). A WFPB diet is fundamentally and foremost based on a strong health ethos. The WFPB is devoid of any animal products but is also very health conscious and avoids any unhealthy, high fat (such as coconut) or processed plant foods. Essentially WFPB comprises the 4 principal food groups: Whole grains, Vegetables, Legumes and Fruits and plus a small quantity of nuts and seeds. Processed food is avoided or at least, where possible, kept to an absolute minimum and adheres to the adage, *'nothing good removed, nothing bad added'*. A variation of the diet also minimizes or completely excludes salt, sugar and oil and avocado and reduces seeds and nuts to a bare minimum for essential fats only.

It is encouraging to observe that each year there is a significant and growing number of people adopting this new *Plant-Based* way of life around the globe, as it is bringing considerable benefits to people's lifestyles and health, the environment, forests, land, the sea, animals and other organisms and hopefully leading to a reduction in animal suffering. The world would be in a far better place in so many ways, if meat production and the staggering amount of land used to graze livestock and the ensuing pollution of the rivers and seas was minimised. Look at the destruction of all the animal and plant species wrought on our Earth since 1960. Section 15.14 provides some further detail. [48, 49, 50, 51, 52, 53, 54, 55, 56]

WFPB is not to be confused with a Vegan diet. Many people adopt a Vegan diet primarily because of animal welfare concerns, then concerns for the environment, then for health reasons, in that order. There is nothing wrong with this whatsoever, as, in the opinion of the author, they are clearly right in addressing these cruel and highly relevant and concerning issues for animals and the environment. However, a WFPB diet also excludes any relatively unhealthy ingredients that may be part of some vegan diets, such as vegan cheese, coconut oil, vegan cakes or meat substitute meals etc. The ethos is to eat adequate, sufficient portions of 'complex carbohydrates' or 'starches, obtaining daily calories from whole grains, legumes and root vegetables, with these making up around half of the plate.

You can load up the rest with salads and vegetables. In sampling, then adopting this new way of life, the knock on benefits are considerable, including you own health, lowering pressure on an overstretched, health industry and a considerable reduction in environmental damage to all species throughout our damaged Earth. Furthermore, a reduction in greenhouses gasses towards lowering carbon emissions and last but by no means least, a reduction in animal suffering, due to people's insatiable desire to have meat on their plate, though, as this book shows, meat is not necessary to maintain good health, but instead, is continually marketed as healthy for human taste buds, convention and habits. [57] All that said, even relatively unhealthy Vegan ingredients are far better for the environment and the animals than meat and dairy products.

The author has fully researched and reviewed numerous studies on the WFPB way of life over the past 20 years, including scores of scientific papers and evidence and has concluded that adopting and following a WFBP way of life is the right approach for him. From a nutrition basis alone, there have been countless studies and overwhelmingly robust science based evidence over the past 100 years or more, that a plant-based diet and even healthier a WFPB diet is the optimal diet for a longer, healthier life, let alone helping to achieve weight loss goals.

A WFPB diet will reduce the risk of cardiovascular diseases, diabetes, obesity, neurodegenerative diseases, autoimmune disease and other serious diseases. There are also several issues with excess dietary protein in more meat oriented diets. [56] Readers should review the wealth of information and research health risks associated with meat and decide for themselves and the author completely respects the rights of all individuals to be responsible for their own nutrition choices and select and eat and drink what they enjoy, whether animal or plant or both and it is, of course, entirely up to each person to decide what is right for them as individuals, the animals and the environment.

The phrase 'High Biological Value' (HBV), referring to animal based protein sources and 'Low Biological Value' (LBV), referring to plant-based protein sources, is still used in the UK in 2024.

These terms are considerably outdated and flawed and perpetuate the myth that people adopting a Whole Food Plant Based diet are and will be deficient in protein and indeed other essential nutrients.

Section11.19 shows that all 20 amino acids are available from a Whole Food Plant Based diet, simply by eating sufficient amounts of food each day, with the added benefit of all the phytonutrients, antioxidants, vitamins and minerals, and plenty of fibre (absent from meat) and moreover, with none of the health risks associated with eating meat.

Animal protein is associated with higher odds of developing chronic disease with increasing age, whereas plant protein reduces this risk. Plant protein also benefits mental health. [54, 56, 57, 60, 63]

The only 2 supplements that should be required on a healthy, wholesome Whole Food Plant Based diet are Vitamin B12 and possibly DHA, for Omega 3 from algae, that is, only if Omega 3 is shown to be deficient by tests. On a vegetarian, vegan, or plant-based diet, it is impossible to eat sufficient daily calories (EI) to maintain your weight and be protein deficient. [54] One of the quickest ways to age is to follow a high fat, high protein, low carbohydrate diet (Biogerontology).

Too much protein disrupts healthy cell maintenance (Section 11.4). Animal protein builds up (organic, grass fed or not) in the body, accelerates ageing and plays a role in Alzheimer's disease, heart failure, blood vessel ageing and a higher risk of other diseases. [56]

Animal protein amino acid profiles differ markedly from the plant protein amino acid profile. Animal protein contains higher amounts of leucine and methionine. High amounts of these amino activate the rate limiting step for biochemical ageing pathways; (mTOR, [5] IGF-1, [6] pKRAS, [7] Sirtuins [8]).

Free sugars and dairy milk (galactose) and whey protein and muscle growth supplements, activate harmful, inflammatory, pKRAS and IGF-1 and growth hormone paths. Vegetable protein from less branched chain amino acids (BCFA) does not accelerate these pathways and is healthier. [63]

Excessive caloric, dietary or high fat diets, typically with high levels of meat and dairy and free sugars, leads to insulin resistance, increasing the release of insulin like growth factor IGF-1 and therefore, indirectly leads to the activation of the mTOR pathway. [35] The only medical condition widely recommended to adopt a ketogenic diet (a physiological emergency status, emulating starvation, high fat diet minimizing carbohydrates) is for refractory seizure epilepsy. Relatively healthy diets can of course include low contributions of protein from meat and higher intakes of plant protein from vegetables, legumes, nuts and seeds.

The higher the animal in the food chain the higher the protein ratio, the higher risk of some diseases. [57] Therefore, it is probably even better to go one step further and remove the fibre lacking meat proteins, replacing these with plant-based proteins such as legumes, grains, fruit and more vegetables. Table 9.1 lists 120 foods both in terms of nutrition and overall health. This builds on the knowledge discussed in Chapter 3 regarding caloric density and shown in Table 3.1. Table 9.1 now lists these foods in 3 columns as; (A); foods recommended as highly nutritious and health promoting, (B); foods that should be consumed in moderation and (C); foods that should be avoided where possible except for very occasional 'treats'.

The criteria determining these recommendations is based on overall benefit to good health, whole food, home prepared, promoting weight loss, nutrient density, caloric density, plant-based or animal-based, saturated fat, unsaturated fat and total fat content, cholesterol, salt, processed and ultra-processed food. Table 9.1 can be used as a guide to follow healthy nutrition for weight loss. To reiterate, every individual should choose what they feel is right for them.

Table 9.1 List of Foods Supporting Weight Loss and improved health and Foods to avoid

Very Healthy Foods to Include	Healthy Foods in Moderation	Unhealthy Foods to Avoid
A	B	C
All Vegetables incl. Watercress	Flax Seeds	Salami
Celery/Lettuce/Cucumber	Sesame Seeds	Bacon
Courgette (Zucchini)	Chia Seeds	Sausages
Bamboo Shoots	Sunflower Seeds	Ham
Cabbage/Sprouts	Pumpkin Seeds	Beef
Tomatoes/Canned Tomatoes	Walnuts	Chicken
Turnip	Pecan Nuts	Pork
Rocket (Arugula)	Brazil Nuts (one per week)	Lamb
Cauliflower	Macadamia Nits	All Other Meat
Broccoli	Cashew Nuts	Cheese (Includes Vegan Cheese)
Mushrooms	Dates	Mayonnaise
Radish	Dried Fruit	Butter/Margarine
Spinach	Salmon	All Dairy Products Milk/Cream
Kale	Mackerel	Fried Fish, Fried anything
Beetroot	Kippers	All Pastries/Croissants
Bell Peppers	Sardines	Pizza
All Onions/Garlic/Herbs/Spice	Tuna (Canned)	White Flour
Carrots/Parsnips	Avocado	Maple Syrup/Honey/Treacle
Chilli Peppers	Avocado Oil	White/Wholemeal Bread
Potatoes/Sweet Potatoes (Baked)	Olive Oil	White Pasta
Potatoes (Boiled/Mashed)	White Rice	Table Sugar Sucrose
Garden Peas/Green Beans	Baked Beans (Low Salt)	Peanuts
Aubergine (Egg Plant)	Olives	All Crisps & Cereal/Energy bars
Corn on the Cob	Eggs	Doughnuts
Corn (Tinned)		All Biscuits & All Crackers
All Legumes incl. Lentils		Artificial Sweeteners
Chick Peas, Soy Beans		All Other Cereals (See A)
Black Beans, Kidney Beans		Chocolate Bars/Chocolates
Houmous		Crackers (Plain)
Quinoa		Orange Juice/All Soft Drinks
Quorn/Tempeh/Tofu		Cola Drinks/Energy Drinks
Oats (Steel Cut) for Porridge		All Sugar Free Drinks
Wheat Cereals (Whole grain)		All Cakes
Quinoa		Ice Cream
Wholegrain Rice		All Crisps (Any Flavour)
Whole Grain Pasta		Peanut Butter
Wholegrain Bread (Rye)		Ketchup/Brown Sauce
Sour Dough Bread		Coconut Oil/All Coconut
All Fruit Including		Sunflower Oil
Apple, Pear, Oranges, Grapes		All Other Oils
Raspberries, Strawberries		Noodles
Blueberries, Blackberries, Melon		Hamburgers/Kebabs
Bananas, Pineapple, Lemon		Take Away Meals
Oat Milk/Almond Milk		Ready Cook Meals

CHAPTER

10

Carbohydrate

Nutrition comprises 6 principle dietary components; Carbohydrates, Proteins, Fats, Minerals, Vitamins and Water. Chapters 10 to Chapters 13 explore and describe the 3 dietary macronutrients, carbohydrates, proteins and fats, plus the essential micronutrients; minerals, vitamins and water, perhaps the most essential. All are required for humans to survive, including energy production, body functions and growth and maintenance. The principal functions of these dietary elements are discussed, together with their role in metabolism and healthy weight management. Though not considered as a nutrient, alcohol is also included in Chapter 13, as it can and often does, nevertheless, play a significant part in diet and energy intake (EI).

10.1 What is Carbohydrate

Carbohydrates molecules exist in a very wide ranging, structural diversity and comprise compounds that can be digested or metabolically transformed directly into glucose. The body requires carbohydrate to function well and despite all the hype to the contrary, it can certainly be defined as essential, as it plays a vital role in the provision of glucose for immediate energy needs of many key cells in the body, including the growing fetus and plays a considerable part in driving the body's healthy metabolism efficiently and is essential for DNA synthesis. Carbohydrate contains carbon, hydrogen, and oxygen in the ratio of 1:2:1.The word carbohydrate originally comes from the French term, 'hydrate de carbone', or hydrate of carbon. However, this is now known to incorrect, as the compounds are not exactly hydrates of carbon because they are not crystals but long chains of carbon atom with hydroxide side chains. Carbohydrates are also known as saccharides (Greek for sugar). The word 'carbo' comes from the Latin meaning charcoal (for example carbonara).

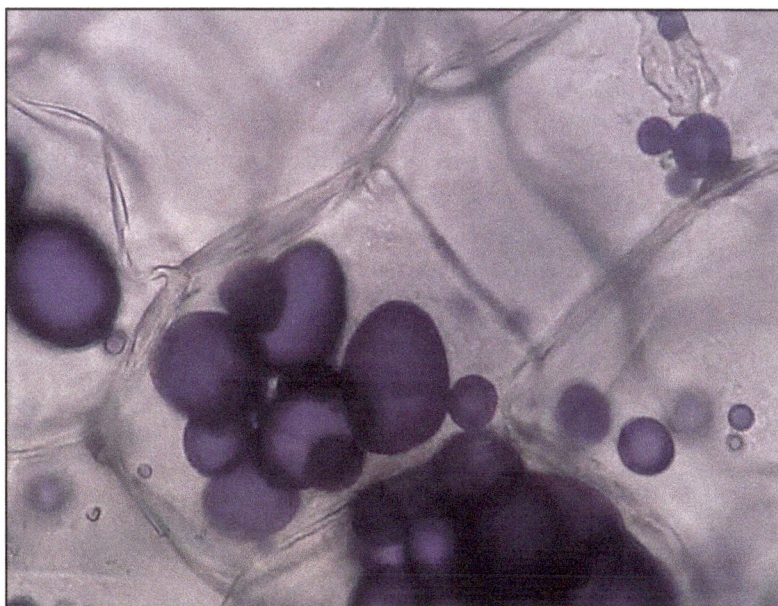

Starch grains in a potato cell (Solanum tuberosum), Optical microscopy, stained with Lugol (25% dilution)
Magnification: 400x
Author Ganimedes July 2019
Wikipedia Creative Commons Attribution-Share Alike 4.0

International

10.2 Dietary Carbohydrate & Weight Management

There is evidence that a healthy, wholesome, high carbohydrate, low fat, plant based diet leads to healthier weight and sustained weight loss for individuals that are overweight. [25, 37, 39, 75, 76, 77, 78, 79, 80]. Contrary to the benefits described for carbohydrate-restricted/ketogenic diets, various trials in humans have shown improved metabolic outcomes, improved lipid profile and insulin sensitivity on high carbohydrate/low fat diets (vs. control diets) [92, 93, 94]

Carbohydrates are an important and essential part of the human diet. There are several reasons why low carbohydrate diets have become more fashionable over the past few years. Firstly, the considerable growth in processed foods containing free sugars that contribute to weight gain due to caloric excess has led to the message that as all carbohydrates are digested and broken down into 'sugar' (actually glucose) and that all carbohydrates are to be avoided. Therefore, the amount of all sources of healthy carbohydrate energy providing foods, such as rice, potatoes, sweet potatoes, peas, corn, wheat, oats and grains is reduced in many of the fad diets. Secondly, the growing fear that carbohydrates are the source of all gluten and that gluten causes the associated coeliac disease, which affects approximately 1% of the population, or non-coeliac gluten sensitivity, which affects a further 2% of the population. Despite this relatively low number of affected people there is a significant growing trend in gluten free products. Thirdly, the common view that reducing fat via 'low fat' products would be beneficial, which was supported by health professionals during the 1980's but seemingly, did not work, therefore people now blame and avoid the carbohydrates instead . But in fact total fat consumption since then has actually risen, mainly due to increased consumption of mono and polyunsaturated fat comprising processed and vegetable oils. (Figures 1.1 and 1.2). Lastly, the recent growing popularity of insulin monitors promoted to all people, where people can monitor glycaemic index and the insulin effect of all foods eaten, hitherto necessary for people with diabetes, caused in part by the significant increase in the number of people with diabetes and pre-diabetes. This has caused people to fear eating even healthy carbohydrates which may cause a completely normal. natural and healthy spike in insulin in levels to complete its function of facilitating energy to the cells. All these factors have led to the growing trend in fad diets, most of which generally vilify all carbohydrates. Such diets necessarily lead to increases in the other 2 macronutrients. on high fat or high protein or both high fat and high protein intake. (Section 1.10). For managing both weight and overall health, wholesome carbohydrates, such as rice, barley, oats, grain, potatoes, sweet potatoes, quinoa and beets (beetroot) should, ideally represent approximately 60-75% of total dietary intake. Free sugars and processed food should be minimised as far as possible.

10.3 The Function of Carbohydrate in Human Diets

Glucose from dietary carbohydrates, is the predominant key energy source for virtually all forms of life. Carbohydrates are the body's preferred first choice to burn for energy and the preferred fuel source during increased exercise intensity and exercise duration. Glucose is also the main precursor for the synthesis of every molecule that the cell can make, including DNA. The body uses glucose to provide most of the energy for the human brain and all the energy for RBCs. Fundamentally, the 25 trillion RBCs can-not function without glucose from anaerobic glycolysis and the brain CNS and kidney medulla much prefer a reliable, constant supply of readily available glucose.

The fundamental purpose and major function of carbohydrate is to provide energy to all the cells in the body, through oxidation of glucose and storage of glycogen (up to 50,000 glucose units) in the muscles and liver.

About half of the energy used by muscles and other body tissues is provided from glucose and glycogen. The liver closely regulates blood glucose levels within a narrow range to ensure the body has both safe and sufficient glucose for energy for vital tissues. Carbohydrate provides fibre which is essential for a healthy microbiome and gut.

Fatty acids from stored triacylglycerol fats are the second hierarchical main provider of energy throughout the day and together with small contributions from other precursors, through gluconeogenesis and ketogenesis, ensure blood glucose levels are maintained and glycogen reserves are not significantly depleted.

Glucose from carbohydrates is required to synthesise many key substances in the human body, including 5 carbon sugars such as ribose, which is an important component of DNA, non-essential amino acids, cholesterol, complex lipids for membranes, essential coenzymes; ATP, ADP, ATP, FAD, NAD^+, $NADP^+$ and the backbone of RNA. It also provides carbon atoms for many essential proteins, lipids, porphyrins (haemoglobin) purines, pyrimidines, neurotransmitters and many more.

The energy from carbohydrate is derived from the chemical bonds between the carbon atoms. These high energy bonds were produced from light energy (photons) from sunlight in the process of photosynthesis (Section 15.1). The chemical energy is captured in the cells via the step by step, incrementally controlled, burning process of cellular respiration (Chapter 15) to release energy for the body efficiently. This process comprises glycolysis, the TCA cycle and OXPHOS and is carefully regulated by enzymes and hormones.

The body converts carbohydrates into glucose for immediate energy and into glycogen as stored energy. Because many foods are high in carbohydrates, it is erroneously rumoured that all carbohydrates are "fattening", (whereas fat isn't...?). Section 1.10 showed this assumption to be incorrect and moreover, either choosing a high-carbohydrate, high-fibre, low-fat diet or, low carbohydrate, high fat, low fibre diet helps can reduce weight, with the caveat that the latter of these 2 diet patterns is less healthy overall. [25] The body relies on glucose; it is not logical, nor common sense, to vilify and demonise the very substance that is so fundamental to the human body and for that matter all animals and plants.

The 'anti carb' brigade is mistaken and wrong. Healthy carbohydrates are found in starches like potatoes, corn, oats, grain foods, vegetables, fruits, beans, peas and lentils. These provide healthy carbohydrates, protein with a little fat, plus many other beneficial nutrients, minerals and vitamins and critically, fibre. Healthy, nutritious complex starchy carbohydrates are a crucial source of energy. It's the excessive consumption of refined carbohydrates (such as sugar based snacks, pastries, drinks and free sugars), that has, for most, led to this misunderstanding, together with excessive and unhealthy fats and excess protein, in other words, over nutrition, that contributes to weight gain and health issues and it is not due solely to eating carbohydrates and especially not healthy, complex carbohydrates.

10.4   Carbohydrate Deficiency

The only reported deficiency in carbohydrate, that the author is aware of, is from people who follow extreme versions of 'low carb' or 'keto' type diets, where carbohydrate is minimised to very low levels below 50 g/d and sometimes even lower than 25 g/d, at around 5-8% of daily energy intake (EI). This level would mean that dietary protein and fat constitutes 92-98% of total dietary food EI. High dietary fat may lead to many diseases, including cardiovascular disease, diabetes and atherosclerosis. [40]   Symptoms of such carbohydrate deficiency are hypoglycaemia, impaired cognitive abilities, headaches, fatigue, brain fog, heart arrythmias, kidney damage, constipation, muscle cramps, insufficient Thyroid Stimulating Hormone (T3), causing metabolic slow down, muscle loss, prevention of muscle gain, and plus raised levels of cortisol leading to disruption of cortisol and testosterone ratios, significant interference with female hormonal balance, mood swings. Prolonged hypoglycaemia (low blood glucose) can lead to brain damage.

10.5 Classification of Carbohydrates

Carbohydrates are often referred to as CHO, as a reference to the fact that the compounds are comprised of carbon, hydrogen and oxygen atoms. There are 3 principal subcomponents of dietary carbohydrates (Figure 10.1). In fact, all 3 macronutrients are CHO, with N (nitrogen) included in and differentiating protein from the other two macronutrients. Carbohydrates are grouped as follows:

1) Sugars - either monosaccharide sugars or two joined monosaccharides, i.e., Disaccharides
2) Oligosaccharides - typically comprising from 3 to 9 monosaccharides
3) Polysaccharides - long chain polymeric carbohydrates, with 10 or more monosaccharides

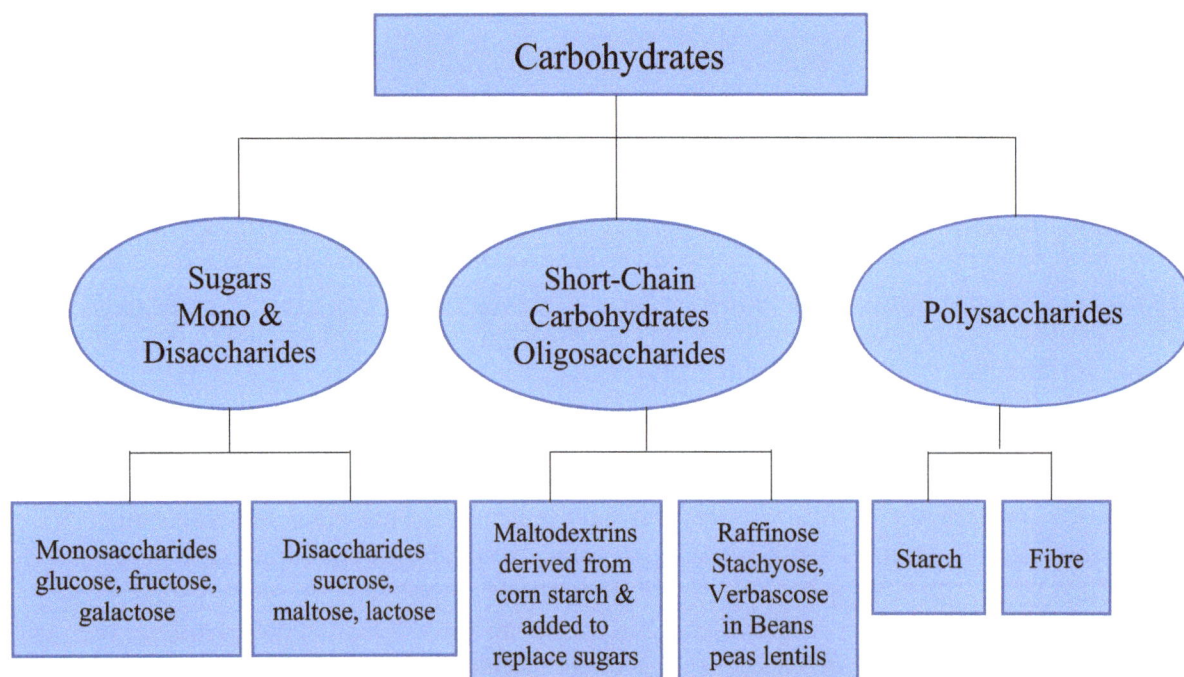

Figure 10.1 Principal make up of dietary carbohydrates (CHO), fibre, starch and sugars

10.6 Sugar - Monosaccharides

Glucose - Multi-units make polymers, glycogen, starch and cellulose. The fundamental and the body's preferred energy source.

Fructose - 'Fruit sugar', found in fruit, vegetables such as mushrooms, onions and red peppers, artichokes and asparagus, honey and sugar cane. It is metabolised in the liver and does not affect insulin levels. The most harmful form of fructose is produced commercially from corn or sucrose into a crystalline form and used in packaged foods to make it sweeter and more attractive to the taste, and also as high fructose corn syrup, used for food browning, structure, moisture (highly soluble) texture, stabilising and to improve shelf life.

Galactose - 'Milk sugar', is found in animal milk and is only found in nature when it links with glucose to form the disaccharide lactose.

10.7 Sugar - Disaccharides

Maltose - Made from 2 units of glucose. Occurs naturally in malted cereals and also manufactured into high maltose syrups, molasses.

Sucrose - Made from glucose and fructose which is commonly found in fruits, vegetables and nuts and sugar cane and sugar beet. Sucrose is also synthesised, or 'man made' from sugar cane and beets, known as granulated table sugar.

Lactose - Made from glucose and galactose. It is found in the milk of mammals. Approximately 65 percent of the human population has a reduced ability to digest lactose after infancy, so called lactose intolerant. Sugars are most commonly found in combination with other sugars, for example honey is 50% fructose, 44% glucose, 4% galactose and 2% maltose.
In maple syrup, <1% is fructose, 3% is glucose and 96% is sucrose. Obtaining sugar from natural plant-based sugars has the benefit of fibre, vitamins and other phytonutrients, whereas manufactured, synthesised free sugars contain no nutrients and can lead to dietary, obesity and other health issues.

10.8 Free Sugars

Free sugars are so named because unlike the sugars found in fruit, vegetables and milk, these sugars are not found inside the cells of the food we eat. These sugars do not come with extra nutrients such as fibre, vitamins or phytochemicals. Free sugars are either any sugar added to a food or drink or, naturally occurring in honey, syrup and fruit juice. It is easier to consume extra sugars without realising. Free sugars should make up less than 10% of total dietary intake (EI).

10.9 Oligosaccharides - Complex Carbohydrates

Galacto-oligosaccharides are short chains of galactose molecules found in soybeans and other legumes, typically comprising from 3 to 9 monosaccharides. (Some scientists define oligosaccharides as 2 to 10 monosaccharides and therefore including disaccharides).

10.10   Polysaccharides - Complex Carbohydrates

Polysaccharides are carbohydrate polymers consisting of >10 monosaccharides, commonly ranging from tens to hundreds to several thousand monosaccharide units. All of the common polysaccharides contain glucose as the monosaccharide unit. These carbohydrates have long chain branches and are the most abundant carbohydrates. Polysaccharides are synthesized by plants (cellulose and starch), and animals (glycogen), to be stored for food, structural support, or metabolized for energy. They are found in food and comprise starches and fibre.

10.11   Starch

Starch is a complex carbohydrate polysaccharide stored in plants and digested as a whole food source of carbohydrates by animals and humans. Digestion occurs in the duodenum and the glucose is absorbed in the jejunum and ileum. Starch foods are foods high in long chain digestible complex carbohydrates. Examples; potato, sweet potato, yams, corn, oats barley, pasta, rice, whole grains, legumes green peas and red kidney beans. (Figure 10.2). Starch can be further separated into 2 fractions; amylose and amylopectin. Natural starches are mixtures of amylopectin (70-80%) and amylose (20-30%). Amylopectin is the most common carbohydrate eaten by humans and is contained in many staple foods. The major sources of amylopectin of starch intake worldwide are the cereals such as rice, wheat, corn, and potatoes. Amylopectin is highly branched, formed of 2,000 to 200,000 glucose units. Amylopectin is more easily digested by enzymes in contrast to amylose, which tends to form helices that contain hydrogen bonding and is therefore an important form of resistant starch. Amylose takes up less space and is the preferred starch for storage in plants. Amylose, serves as an energy source in the human diet. Unlike dietary fibres, starches are fully digestible and provide energy calories to the body. By refrigerating cooked potato, sweet potato, rice and corn after cooking overnight and freezing whole grain bread and defrosting before eating, the amount of resistant starch will be significantly increased which will slow absorption and digestion and reduce insulin spikes.

10.12   Dietary Fibre

Fibre, also known as roughage, is a complex carbohydrate and is the portion of plant derived food that can-not be completely broken down by human digestive enzymes. Most carbohydrates and starches are broken down into molecules of glucose for energy. However, fibres cannot be broken down into sugar molecules, and the fibre passes through the body undigested. Fibre therefore, helps regulate the body's use of sugars, helping to keep hunger and blood sugar in balance. This complex carbohydrate is found in wholegrains, cereals, fruits and vegetables, made up of the ingestible parts or compounds of plants that pass through the gut relatively unchanged. Dietary fibre consists of non-starch polysaccharides and other plant components such as cellulose, resistant starch such as barley, legumes, potatoes, resistant dextrins, inulin, lignins, chitins, pectins and beta-glucans. Cellulose and hemicellulose found in cereal grains, provide bulk to the diet and promote regular bowel movements. Lignins are found in wheat and corn bran, nuts, flaxseeds, vegetables, and unripe bananas that trigger mucus secretion in the colon, add bulk to stools and have a laxative effect. Pectins are soluble, highly fermentable fibre found in apples, berries, and other fruits.

They have gelling properties and because of this property, they may slow digestion and help normalize blood sugar and cholesterol levels. Fibre helps protect against leaky gut. [49] Hence, overall, fibre is an essential component of a healthier diet. A diet high in regular fibre consumption from carbohydrates is generally associated with supporting better health and lowering the risk of several diseases. It keeps the digestive system healthy and can help reduce cholesterol. It also facilitates egestion of harmful oestrogenic chemicals and maintains the blood brain barrier and protects against dementia. It is recommended to gradually add more fibre to the diet over a few weeks. Recommended levels of dietary fibre for males is a minimum of 30 g/d and for females, a minimum of 21g/d. There are two main types of dietary fibre: soluble and insoluble. These are processed in different ways, and each carries different health benefits. Examples of healthy carbohydrates containing fibre are fruit, celery, leafy green vegetables (Figure 10.2).

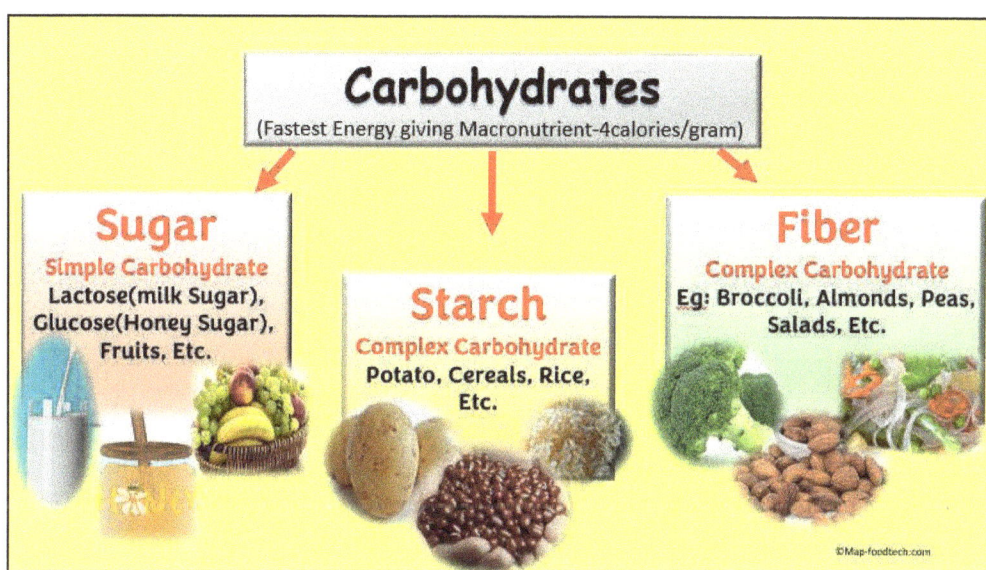

Figure 10.2 Food examples components of carbohydrates (CHO), fiber, starch and sugars. Source Map-foodtech.com

10.13   Soluble Fibre

Soluble fibre can help lower blood cholesterol and stabilise glucose levels. It can also lower fat absorption due to its gelling properties. It is found in oats, peas, beans and legumes such as lentils, apples, berries, citrus fruits, carrots, barley, cucumber, flax seeds, chia seeds, strawberries, coffee. Some fibre rich soluble and fermentable foods feed the gut microbiome and enable the bacteria to thrive for longer periods and hence promote the growth and health of the bacteria to benefit the host human body.

10.14   Insoluble Fibre

Insoluble fibre promotes the movement of material through the gut and increases stool bulk. It draws fluid into the gut and sticks to other by-products of digestion that are ready to be formed into the stool. Its presence therefore, speeds up the movement and processing of waste, helping prevent gastrointestinal blockage and constipation or reduced bowel movements. Examples of insoluble fibre are; brown and whole grain rice, whole-wheat flour, quinoa, wheat bran, nuts, beans, kale, cauliflower, green beans, almonds, potatoes.

10.15   Carbohydrate Promotes Gut Flora and Gut Health

Starchy carbohydrates such as potatoes and rice in resistant starch have been shown to promote and affect the abundance and composition of the intestinal flora in the body. Estimates of the total number of gut micro-organisms are in the order of 38 trillion, compared with around 30 trillion human cells, thus, these symbiont friends are mightily important for the human gut and nutritional health. Fibre from carbohydrates is essential for a healthier gut microbiome and the bacterial production of Short Chain Fatty Acids (SCFA). SCFAs are produced when the gut bacteria ferment fibre and are very important for health.

Although these molecules are types of fatty acids, they are included in this discussion of carbohydrates, rather than included in Chapter 12 on fats because they can only be formed by eating carbohydrates and only then synthesised by other organisms, i.e., the gut microbiome. They are composed of a straight hydrocarbon chain, terminating in a carboxylic acid group, with a polar hydrophilic end and a straight hydrophobic end. They play a key role in many gut health processes. The most commonly abundant SCFA are propionate, acetate and butyrate. Most SCFA produced are used for energy by the human cells lining the gut. The remaining SCFA enter the blood stream and then go on to promote substrate energy metabolism and provide approximately 5-10% of daily energy IE calorie requirements to the rest of the human body. The SCFA help keep the epithelial cells close together, reducing the risk that unwanted compounds may leak into the bloodstream, called 'leaky gut'. SCFA also help the gut lining intestinal epithelial cells, aiding food digestion significantly, absorbing nutrients and helping protect against infections. Epithelial cells also pump hormones into the bloodstream to communicate with the nervous system. Some SFCA bind to receptors and trigger release of 2 hormones, GLP-1 and peptide YY and also lower blood glucose by stimulating the release of insulin and slowing the release of glucagon. There are many studies demonstrating the health benefits of short chain fatty acids (SCFA). In the fasted state, free fatty acids (FFA) from adipose tissue circulate in the plasma in association with the protein albumin (the most abundant protein circulating in the blood). For the purpose of clarification, the term Free Fatty Acid (FFA) is the same as Non Esterified Fatty Acid (NEFA) and refers to a fatty acid which is not esterified with glycerol to form a glyceride or triacylglycerol (TAG). It is then free either to be oxidised or to be re-esterified. The FFA are then taken up by the liver or muscles and oxidised via acetyl-CoA. In the liver, some of the acetyl-CoA is diverted to produce ketone bodies. The liver continues to synthesise VLDL and HDL and releases these into circulation. This action is reduced in prolonged starvation. The LDL fraction is the major carrier (around 60%) of cholesterol binding. The FFA is activated by coenzyme A, yielding adenosine monophosphate (AMP). The (β)-oxidation of fatty acids occurs within the mitochondria. SCFA can pass into the mitochondrion matrix directly. Long chain fatty acids (LCFA) can-not cross this membrane and require carnitine for transport. SCFA can also regulate and improve the function of insulin and improved response to carbohydrate in meals. There is also some evidence that SCFA play a party in healthy weight maintenance, due to GLP-1 and Peptide YY increasing satiety, plus the action of SFCA on fat cells promotes the release of the hormone leptin from adipose tissue, which reduces appetite. SCFA also influence glycolysis (the cellular break down of glucose for energy), gluconeogenesis and glycogenesis, (glycogen synthesis from the conversion of glucose to glycogen).

10.16 Cellulose

Cellulose is a type of insoluble fibre that makes up the rigid cell walls of most plants and is the most abundant of all natural organic compounds. Cellulose is a linear, polysaccharide polymer with many glucose monosaccharide units. The acetal linkage is beta, which makes it different from starch. Humans are unable to digest cellulose because the appropriate enzymes to breakdown the beta acetal linkages are lacking.

Animals such as cows (ruminants), horses and termites have the necessary symbiotic bacteria to digest cellulose in their GI tract but no animal can do this alone without these bacteria. Indigestible cellulose is the fibre which aids in the smooth working of the intestinal tract. It should be included as part of a healthy diet.

Table 10.1 Carbohydrate component types in common foods

Type of Carbohydrate Per 100g	Apple	Oats	Kale	Potato
Sugars (Glucose, Fructose)	11.8	0.99	0.5	0.3
Starch	0.9	56.61	0.1	15.2
Fibre	2.1	10.1	3.8	1.7
Total Carbohydrate	14.8	67.7	4.4	17.2

10.17 Glycogen

Glycogen is stored by humans in the liver and muscle, as the main source of stored carbohydrate for energy reserves and body needs for approximately 24 hours (Figure 10.3).

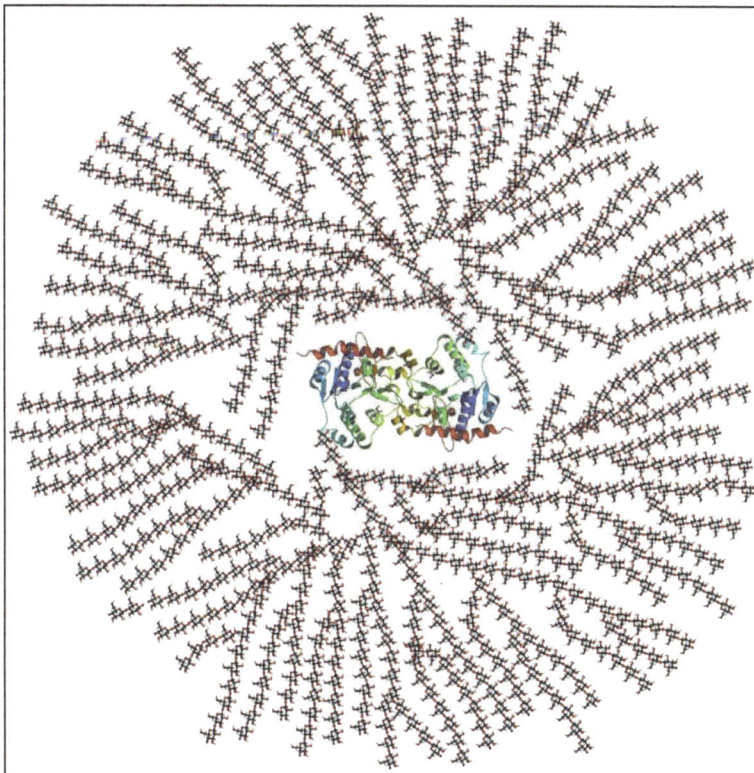

Figure 10.3 Schematic two dimensional cross-sectional view of glycogen: A core protein of glycogenin is surrounded by branches of glucose units. The entire globular granule may contain around 30,000 glucose units. Source Wikipedia. Mikael Häggström, (2014). "Medical gallery of Mikael Häggström 2014". WikiJournal of Medicine 1 (2). DOI:10.15347/wjm/2014.008. ISSN 2002-4436. Public Domain.

10.18   Glucose the Preferred Accessible Energy Fuel

Glucose is absolutely central to the body's entire energy consumption and the body's first choice to provide a rapid, preferable source of energy. Carbohydrates and also proteins ultimately break down into glucose, providing the primary metabolic fuel of humans. Glucose is the principal energy substrate for the placenta and the fetus and is essential for normal foetal metabolism and growth. Both glycogen and glucose play a key role in regulating calcium levels in muscle ensuring efficient contraction. As discussed in Section 1.2, there is a great deal of misinformation continually reported on social media and the internet and misleading portrayals, effectively vilifying and demonising carbohydrates. The proponents of so called 'low carb', 'keto', 'paleo' or even ….'carnivore' diets often claim that high-protein and/or high-fat diets, with considerable restriction on carbohydrates, is the best way to lose weight and maintain good health. This has been shown to be ineffective for long term sustainable weight loss.

Humans developed as starch, plants, seeds and nut feeders and have never been carnivores. Any meat ancient humans did manage to catch was, most probably, a small part of the diet after humans discovered fire and cooking. Basic anatomy and physiology of the human body, including sweating, stomach acidity pH, cholesterol maintenance, the inability to synthesise vitamin C, the mouth, chewing, teeth, jaw, gut, male seminal vesicles, placenta shape, urine concentration, colour vision, sleep duration, hands, nails instead of claws and running ability all confirm this.

On the contrary, there is an abundance of evidence that clearly confirm that a diet rich in wholesome, complex, carbohydrates is beneficial to health and fitness of all humans, including athletes. Note, the best endurance runners throughout history, across the globe, dominating sporting events for decades, largely from Kenya and Ethiopia. The carbohydrate in their diet is >75% of total daily caloric intake (EI). This does not mean that one should ingest refined carbohydrates, such as fructose or free sugars found in many processed foods, white bread, fast food and cakes, pastries, syrups and biscuits, orange juice and energy and cola energy drinks. Instead, concentrate on the wealth of all the healthy, complex carbohydrates found in starches, grains legumes and vegetables.

10.19   The Preservation of Glucose

The body goes to elaborate lengths to preserve glucose. Both glucose and fatty acids are the body's evolved, main energy substrates. Glucose is required by all cells as the most accessible and least expensive easily fuel for all cellular activities. Glucose is essential for (RBCs) and the preferred fuel for the brain, CNS and other key tissues. Glucose is the major form of energy for muscular work in glycolysis, via anaerobic breakdown of glycogen to lactate. The body preserves glucose, preferentially, by utilising plentiful adipose (fat) rich as an energy source for other constantly energy demanding organs such as the heart. The heart muscles normally utilise 60% of their energy requirement from fatty acid (fats) (β)-oxidation (glucose 40%) and up to 90% when glucose is low. Similarly, the kidney cortex regularly uses fats as fuel, in order to preserve and reserve, much lower stocks of glycogen glucose stores for the brain, and other obligate glucose reliant cells. On top of this, the ingeniously evolved body (i.e., the liver) carefully selects non-carbohydrate fuel substrates such as glycerol, lactate, amino acids and propionate for gluconeogenesis. Furthermore, if absolutely necessary, during starvation, the body and brain uses a significantly higher amount of ketone bodies (emergency back-up chemicals for fuel converted from fats) to protect RBCs. This will be discussed further in Sections 14.14 and 14.39

10.20 Glucose is Essential for Efficient Fat Burning

Glucose is essential for the efficient (β)-oxidation of fats for fuel, via oxaloacetate to acetyl-CoA, leading to citrate and the Tricarboxylic Acid (TCA) cycle. Appropriate levels of fat, glycogen and glucose must be present for the body to efficiently burn fat as energy. Without a sufficient quantity of glucose, the body will increase gluconeogenesis (the formation of glucose from non-carbohydrate precursors), divert oxaloacetate and burn other substrates such as lactate, glycerol and proteins, reducing fatty acid beta (β)-oxidation. An efficient body, managing weight, anabolic and catabolic processes and cellular functions, requires a sufficient and steady rate of dietary carbohydrate as glycogen and glucose then pyruvate, to ensure the metabolic processes, including normal fat burning, are optimally regulated.

10.21 Carbohydrate is Both Protein and Fat Sparing

Carbohydrates are also more readily available for energy supply compared to fats which makes them more vital for preserving protein. Carbohydrates have a key protein sparing action via a balance between glucose preservation and structural protein preservation, through reduction of anaplerotic gluconeogenesis. This ensures protein is retained for structural functions, not fuel (Table 14.3 and Figures 14.6 and 14.7). One of these actions is through the liver, directing the mechanism to use other non-protein substrates in gluconeogenesis. Also, preventing too much gluconeogenic use of the principle glucogenic amino acid alanine, therefore, protecting muscle and other protein structure from long term depletion. Carbohydrates stimulate the release of insulin, which plays a role in promoting protein synthesis and inhibiting protein breakdown. When insulin levels are elevated, it helps to drive amino acids into muscle cells for repair and growth, which can be beneficial for preserving muscle mass. As the primary source of energy, especially during high intense activity, consuming sufficient carbohydrates helps replenish glycogen stores, which spares protein from being used as an energy source during physical activity. In the same way, when the body is in the fed state and particularly after a large meal, the body will hold on to its adipose fat tissue, supress lipolysis and also add further dietary fat to the adipose reserves via lipogenesis, reforming triacylglycerol from the constituent fatty acids and glycerol, broken down during digestion.

10.22 Carbon and Nucleic Acid Structure

Carbohydrates provides essential pentose sugars such as ribose and carbon atoms for many essential proteins, lipids and the structural framework of nucleic acids (DNA). The 5 carbon monosaccharide ribose is an important component of essential coenzymes; ATP, ADP, ATP, FAD, NAD^+, NADP and the backbone of RNA. Deoxyribose is a component of DNA. Carbohydrates are key molecules in the synthesis of many other essential molecules and proteins from pyrimidines to porphyrins, purines and many more.

CHAPTER

11

Protein

11.1 What is Protein

Dietary proteins are classes of nitrogenous organic compounds, composed of one or more long chains of amino acids, essential for body tissue growth, repair structure, skeletal muscle, bones and tissue and cell maintenance. Protein is a vital, major component of DNA, enzymes, hormones, cell membranes and antibodies. They transport and synthesise other proteins and also enable other metabolites to be carried through system circulation. They are very large molecules and perform the majority of the complex work within each cell, including the final release of cellular energy from food. The word protein is derived from the Greek 'proteios' meaning first place or primary and first adapted to emphasize the key importance of these molecules in the processes of all living things on earth.

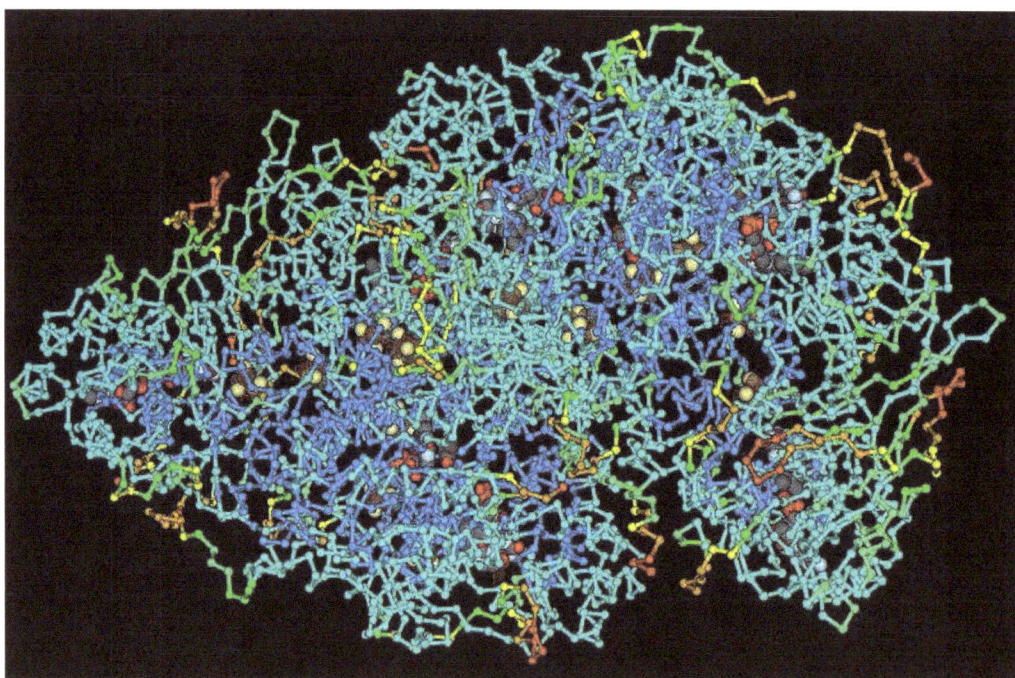

Chemical Protein structure of a polypeptide macromolecule Source: enwikipedia to commons by user Maksim c2006

It is estimated that the human body contains between 80,000 and different 400,000 proteins. Each of these comprises a chain of 50 to 2,000 amino acids, which has a specific composition and a unique large molecular weight in the range from 6,000 to 3,000,000 Daltons (one Dalton is the weight of one hydrogen atom) and its own sequence of different amino acids along the polymer chain. Most proteins have a molecular weight ranging from 10,000 to 100,000 Daltons. Titin, is the third most abundant protein found in muscle tissue and is the largest protein in the human body with a molecular weight of 3 million Dalton and composed of 27,000 amino acids.

An adult human contains approximately 0.5 kilograms of this one protein alone from a total body protein of around 9.5 kg. A single cell can contain many thousands of proteins, each with a unique function. Proteins are found in all parts of the body in the following proportions:

Skeletal Muscle	40% of total body protein
Skin and Blood	35% of total body protein
Organs	25% of total body protein

11.2   Dietary Protein & Weight Management

There is evidence that a high protein diet (HPD), for example where protein is 18-25% of total energy intake leads to weight loss and sustained weight loss over time. [81,] Many clinical trials have shown that consuming more protein than the recommended dietary allowance (10%) induces weight loss and improves body composition regardless of total energy intake. HPD has also observed to have long-term weight-loss effects and to prevent weight regain following initial weight loss. [82, 83, 84] There is also evidence that HPD, such as ketogenic, paleo or carnivore diets may lead to kidney problems, metabolic issues bone deterioration and a lower lifespan (Section 11.4) [85, 86,87, 89] . Longer term clinical trials are needed to fully identify these risks for those considering adopting such diets in the long term (>6 months).

The author recommends adherence to current guidelines of around 10-15% of DI or around a minimum of 50 gram of protein/d until and if these guidelines are revised. Low protein, high fat diets lead to several metabolic disorders, especially compared to low protein high carbohydrate diets and a low protein and high fibre diet can reduce and prevent the development of insulin resistance and type 2 diabetes.

This reinforces the importance of the holistic macronutrient approach to a balanced diet as discussed in Section 1.4. On balance the evidence suggests a diet based mainly on plants and lower consumption of animal protein in the order of 60%-75% carbohydrate, 10%-15% protein and 10%-30% fat (<10% total saturated fat).

For managing weight and overall health, protein should, ideally represent approximately 10-15% of total dietary intake with the majority of protein derived from plant based sources such as legumes such as kidney beans and lentils, broccoli, oats and nuts.

11.3   Protein Deficiency

Protein deficiency is extremely rare in the Western world. The vast majority of people are eating plenty of protein and some to excess. The current guideline is around 0.8 g/kg/d body weight, which works out to around 48 g/d to 56 g/d (females and males respectively) and 9-10% of total daily intake EI. This data was based on experiments conducted fifty years ago during the 1970's and assumes that people are eating meat, eggs and dairy (cheese). However, most people are eating more protein than this, nearer a minimum of 13% of total daily intake EI and equates to 72 g/d total and an average of 1.2 g/kg/d for females and 87 g/d total and 1.24 g/kg/d for males.

There is a difference in the distribution profile of essential amino acids in meat, eggs and dairy compared with the amino acids in plant foods. On these plant-based diets, more protein intake is required than the average guideline, increased from 0.8 g/kg/d to around 1.2 g/kg/d. This equates to a total daily amount of protein at 72 g/d or 14% total daily EI for females and 84 g/d and 13% of total daily EI for males. This is easily and unnoticeably achieved on a whole food plant based diet. Deficiency on a plant-based diet is therefore unlikely, as the majority of people are in fact eating protein at around 1.2 g/kg/d.

Therefore, true protein deficiency is rare, especially across Western populations consuming high quantities of meat and cheese. Extreme and poorly designed vegan diets can lead to consumption of lower amounts than the guidelines, if the individual is not consuming sufficient daily calories, where EI is significantly less than TEE.

However, there is plenty of sufficient protein in a healthy plant-based diet, providing sufficient total calories are consumed and even on a weight loss plant-based diet. The main symptoms of significant protein deficiency are swollen skin, loss of hair, poor nails, loss of muscle mass, weakened bones, stunted growth in children, lower immunity, increased appetite leading to intake of excess fat and carbohydrates and build-up of fat in the liver, i.e., fatty liver disease.

There is recent compelling evidence that protein should be increased to a range from 1.2g/kg/d to 1.6g/kg/d to improve healthy ageing for people over 65 and that this should be from plant sources, combined with exercise and some resistance training 2 to 3 times per week. [63]

11.4   Protein Excess

As ever, the watchword is dietary sufficiency, i.e., dietary deficit or dietary excess. Whatever macronutrients are eaten, when the EI is in excess of TEE, the body will not burn fat sufficiently to diminish fat stores and fat stores will grow. When EI is equal to TEE, body fat will remain at the same level. However, when EI < TEE, the body will use up glycogen then start to increase fat burning, at a net rate in excess of fat storage. This is no different with excess dietary protein.

Unlike carbohydrate and fat, protein does not have a specialized storage system to be used later for energy. Protein is not stored. If a person is consuming a high protein diet and also eating more calories than their body needs, the extra amino acids will be broken down (removing nitrogen) and these will be used as fuel in gluconeogenesis (Section 14.16), or lost in urine or transformed into glucose or glycogen and eventually a small amount of fat via the highly bioenergetic cost in the process of de novo lipogenesis (DNL) Section 14.40), however, the key is that with protein excess, less fat will be burnt than stored.

Excess dietary protein is metabolized by the liver as urea, cleaving off nitrogen and excreted in the urine. Protein can-not be stored for energy and has the highest thermic effect up to 35% of consumed energy if used as fuel, hence wasteful in terms of heat, when used as fuel for the body's energy needs in times of deliberate carbohydrate deprivation or genuine shortage. Eating a high-protein diet, or supplementing, will not therefore, necessarily help the body build more muscle protein.

On the contrary, it is movement through exercise and resistance training that builds protein and muscle strength rather than from attempts to eat additional, excessive protein in the diet. Many high protein diets (>2.0 g/kg body weight) or costly protein powder supplements including whey powder) are effectively flushed down the toilet on urination, as the body excretes much of the excess protein.

On average, most people require in the order of approximately 10% -15% of total daily energy intake (EI) as protein. That is in the order of 0.8 g to 1.2 g/kg body weight (50 g to 62 g to 75 g to 95 g, females-males) of dietary protein per day. People in the UK, who are taking at least 10% of their energy as protein are not protein deficient. Although some studies show that excess dietary protein (25% of DI) does not lead to harmful effects on the kidneys and bones, [81] Despite these studies, many scientists consider that excess dietary protein can also lead to health issues and damage tissues, especially the kidney.

The kidney has to deal with the extra urea which results in higher demand for water, and potential dehydration. Further problems include hyper urea cycle disorder, aminoacidemia (excess amino acid in blood), hyperammonaemia (excess ammonia), hyperinsulinemia (excess insulin), calcium loss and overreaction within the immune system.

If protein levels exceed approximately 16 % protein as a proportion of total daily dietary intake (EI), this may produce a negative calcium balance, leading to loss of calcium in the urine and excess protein is the principal cause of osteoporosis. [56] [85, 86, 87] More long-term clinical trials are required to identify a safe upper limit of high protein diets.

The vast majority of protein (75%) is used in the liver for synthesis of other body proteins, hormones and enzymes etc., plus a further 25% for muscle turnover and synthesis. Conversion of protein to significant quantity of fuel energy via gluconcogenesis is physiologically, bioenergetically expensive and intensive and only occurs when there are insufficient carbohydrates or fat for energy needs and during starvation, first using dietary protein, then catabolising structural protein in the organs and muscles for amino acid conversion to energy.

Low carbohydrate, or carbohydrate deprivation diets cause the body to convert protein amino acids to glucose via gluconeogenesis, to ensure blood glucose levels are both safe sufficient and glycogen stores are topped up.

Consult a doctor if you think you require high protein or supplementation. High and continued excess protein intake can lead to liver and kidney issues, where the liver is unable to convert nitrogen into urea quickly enough.

11.5 Protein Renewal

It is necessary to synthesise approximately 300 g of new protein each day, irrespective of age. This total amount is replaced or 'turned over' in the human body each day. Protein dietary intake is around 70 g/d. Amino acids for synthesis (300 g/d) mostly come from endogenous breakdown (300 g/d). Thus, net intake and excretion is the same (70 g/d).

Although skeletal muscle is around 40% of total body weight and contains between 50-75% of body protein and has the highest tissue mass specific metabolic rate in the (male) body (Table 14.2), only 25% of the c300 g daily turnover, i.e., 75 g/d is for muscle turnover. The majority, i.e.,75% of total body protein turnover, or 225 g/d is combined in the liver, gut and kidneys.

The rate of different tissue turnover varies considerably throughout the body with some proteins turned over on an hourly basis in the liver, whereas some muscle turnover may be in the order of 40 days and for example, collagen every 3 years. All body protein is replaced around 4 times per year.

The maximum net rate of muscle growth from exercise is approximately 5 g/d, due to muscle turnover. This means making a total of around 30 g/d, to derive a net of 5 g/d additional lean mass. Essential amino acids are continually oxidised to supplement energy and the amount of these amino acids within the diet is equal to this oxidation and excretion.

During rest and sleep, skeletal muscles supply the liver with amino acids, as the liver is constantly synthesising and processing amino acids for new proteins via transamination and deamination or gluconeogenesis (Section 14.16). Hence, during the fasted state (catabolic), muscle tissue is in degradation mode and during the fed state (anabolic) muscle tissue is in the positive build mode.

11.6 Multitasking Functions

The collection of proteins within a cell determines its health and function. Proteins are responsible for nearly every task of cellular life, including cell shape and inner organization, product manufacture and maintenance. Proteins also receive signals from outside the cell and mobilize intracellular response.

They are the workhorse macromolecules of the cell and are as diverse as the functions they serve. Proteins can be large or small, mostly hydrophilic (water loving) or mostly hydrophobic (water hating).

They exist alone or as part of a multi-unit structure, and change shape frequently or remain virtually immobile. All of these differences arise from the unique amino acid sequences that make up proteins. Fully folded proteins also have distinct surface characteristics that determine which other molecules they interact with.

When proteins bind with other molecules, their conformation can change in subtle or dramatic ways.

11.7 Enzymes

Enzymes are proteins that conduct specific chemical reactions. The release of energy relies on enzymes. An enzyme's function is to provide a site for a chemical reaction and to lower the amount of energy and time for that chemical reaction to occur (catalysis). On average, more than

100 chemical reactions occur in cells every second, and most of them require enzymes. The liver alone contains over 1,000 enzyme systems.

Enzymes are specific and will use only particular substrates that fit into their active site, similar to the way a lock can be opened only with a specific key. Fortunately, an enzyme can fulfil its role as a catalyst over and over again, although eventually it is destroyed and rebuilt. All bodily functions, including the breakdown of nutrients in the stomach and small intestine, the transformation of nutrients into molecules a cell can use, and building all macromolecules, including protein itself, involve enzymes.

11.8 Hormones

Proteins are responsible for hormone synthesis. Hormones are the chemical messengers produced by the endocrine glands. When an endocrine gland is stimulated, it releases a hormone. The hormone is then transported in the blood to its target cell, where it communicates a message to initiate a specific reaction or cellular process. For instance, after a meal is eaten, blood glucose levels rise. In response to the increased blood glucose, normally the pancreas beta cells release the hormone insulin. Insulin is the principle anabolic hormone and informs the cells of the body that glucose is available and instructs the cells to take glucose and other nutrients up from the blood for energy or store it, or use it for building macromolecules. A major function of hormones is to turn enzymes on and off, so some proteins can even regulate the actions of other proteins. While not all hormones are made from proteins, many of them are. Section 14.3 discusses hormones in further detail.

11.9 Collagen

By far the most abundant protein in the human body is collagen which makes up about 6% of total body weight. Collagen also makes up 30% of bone tissue and comprises large amounts of tendons, ligaments, cartilage, skin, and muscle. Collagen is a strong, fibrous protein made up of mostly glycine and proline amino acids. Within its quaternary structure, three protein strands twist around each other similar to a rope. These collagen ropes then overlap with others. Collagen makes bones strong and flexible.

Collagen fibre in the skin's dermis provide the skin with structure, and the accompanying elastin protein fibrils make it flexible. Collagen and elastin proteins keeps the flexibility and shape of the skin.

Smooth-muscle cells that secrete collagen and elastin proteins surround blood vessels, providing the vessels with structure and the ability to stretch back after blood is pumped through them. Another strong, fibrous protein is keratin, an important component of skin, hair, and nails.

11.10 Transport

Proteins also play vital roles in transporting substances around the body. For example, the most abundant protein in blood is albumin, very distinctively butterfly-shaped. This protein has important roles in a multitude of ways in the body. For example, it chemically binds to hormones, fatty acids, some vitamins, essential minerals, and drugs, and transports them throughout the circulatory system.

Each red blood cell contains millions of protein haemoglobin molecules that bind oxygen in the lungs and transport it to all the tissues in the body. A cell's plasma membrane is usually not permeable to large polar molecules, to enable transport of the required nutrients and molecules into the cell, many transport proteins exist in the cell membrane. Some of these proteins are channels that allow particular molecules to move in and out of cells. Others act as one-way taxis and require energy to function.

11.11   Fluid Balance

Adequate protein intake enables the basic biological processes of the body to maintain homeostasis (constant or stable conditions) in a changing environment. One aspect of this is fluid balance is keeping water distributed properly in the different compartments of the body.

If too much water suddenly moves from the blood into a tissue, the results are swelling and, potentially, cell death. Water always flows from an area of high water concentration to an area of low water concentration. As a result, water moves toward areas that have higher concentrations of other solutes such as proteins and glucose.

To keep the water evenly distributed between blood and cells, proteins continuously circulate at high concentrations in the blood. The presence of albumin in the blood makes the protein concentration in the blood similar to that in cells. Therefore, to preserve homeostasis, fluid exchange between the blood and cells is minimized, rather than extreme.

11.12   Acid Base Balance

Protein is also essential in maintaining proper pH balance (the measure of how acidic or basic a substance is) in the blood. Blood pH is maintained between 7.35 and 7.45, which is slightly basic (alkaline). Even a slight change in blood pH can affect body functions. The body has several systems that hold the blood pH within the normal range to prevent this from happening. One of these is the circulating, albumin.

Albumin is slightly acidic, and because it is negatively charged it balances the many positively charged molecules circulating in the blood, such as hydrogen protons (H^+), calcium, potassium, and magnesium. Albumin acts as a buffer against abrupt changes in the concentrations of these molecules, thereby balancing blood pH and maintaining homeostasis. The protein haemoglobin also participates in acid-base balance by binding hydrogen protons.

11.13   Immunity

Proteins also play important roles in the body's immune system. The strong collagen fibre in skin provides it with structure and support, but it also serves as a barricade against harmful substances. The immune system's attack and destroy functions are dependent on enzymes and antibodies, which are also proteins. For example, the enzyme lysosome is secreted in the saliva and attacks the walls of bacteria, causing them to rupture.

Certain proteins, circulating in the blood, can be directed to build a molecular weapon that damages the cellular membranes of foreign invaders. The antibodies secreted by white blood cells survey the entire circulatory system, searching for harmful bacteria and viruses to surround and destroy. Antibodies also trigger other factors in the immune system to seek and destroy unwanted intruders.

11.14 Another Back-up Energy Source

Most of the amino acids in proteins can be disassembled in the liver and used to make energy. The complete oxidation of amino acids generates:

$$Energy + CO_2 + HCO^{3-} + NH3 + NH4^+$$

About 5 -10% of endogenous proteins are catabolized each day to make cellular energy via the process of gluconeogenesis and proteins represent a theoretical 3% of average daily total energy storage reserves in the body (Section 14.16) in the fed state.

The amount of protein degradation in this pathway increases during fasting and starvation to preserve and maintain glucose levels. The liver is able to break down amino acids to the carbon skeleton, which can then be fed into the TCA (or Krebs) cycle.

During this process, the amino groups are cleaved and lost to urea which then enters the blood. These glucogenic amino acids can also be used to synthesize glucose for energy, through gluconeogenesis. If a person's diet does not contain enough carbohydrates and fats, their body will use more amino acids as a back-up reserve to make energy for the body.

This can compromise the synthesis of new proteins and destroy muscle proteins if caloric intake is also too low.

11.15 Metabolic Messengers

Other proteins work as monitors, changing their shape and activity in response to metabolic signals or messages from outside the cell. Cells also secrete various proteins that become part of the extracellular matrix or are involved in intercellular communication.

Proteins are sometimes altered after translation and folding are complete. In such cases, so-called transferase enzymes add small modifier groups, such as phosphates or carboxyl groups, to the protein.

These modifications often shift protein conformation and act as molecular switches that turn the activity of a protein on or off. Many post-translational modifications are reversible, although different enzymes catalyse the reverse reactions. For example, enzymes called kinases add phosphate groups to proteins, but enzymes called phosphatases are required to remove these phosphate groups.

11.16 Cell Cytoplasm Organisation

The cytoplasm of cells is highly structured due to proteins. Particularly in eukaryotic (complex organisms) cells, which tend to be larger and need more mechanical support than prokaryotic (single cell organisms) cells, an extensive network of filaments (microtubules, actin filaments, and intermediate filaments) can be detected with microscopic methods. Microtubules play a major role in organizing the cytoplasm and in the distribution of organelles (cell sub components). They also form the mitotic spindle during cell division. Actin filaments are involved in various forms of cell movement, including cell locomotion, contraction of muscle cells, and cell division. Intermediate filaments are strong fibre that serve as architectural supports inside cell.

11.17 Amino Acids

Amino acids are the sub units that make up proteins. A protein is made up of one or more linear chains of amino acids, each of which is called a polypeptide. There are 20 types of amino acids commonly found in proteins in humans (Table 11.1).

Table 11.1 List of essential (required within the diet) and non-essential (not required within the diet) amino acids

Essential Amino Acids (9)	Non-Essential Amino Acids (11)
Histidine	Alanine
Isoleucine	Arginine
Leucine	Asparagine
Lysine	Aspartic Acid
Methionine	Cysteine
Phenylalanine	Glutamine
Threonine	Glutamic Acid
Tryptophan	Glycine
Valine	Proline
	Serine
	Tyrosine

11.18 Essential Amino Acids

Table 11.1 also lists the 9 essential amino acids that the body is unable to synthesise and these must be therefore be obtained exogenously via diet. The remaining 11 non-essential amino acids can be synthesised endogenously via transamination, from free ammonia, or from modification of carbon skeletons of existing amino acids in liver hepatic cells.

11.19 Plant Have all Amino Acids

Note Table 11.2. Contrary to common belief and misinformation, plants contain all 20 amino acids, including all 9 essential amino acids. It is certainly true that the proportions of each amino acid differ in meat and plants for example, legumes have proportionately less methionine and cysteine than meat and grains have less lysine than meat. However, a Whole Food Plant Based Diet (WFPB) (Section 9.4) provides a healthy nutrient rainbow, that comprises a wide variety of nutrients and contains all amino acids, plus sufficient carbohydrates and fats, fibre, macronutrients, micronutrients, antioxidants, phytonutrients, minerals, and vitamins. There is no need to complete the amino acid profile in every meal. [54, 60, 63] A sufficiently caloric plant-based diet is likely to guarantee the sufficient daily protein requirements. Grams per calorie, plant foods have high amounts of protein. They contain almost as much protein as animal foods. [60, 63] It is not possible to be deficient in protein on a healthy plant-based diet, unless an individual is starving, or not eating a sufficiently balanced diet with sufficient overall calories.

What usually occurs on a WFBP diet is that because of the highly nutritious food and the high nutrient and low caloric density ratio, people are obtaining enough protein and amino acids (1.2g/kg/d) from vegetables, grains and legumes, with sufficient quantities of lysine, methionine and cysteine. As discussed in Section 11.3, most people are in fact eating on average around 1.2 g/kg/d, at around 14% of daily calories from protein and which is perfectly adequate on any diet. This can be increased up to 1.6 g/kg/d for people over the age of 65. [63] For example, if an individual is on a healthy weight loss plan, then sufficient protein will be provided from vegetables, legumes, grains, seeds and nuts as shown in Table 11.2. This lists the foods that specifically contain the essential amino acids but recall that plants contain all 20 amino acids. It is evident that protein insufficiency on a WFPB diet is highly unlikely. For those wishing to explore adopting a plant-based diet, humans do not require extra protein or modern-day synthesised supplements, when their diet comprises a wide range of healthy plant-based foods, including lentils, beans, grains and vegetables and foods with fibre. [60, 63]

Table 11.2 List of essential amino acids and examples of plant sources

Essential Amino Acids (9)	Examples of Plant Sources
Histidine	Oats,Quinoa,Hemp,WG Rice,Cauliflower,Pumpkin, Potatoes,Legumes, Wheat,Almond, Chia
Isoleucine	Oats, Quinoa, Hemp, Lentils, Beans, WG Rice, Cabbage, Spinach, Blueberries, Apples, Chia
Leucine	Peas, Quinoa, Hemp, WG Rice, Watercress, Kidney Beans, Bananas, Olives, Apples, Sesame
Lysine	Oats, Quinoa, Hemp, Beans, Watercress, Parsley, Lentils, Chickpeas, Almonds, Avocado
Methionine	Oats, Quinoa, Hemp, Sweet Potatoes, Spinach, Wheat, WG Rice, Peas, Onions, Chia,
Phenylalanine	Quinoa, Hemp, Pumpkin, Beans, WG Rice, Avocado, Leafy Greens, Berries, Olives, Almonds
Threonine	Oats, Quinoa, Hemp, Watercress, Pumpkin, Leafy Greens, Almonds, Chia, Soy, Wheat
Tryptophan	Oats, Quinoa, Hemp, Spinach, Sweet Potato, Watercress, Chia, Carrots, Peppers
Valine	Oats, Quinoa, Hemp, Watercress, Leafy Greens, Wheat, Pumpkin, Avocado, Sunflower Seeds

Sources of leafy greens & cruciferous vegetables: kale, cauliflower, sprouts, spinach, broccoli, arugula, watercress, cabbage, Swiss chard

11.20   Branch Chain Amino Acids (BCAA)

The Branch Chain amino acids (BCAA) are three of the essential amino acids; valine, leucine and isoleucine. There is a popular movement currently regarding supplementation of these BCAA. Some people, particularly body builders and athletes, may take BCAA supplements and some reports suggest these may possibly reduce fatigue and improve endurance, improve muscle repair, lower creatinine levels in the body and reduced blood serotonin.

Changes in BCAA metabolism are common in a number of diseases and the BCAA have therapeutic potential due to their proven protein anabolic effects. However, more studies are needed to properly assess the effectiveness of supplementary BCAAs. Unless deficiency in any other minerals or requirements are confirmed from symptoms or medical tests, (or prescribed by the doctor) no supplements, apart from vitamin B12 are recommended.

CHAPTER

12

Fat

12.1 What is Fat

Dietary fat is the fat that comes from food and includes all the lipids of animal and plant tissues that are eaten, generally classified as visible and invisible fats. Visible fats are seen in different forms, either solid, for example, butter or liquid, for example, vegetable oils such as olive oil. Invisible fats are contained within meat, dairy products and processed foods.

The body breaks down dietary fats into smaller particles called fatty acids and glycerol that can enter the bloodstream. The body uses a small amount of these fatty acids for cellular processes and stores the majority as triacylglycerol (TAG) within the adipose (fat) stores, the most dense and largest storage reserve of energy (c>90% total stored energy) in the body.

The word fat in old English *fætt* meant "fat, fatted, plump, obese," a contracted past participle of *fættian*, which meant "to cram, or to stuff." In the 14th century it was also used to describe fat land, i.e., fertile or abundant and also to describe people who were wealthy or affluent, hence the term 'fat cat', still used today.

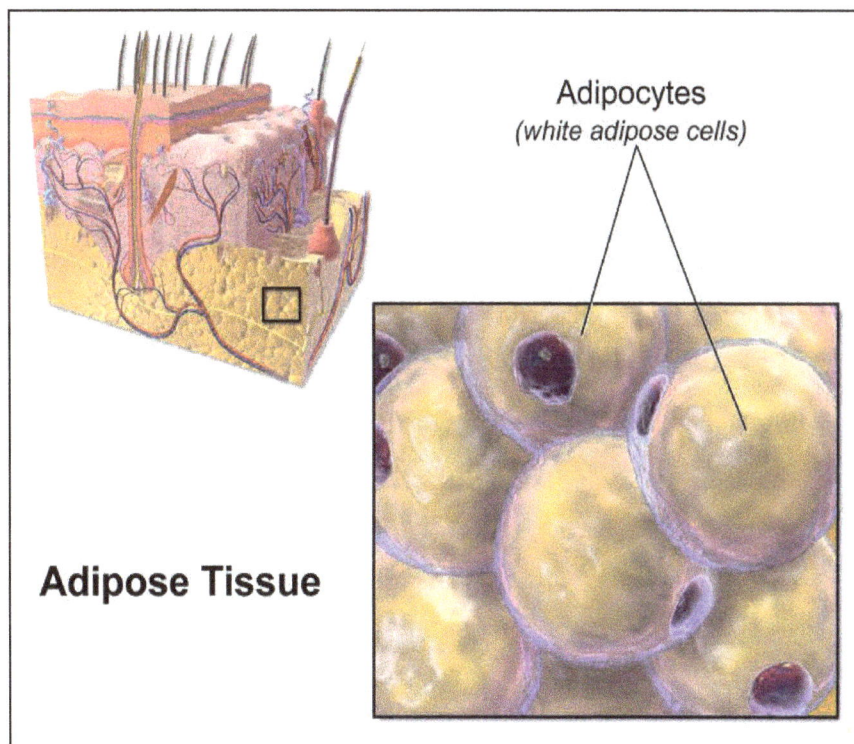

Adipocytes
(white adipose cells)

Adipose Tissue

Adipose Tissue
Blausen.com staff (2014). "Medical gallery of Blausen Medical 2014". WikiJournal of Medicine 1 (2). DOI:10.15347/wjm/2014.010. ISSN 2002-4436. - Own work

12.2 Dietary Fat & Weight Management

As for diets including excess calories from free sugars and processed foods, there is evidence that over consumption of fat and high fat diets lead to weight gain. [22, 34, 40, 90] Fat overfeeding has minimal effects on fat oxidation and total energy expenditure, leading to storage of 90-95% of excess energy. Excess dietary fat leads to greater fat accumulation than does excess dietary carbohydrate. [37] Increased fat intake significantly increases the risk of being overweight and development of obesity. [91] Carnivore, paleo, ketogenic diets and low carbohydrates are typically high in fat ranging from 50-75% fat. The carnivore diet eliminates all fruit vegetables and nuts and is typically 70-80% fat. There is a considerable amount of anecdotal comment including various celebrities claiming significant weight loss. Advocates of high fat and meat, carnivore, ketogenic and paleo diets claim weight loss (which is mostly due to an overall reduction in dietary intake) and increased energy and mental clarity.

Effectively, all these diets signal an altered metabolic state and starvation conditions to the body as carbohydrate and glucose is limited. Therefore the body burns fat but invokes priorities 1 and 2 to preserve status. Research does show that a ketogenic diet is an effective way to jump-start weight loss and improve blood-sugar levels. However, it is hard to maintain, and there remains a lack of statistically significant long-term evidence from clinical studies, that show it to be a sustainable eating pattern for keeping weight off compared with other diets. A weight loss study was conducted in 2003 on 63 adults, restricting either dietary fat (low fat) or carbohydrate (high fat) for 12 months. Although weight loss was 4% greater in the high fat group after 6 months, weight loss due to both diets were similar after 12 months. [95]

Another study in 2003 on 132 severely obese subjects was carried over a 6 month period to establish the weight loss achieved comparing a low fat high carbohydrate or a high fat low carbohydrate. [96] The results showed that the high fat diet group lost significantly more weight; (-5.8 ± 8.6 kg vs. -1.9 ± 4.2 kg) but the study concluded that the findings should be interpreted with caution, given the small magnitude of overall and between-group differences in weight loss in these markedly obese subjects and the short duration of the study and that future studies evaluating long-term cardiovascular outcomes are needed before a carbohydrate-restricted diet could be endorsed. Similar results were found in with 2 other studies, again in 2003 with duration over 3 months and 6 months respectively. [97, 98] However, there are other later studies that demonstrate the opposite, for example, the National Institutes of Health Clinical Center 2021 study. [25] The risk of a high fat diet and cardiovascular and diabetes health issues is well documented and also discussed in Chapter 1. Reviewing all the studies, literature and evidence it can be concluded that in order to maintain or lose weight, calorie (energy intake) adjustment and physical activity appear to be more important than the proportion of fat in the diet. However, an increase in total fat will invariably increase the intake of saturated fat and thus increase serum cholesterol and contribute to the CVD risk of obesity.

For managing weight and overall health, fats, such as chia, flax and hemp, sesame, sunflower seeds, walnuts and avocado should, ideally represent approximately 10-30% of total dietary intake. Animal fats, cheese, eggs, butter and all oils should be minimised as far as possible. Fats should preferably be omega 3 unsaturated fats and saturated fat should be minimised and not exceed 10% of total dietary intake.

12.3   Classification of Fats

Figure 12.1 shows the classification of fats into simple, compound and derived lipids. Note, the triglycerides (also referred to as triacyclglycerols TAGs) are the fats contributing to body fat and excess weight gain, as subcutaneous, visceral and ectopic fat. Triacylglycerol consists of a glycerol backbone with 3 fatty acids attached. Glycerol is a 3 carbon alcohol molecule, and is the structural backbone for the fatty acids and can be used as a gluconeogenic substrate energy substitute during low levels of glucose. Steroid hormones are synthesised from a sterol lipid cholesterol and proteins to form the sex hormones in the gonads and as corticosteroids in the adrenal glands.

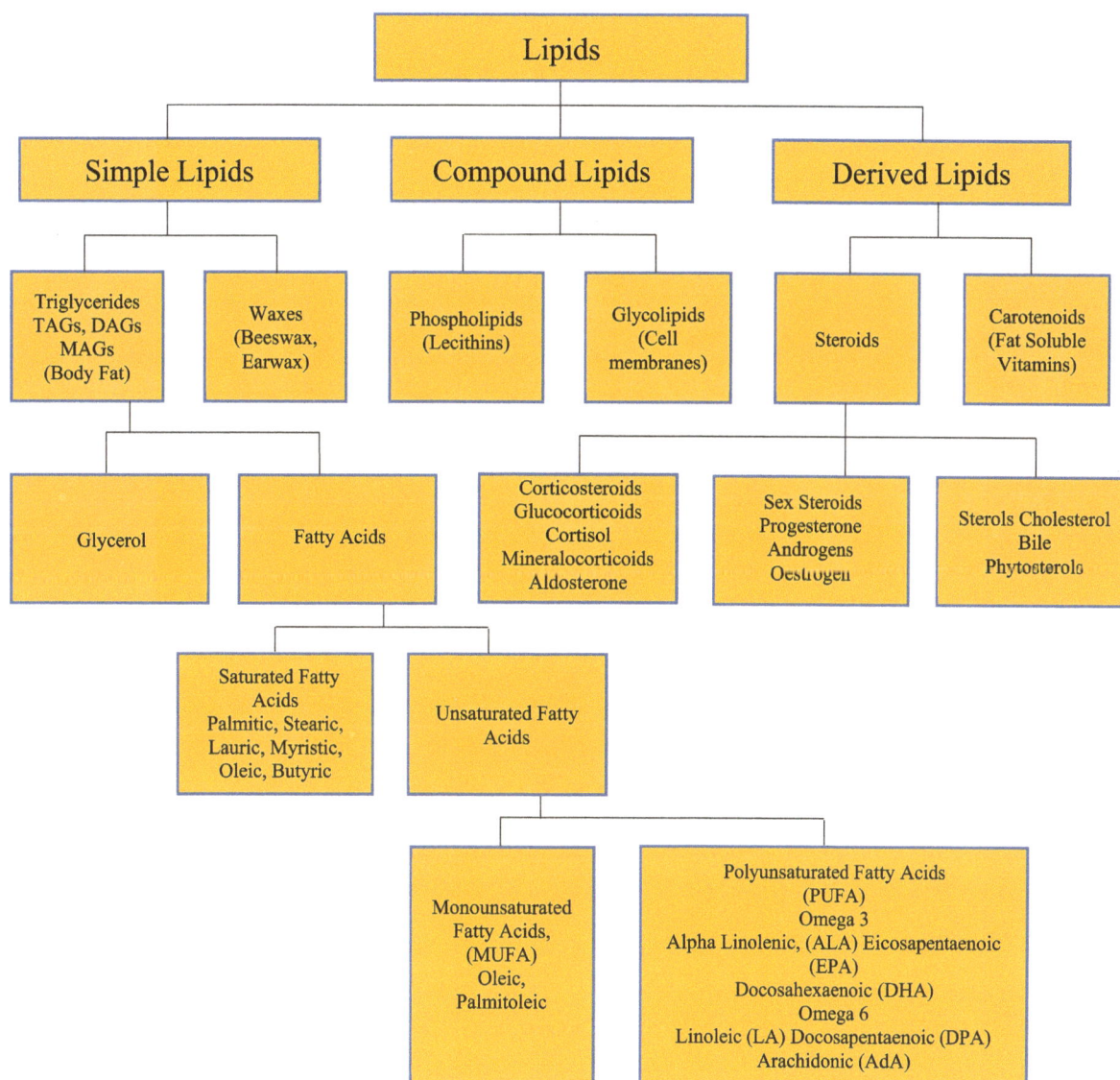

Figure 12.1 Classification of Lipids based on relevance to and health of the human body and available from dietary sources and synthesised in the body.

12.4   Fat Deficiency & Excess

The author has yet to come across a case of fat deficiency in the Western world. As discussed in the abstract of this book and in Chapter 1, a very significant proportion of the population in the west and in westernised countries across the world are either overweight or obese. These populations are generally carrying too much fat, not too little. Section 14.4 shows that even lean people who are not overweight, have considerably large fat reserves, in the order of 12-15 kilograms (26-33 pounds), evolved naturally to cope with a potential extended period of starvation. In this context therefore, the term fat deficiency can therefore be seen as an oxymoron. The functions of fat described in this Chapter are the various key functions and requirements of the body's endogenous fat, principally laid down in the first 6 years of life. For example, the brain is composed of 60% fat by dry weight formed during growth, primarily composed of phospholipids, which are essential for maintaining the integrity and fluidity of cell membranes. However, that does not mean people need to eat significant quantities of fat to constantly replace and recycle fat cells in the brain or elsewhere.

It seems there is still a message of confusion and pseudo-science perpetuated in literature, the internet and social media concerning total fats, saturated fats and the two essential fats (LA and ALA as detailed below) and therefore, in the UK at least, the message that humans need to eat relatively large quantities (c30%) of total fat in the diet. The functions of endogenous fat and the two essential fats are described in Section 12.23. It is certainly necessary to include small amounts of these essential fats in the diet. Counterintuitively, considering the preponderance of obesity, these essential fats are generally lacking in most populations. These can be obtained by eating a little oily fish, flax, chia, seeds, kidney beans, kale, spinach, watercress, nuts and polyunsaturated fats in moderation (See Alpha Linolenic Acid (ALA) Omega 3 below). This requirement will easily be met as part of a wholesome, nutritious and calory sufficient diet. As evidenced by the continually growing overweight and obesity statistics, it is most certainly not necessary, nor 'essential' to eat significant quantities of other fats on top of that. To reiterate, in the UK, the current government guidelines, based on advice from the Scientific Advisory Committee on Nutrition (SACN), still recommend that total fat intake should not make up more than 35% of total daily of Energy calory Intake (EI) and that saturated fat intake should not be greater than 11% of EI from food.

This was based on the 1991 report by The committee on Medical Aspects of Food and Nutrition Policy (COMA) [15]. In 2019, SACN revised the saturated fat limit to *'reducing population averages of SFA from current intake to no more than about 10% of total dietary energy'* [16]. but have not revised the total fat recommendation and I quote; *'This report does not consider total fat in the diet'*. In 2023, NICE lifestyle advice was that *'total fat intake should be 30% or less of total energy intake and saturated fat 7% or less'* [17]. In contrast, in the United States, the latest 2020 guidelines from the Dietary Guidelines for Americans (DGA) [1] do not refer to total dietary fat at all, but broadly agree that *'intake of saturated fat should be limited to less than 10 percent of calories per day'* and that this should be achieved along with *'replacing them with unsaturated fats, particularly polyunsaturated fat'*. Hence, people are left with uncertainty regarding total fat and thus a great deal of people are eating at least 30% total fat and, in some quarters, well over 50% total fat. This, compounded by the current populist demonisation of all carbohydrates (including the healthy carbohydrates), has resulted in the widespread adoption of a high fat, high protein, high free sugars carbohydrate and low healthy carbohydrate diet overall. People are eating too much unsaturated fat (Section 1.8 and Figure 1.2). Hence, dietary fat, together with added free sugars and processed food and over consumption, has caused the growing obesity pandemic. The vast majority of people have plenty of fat stores and need to lose weight, not gain it, hence why this book was written. Clearly the guidelines are not working and although the current UK 2014/15 to 2015/16 mean intake of saturated fats for adults (19-64 years) is only marginally above the SACN 10% guide lines at 11.9%, the obesity issue continues to grow exponentially.

Given all the discussion regarding ultra-processed food, that contains both free sugars and fat, this major obesity and health issue is all too often blamed firstly, on all carbohydrates, including complex starches and also on saturated fats but seldom, if at all, on total dietary fat. This widely held attitude, that 30% fat is acceptable as long as saturated fat is reduced, including endorsement from most of the medical profession and governments, continues to be one of the most surprising and baffling aspects and attitudes within the nutrition and health provider world to date. As for the perplexing reasons why total fat should not be discouraged, it is probable that this love affair with total dietary fat has been driven by the many people who wish to continue enjoying the very attractive taste of plenty of oils, butter and cheese (remember cheese has high salt and fat as well as protein). There is also the influence of Big Farming agriculture, politics, Big Food, then given the thumbs up from the now famous article that appeared on the front cover of Time magazine in June 2014, stating that the scientists were wrong and that butter is not the enemy (processed foods and sugars are). Apart from a blip in the early 1970's, the level of dietary saturated fat did not fall. This has since served to perpetuate the populist, thinking and belief that the fat on your body does not come from the fat you eat but instead 'it's the carbohydrates that are turned to fat'. (Section 14.40). All fat, whatever its source, has more than double the caloric value of carbohydrates and proteins. Therefore, compared with complex carbohydrates and proteins, relatively small amounts (gram/kcal) of exogenous dietary fat contributes significantly to weight gain. When levels of fat are high in the blood, the viscosity of the blood increases significantly, slowing blood flow through to the tissues and major organs including the brain, clearly visible in blood samples blood taken after a meal containing high quantities of fat.

12.5 The Long Term Energy Reserve

Fat is the body's principal, long term fuel reservoir. Most of the body's energy reserves, normally in the order of >90% in a healthy adult, are in stored fats within adipose tissue (Table 14.3). The brain, RBCs and other key tissues prefer to use glucose for energy and the body (principally via the liver) is able to preserve and direct glucose to these tissues and organs, whilst constantly using fat for energy in other tissues, such as the heart, the muscles, the liver, the kidney cortex and adipocytes. The body principally utilises both fats and glucose constantly for energy and this, along with contributions from gluconeogenic precursors lactate and glycerol, ensures sufficient blood glucose levels remain constant. The extra energy from fats helps humans survive longer periods of fasting or starvation when food is scarce, or when eating is not practicable. Regular over feeding (>TEE) of any of the 3 macronutrients reduces fat burning and increases fat storage and therefore, leads to weight gain, health issues and obesity.

12.6   Energy Storage and Efficiency

Utilisation of fat is a much more efficient way to store energy compared with storing energy from carbohydrate or protein. The calorific value of dietary fat is approximately 9 kcal/g, whilst protein and carbohydrate have just 4 kcal/g. Fat stored in tissues contains very little water. In contrast, every gram of glycogen (the storage form for carbohydrate) holds approximately 0.77 grams of water (see Section 14.8). Muscle protein also holds water; 100 grams of 95% lean ground beef contains 21 grams of protein. One pound of fat (454 grams) stored in body adipose tissue, holds about 4,100 calories, which provides around 2 days fuel energy. However, a pound (454 grams) of hydrated glycogen from carbohydrate holds just 17% of this energy i.e., only 680 calories. A typical lean, healthy adult man carries a range of about 24 to 33 pounds (11 kg - 15kg) of stored fat. However, if this energy was stored as glycogen from carbohydrate, the same number of calories would weigh a colossal 144 pounds (66 kg), thus, each lean individual would be almost double their current weight which would be unsustainable as a species that needs to move in times of danger or seeking food and shelter. Paradoxically therefore, using fat as energy storage means that humans significantly weigh less overall than if carbohydrate or protein was used as long term energy storage. Hence, why fat has evolved to be the long term energy store food of choice. Endogenous fat built up during growth provides several key functions described from Sections 12.5 to 12.18.

12.7   Adipocytes and Health Issues

Fat stores are important during illness and nourish the cells and provide the immune system with energy to fight off the infections when illness prevents eating.

12.8   Cell Membrane Structure

Lipids form the structural component of cell membranes helping to control the movement of substances inside and out of cells, via membrane bound enzymes and receptors. The fatty acid composition of membrane lipids greatly influences membrane function. Endogenous fat forms a major component of nerve cells and endogenous fat forms approximately 60% of the brain.

12.9   Endocrine and Metabolic Control

Adipose tissue produces more than 50 different kinds of signalling molecules that act on many types of cells throughout the body, for example, secreting biologically active factors called adipokines. These molecules contribute to a variety of different functions, including regulation of energy balance, appetite, food intake digestion and satiety, inflammatory response and metabolism of steroid hormones.

12.10   Absorption of Fat Soluble Vitamins

A little dietary essential fat (Omega 3 and Omega 6) helps in the absorption of fat soluble vitamins, A,D,E and K, which are hydrophobic and can-not be transported by water. The amount of dietary fats required for this absorption is often over-stated and in fact very small amounts, at around 5 grams are sufficient. This can be obtained from a few sunflower seeds, or one fifth of an avocado, or two or three walnuts. These small additions will significantly increase absorption of the fat soluble vitamins such as carotenoids from coloured vegetables that can be converted into vitamin A and improve the absorption of vitamin K held in lycopene in tomatoes. Bile synthesised in the liver and stored in the gall bladder also has a significant role in vitamin absorption.

12.11   Leptin Supresses Appetite

Fat cells in adipose tissue produce leptin, the signalling hormone that suppresses appetite, increases energy use and encourages the body to burn fat (Section 14.3). In lean people, fat tissue makes lower levels of leptin promoting additional feeding and weight gain. On the other hand, it is possible for overweight individuals to develop leptin resistance, causing increased intake, even with feeding sufficiency and sufficient levels of leptin. A parallel can be drawn with insulin resistance which can also affect overweight people, a perfect storm due principally to over nutrition.

12.12   Components of Hormones

Fat helps make steroid hormones, including the sex hormones oestrogen and testosterone. Endogenous fat can modify these steroid hormones, converting one type into another.

12.13   Cholesterol and Bile

Fat forms cholesterol which is a key component of cellular membranes and the precursor for testosterone and oestrogen and other hormones. Cholesterol levels, especially LDL cholesterol should be closely monitored if an individual is overweight or considered to have potential risks of developing metabolic syndrome.

12.14   Hydrophobic Modifiers of Proteins

Fat modification of proteins is a widespread, essential process where fatty acids, cholesterol, isoprenoids and phospholipids are attached to polypeptides. These hydrophobic groups may affect protein structure, function and stability and therefore play a critical role in cellular systems.

12.15   Precursors of Eicosanoids

Eicosanoids are a class of bio-active lipids derived from 20-carbon polyunsaturated fatty acids, most frequently from Omega-6 arachidonic acid. This fatty acid has a fundamental role in the brain and serves to act as precursor for the production of important signalling molecules like prostaglandins and thromboxanes in the body and for (smooth muscle contraction and blood clotting.

12.16 Maintenance of Healthy Skin and Hair

Healthy essential fats help prevent skin dryness to maintain healthy and hydrated skin. Omega-3 fatty acids (Table 12.1) are found in cells that line the scalp and provide oils that help keep the scalp and hair hydrated. However, conversely, excess belly fat can lead to higher levels in the hormone dihydroteserone (DHT) which can shrink hair follicles and cause hair to fall out.

12.17 Protects and Insulates the Body From Cold

White adipose tissue helps cushion and protect parts of the body, as well as insulate the body from extreme temperatures, maintaining core human body temperature at around 37 degrees C.

12.18 Generation of Heat

Brown fat burns fuel either as fat or glucose molecules by disrupting and reversing protons in cellular OXPHOS in mitochondria to create heat rather than energy. Brown adipose tissue, has an opposing role to white adipose tissue, as it is used to generate heat through non shivering thermogenesis, serving an important defence mechanism to protect new born babies against hypothermia. Brown fat in human babies and hibernating mammals reduces proton gradients and proton pumping in the process of chemiosmosis in cell mitochondria which makes overall less energy from ATP whilst generating more heat for warmth (Section 15.8). In adults, non-shivering thermogenesis becomes secondary to shivering thermogenesis, achieved by the contraction of skeletal muscles. Brown fat also improves insulin sensitivity which helps reduce the risk of Type 2 diabetes. Brown fat in adults is located in the neck, kidneys, suprarenal glands, heart and chest.

12.19 Triacylglycerol

This is the form in which most fats occur in food. Triacylglycerol (TAG) (also commonly known as triglyceride) consists of three fatty acid molecules attached to a glycerol molecule. A given weight of fatty acids is, therefore, equivalent to a larger weight of fat (triacylglycerol). Triglycerides are fat cells that circulate in the bloodstream and are stored in the body's fat cells. A high level of triglycerides in the blood increases the risk of diseases of the heart and blood vessels.

12.20 Fatty Acids

The fatty acid molecule consisting mainly of a carbon chain (of variable length) with hydrogen atoms attached. There are approximately 16 different fatty acids that make up the bulk of the fatty acids in food. The most common fatty acids in humans are the 2 saturated fatty acids; palmitate (palmitic acid), a 16 carbon chain saturated fatty acid and stearate (stearic acid), an 18 carbon chain saturated fatty acid. Fatty acids are the route to providing energy from endogenous fats via lipolysis then beta (β)-oxidation for many tissues of the body such as the heart and skeletal muscle.

12.21   Saturated Fats and Unsaturated Fats

Saturated fats are generally solid at room temperature, named and are more saturated with hydrogen atoms. Unlike unsaturated fatty acids they contain only single bonds. They are found in foods such as meat, dairy like cheese, butter, whole milk, cream and ice cream, coconut, some vegetable oils and coconut oil. Saturated fats raise both high-density lipoprotein (HDL aka good cholesterol) and low-density lipoprotein (LDL aka bad cholesterol). [58] Unsaturated fatty acids comprise carbon atom chains containing one or more double bonds.

This reduces the number of hydrogen atoms attached and hence are normally liquid at room temperature. Based on current evidence, unsaturated fats are generally considered healthier because research shows that replacing saturated fats with unsaturated fats may decrease heart disease risk. These fatty acids are either monounsaturated or polyunsaturated and discussed in Section 12.22. [33]

12.22   Monounsaturated and Polyunsaturated Fats

Monounsaturated fatty acids (MUFA) and Polyunsaturated fatty acids (PUFA) are both considered healthier types of unsaturated fats that promote good cholesterol levels and other heart-health benefits. Note figure 12.2. Monounsaturated fats have only one carbon to carbon double bond in the structure. Monounsaturated fats can be made in the body or can be eaten in the diet and are found in high concentrations in foods such as almonds, hazel nuts, pecans, pumpkin seeds, sesame seeds, avocado and in olive oil (11% MUFA). However, about half the fats in these foods are saturated and half monounsaturated. They are considered healthy as they may lower bad cholesterol and raise good cholesterol. They also may improve the control of blood sugar levels. Figure 12.2 shows sources. Replacing saturated fats with monounsaturated fats in the diet may lower the level of bad cholesterol and triglycerides in the blood. [33]

Polyunsaturated fatty acids have more than one carbon to carbon double bond in the structure and like monounsaturated fatty acids, these fatty acids can also help lower unhealthy LDL cholesterol. They can-not be made in the body and must be obtained from diet and the healthier varieties are found in high concentrations in foods such as soybean, oily fish, flax seeds, chia seeds, walnuts. They are essential for the body's functions and have a key role in cell membranes and blood clotting. Other sources of polyunsaturated fats include corn oil, safflower oil, sunflower oil, grapeseed oil and are not essential and should be consumed minimally if at all. Generally, the cooking oils are the least healthy polyunsaturated Omega 6 sources. The Oil in fish is used to protect the fish from very cold temperatures. Humans do not live in water of course. Regarding oil generally, whether in melted butter or coconut oil, containing mostly saturated fat, or oils rich in polyunsaturated fat, such as sunflower oil or corn oil, or oils rich in monounsaturated fat such as olive oil, canola oil or sesame oil, the author recommends moderation and to minimise these and use very sparingly, whether used for cooking or as dressings or for baking or frying. This holds even with current advice that these are healthier than saturated fat. There are 2 main types of healthy and essential polyunsaturated fatty acids, Omega 3 ALA and Omega 6 LA. These are the only 2 essential fats in the diet and discussed in Section 12.23 and 12.24. Plants and fish are the best sources of these essential fatty acids.

12.23 Essential Fats

Human cells can-not incorporate a double bond beyond the 9th carbon atom[9] carbon position in the fat molecules because they lack the enzymes [12] and [15] desaturase enzymes. However, plant cells do have the desaturase enzymes and can thus synthesise these fats. The two essential fatty acids are; Alpha Linolenic Acid (ALA), an Omega 3 fatty acid and Linoleic Acid (LA), an Omega 6 fatty acid. This is the reason why both Alpha Linolenic Acid (ALA) (18:3 [9,12,15]), with 3 double bonds and Linoleic Acid (LA) (18:2 [9,12]), with 2 double bonds, are essential fatty acids. These essential fatty acids are both polyunsaturated fats (PUFAs). with 18 carbon chains and are precursors to vitamins and various cofactors in metabolism. Some earlier research indicated that Omega-3 insufficiency or inadequacy is common in people of all ages across the world. However, Omega-3 supplements are not recommended in the UK general population because evidence of benefits remains inconclusive.

12.24 Alpha Linolenic Acid (ALA) Omega 3

Alpha Linolenic Acid is a polyunsaturated fat. ALA is a precursor of Eicosapentaenoic Acid (EPA) which is good for heart health and Docosahexaenoic Acid (DHA) that supports brain function. ALA has 3 double bonds, at the 9th, 12th and 15th carbon atoms. Recent research also shows that Omega 3 fatty acids have a role in the primary prevention of coronary heart disease (CHD). Good sources of Omega 3 ALA are mango, walnuts, flax, hemp, mustard and chia seeds, salmon and other oily fish, kidney beans and green vegetables such as kale, spinach and watercress and seaweeds, the ultimate source of Omega-3. With a healthy diet, an Omega-3 supplement, including fish oil should not be required. Most health benefits are attributed to DHA and EPA. A DHA algal based supplement should only be added if Omega 3 levels are actually considered, or preferably confirmed to be deficient from special tests for each individual and then recommended by your doctor.

12.25 Linoleic Acid (LA) Omega 6

Linoleic Acid is also a polyunsaturated fat. The main function of LA is the maintenance of skin epidermis. Symptoms of deficiency are very rare. LA has 2 double bonds, at the 9th and 12th carbon atoms from the carbonyl function group. Mitochondria are dependent on linoleic Acid. There is approximately 11 times the amount of linoleic acid in the body than alpha linoleic acid. Good sources of LA are peanut butter, almonds, sunflower seeds, eggs and red meat, all vegetables, fruits and grains. Also in vegetable oils, olive oil corn oil, safflower oil nuts.

12.26 Omega 6 to Omega 3 Fatty Acid Ratio

Ensure the ratio to Omega 6 to Omega 3 (OM6 to OM3) is kept in the healthy range of 1:1 up to a maximum of 4:1. Table 12.1 shows the ratio of OM6 to OM3 fatty acids in some foods. Current United States guidelines for the level of Omega 3 fatty acids are 1.1 g/d for an adult female to 1.6 g/d for an adult male. Recommended ratio ranges are OM6 to OM3; of 1:1 to 2:1, with a max' ratio of 4:1. If the ratio of OM6 to OM3 is too high i.e., Omega 6 is high and Omega 3 is low, this can contribute to inflammation and pathogenesis of many chronic diseases. Humans evolved with a ratio close to 1:1. Current western diets are around 15:1, mainly due to vegetable oils, mayonnaise, fried foods, processed foods. There are no specific RDA Omega 6 guidelines in the UK except the recommendation to eat oily fish twice per week. It is important to note that based on current evidence, it is the relative conversion of ALA into DHA and EPA that determines the health benefits. The conversion rates vary from person to person with the average in the order of 6% ALA converted to EPA and 3.8% ALA converted to DHA. Note table 12.1, canned tuna is 0.1:1 OM6 to OM3 i.e., the other way round. The author is not aware of science consensus, or any long term issues with eating foods with a much higher ratio of OM3 to OM6. The total weight of Omega 3 is more relevant which is not high in tuna. The other foods listed have a healthy ratio, apart from olive oil and avocado oil with 13:1 OM6 to OM3. Note that flax and chia seeds have the highest Omega 3.

The RDI (US) is a guide only and not a true reflection of the actual conversion to EPA or DHA. Note that data for DHA and EPA amounts are largely not available as measured or reported apart from fish. On this data it is not possible to confirm that these plant based sources do convert significantly to EPA and DHA but eating reasonable quantities will certainly help as long as in moderation (1 table spoon per day) due to the total fat content.

Table 12.1 Omega 6 (LA) and Omega 3 (ALA) healthy ratio range in the brain is 1:1 - 2:1 (Source My Food Data Nutrition Facts)

Omega 3 (ALA) & Omega 6 (LA) Source	Om 6:Om 3 Ratio	Omega 3 Weight (g/100g)	Omega 6 (% RDI) (US)	DHA	EPA
Flax Seeds	0.3:1	22.81	1426%		
Chia Seeds	0.3:1	17.83	1114%		
Walnuts	4:1	9.08	568%		
Salmon (Atlantic)	0.1:1	2.21	138%	1.429	0.411
Avocado Oil	13:1	0.96	60%		
Tuna (can in water)	0.1:1	0.88	55%	0.629	0.233
Olive Oil	13:1	0.76	48%		
Sprouts	0.5:1	0.17	11%		
Kidney Beans	0.6:1	0.17	11%		

12.27 Dietary Fat and Adipocytes

If energy caloric excess continues, adipocytes (fat cells) are enlarged but the total number of cells remains similar. The number of adipocytes increases at its fastest rate in late puberty. Obese people have more and larger adipocytes compared with lean people and if body fat is lost, the number of cells does not decrease. The fatty acid chemical profile of stored fat resembles the fatty acid profile of dietary exogenous fat and not that of fat synthesised in the liver from carbohydrates i.e., de novo lipogenesis (DNL). Section 14.40 describes how DNL conversion of glucose or amino acids to fat is normally insignificant and also highly inefficient, requiring considerable metabolic energy. [36, 38]

12.28 Fat Levels in Food

Tables 12.2 and 12.3 show the level of cholesterol and the percentage levels of total fat, trans fat (TFA), saturated fat (SFA), monounsaturated (MUFA) and polyunsaturated fat (PUFA) in various foods, derived from animal and plant-based sources.[18] These are listed in order of health benefit with trans-fat as the unhealthiest fat and polyunsaturated the least unhealthy fat. [33] Tables

12.2 and 12.3 show the percentage of fat in any given weight. Note that this is not the percentage of calories in each food but the proportion of different fat. For example, butter is 77% fat, of which 51% of that fat is saturated fat, i.e., saturated fat represents 66% of all the fat found in butter, whereas the beef steak in this example is 20% fat of which 4.3% is saturated fat. As a further example, this time in terms of total calories, cheese is 74% fat, 23% protein and 3% carbohydrate and beef steak is 59% fat, 41% protein and 0% carbohydrate. This difference in percentage terms is due to the fact that fat, at approximately 9 kcal/gram, contains more than double the number of calories than both protein or carbohydrate, each at approximately 4 kcal/gram. The difference between total fat and 100% of the weight of the food in Table 12.3 (for example the missing 2% for olive oil), comprises other nutrients, such as protein or carbohydrates or other contents of the food, such as additives. At the current time and considering current research, the healthiest foods regarding fat type and overall health are shown in blue font and those least healthy for either cholesterol of fat type content in red font.

Table 12.2   The percentage of fat and cholesterol level in animal sourced food (percentage values rounded down)

Food	Cholesterol mg	Total Fat %	TFA %	SFA %	MUFA %	PUFA %
Butter	31	86%	3.6%	56.1%	22.4%	3.9%
Cheese (Cheddar)	28	33%	1.1%	21.4%	9.3%	1.4%
Beef, steak	74	20%	0.0%	4.3%	4.5%	0.4%
Chicken	80	13%	0.0%	3.6%	5.2%	2.8%
Pork	75	17%	0.0%	6.5%	8.3%	2%
Sausages	40	28%	0.0%	8.9%	11.5%	5.2%
Bacon	34	34%	0.3%	11.7%	15.4%	6.0%
Ham	86	5.1%	0.0%	0.97%	1.94%	0.6%
Salami	11	32%	1.0%	11.2%	14.3%	5.1%
Salmon	143	5.5%	0.0%	0.9%	2.1%	2.5%
Tuna (can)	72	2.7%	0.0%	0.8%	0.76%	1.1%
Mackerel	66	7.5%	0.0%	3.2%	5.4%	3.3%
Crab	114	0.8%	0.0%	0.2%	0.3%	0.3%
Squid	66	1.4%	0.0%	0.6%	0.0%	0.8%
Milk	20	3.5%	0.0%	2.2%	1%	0.1%
Eggs	186	9.6%	0.0%	3.3%	3.6%	2.7%

Table 12.3   The percentage of fat and cholesterol level in plant sourced food (percentage values rounded down)

Food	Cholesterol mg	Total Fat %	TFA %	SFA %	MUFA %	PUFA %
Olive Oil	0	100%	0.0%	14.0%	73.0%	11.0%
Avocado Oil	0	100%	0.0%	11.4%	70.7%	13.6%
Sunflower Oil	0	100%	4.8%	6%	85.7%	3.5%
Coconut Oil	0	93%	4.6%	78.5%	6.4%	0.2%
Walnuts	0	65%	0.0%	5%	10.0%	50.0%
Flax Seeds	0	43%	0.0%	5%	9.0%	29.0%
Coconut	0	31%	0.0%	29.5%	1.33	0.44%
Chia Seeds	0	30%	0.1%	3.3%	2.3%	24%
Hemp Seeds	0	41%	0.0%	3.6%	5.3%	32.1%
Oats	0	6%	0.0%	1.3%	2.5%	2.5%
Avocado	0	14%	0.0%	2.5%	9.6%	2.0%
Chick Peas	0	3%	0.0%	0.0%	0.0%	3%
Brown Rice	0	2%	0.0%	0.4%	0.8%	0.8%
Water Melon	0	0.3%	0.0%	0.1%	0.1%	0.1%
Spinach	0	0.3%	0.0%	0.1%	0.0%	0.2%
Apple	0	0.2%	0.0%	0.1%	0.0%	0.1%
Potato	0	0.1%	0.0%	0.05%	0.0%	0.05%
Soya Milk	0	1.6%	0.0%	0.2%	0.0%	1.4%

This is not necessarily a recommendation to eat more of the items in blue font but a guide on what the current weight of evidence of healthier foods containing fats suggests. This general consensus may change as further evidence emerges. Olive oil is 100% fat, of which 14% is saturated fat, 73% is monounsaturated fat and 11% is polyunsaturated fat. Cholesterol levels, generally found in most animal products, are also listed. The most harmful fat in terms of health, are derived from trans-fat and saturated fat. The highest cholesterol foods and highest in trans-fat or saturated fat component values are shown in red font. Foods high in MUFA such as sunflower (and other cooking oils) and coconut and its extracted oil are not generally considered the healthiest and these values are also shown in red font.

Polyunsaturated fats are rich in Omega-3 (essential ALA), (EPA), (DHA) and Omega-6 fatty acids (essential LA) and Arachidonic acid (ARA). Monounsaturated fats are rich in Omega-9 fatty acids, for example oleic acid in olive oil and avocado oil. Human breast milk is also a rich source of oleic acid. Furthermore, some oils such as sunflower may produce more aldehydes and other end products on heating which may be harmful to health and care should be taken to check smoke points if frying, sauteing or baking with oils. Avocado oil has one of the highest smoke points at 271 degrees centigrade and for example, when used for baking, avocado oil is safer than using the highest quality extra virgin olive oil, with a significantly lower smoke point of 207 degrees centigrade. Avocado oil is around 71% monounsaturated fat; mainly oleic and palmitoleic acids; 14% essential polyunsaturated fat; Linoleic acid (LA) and 11% saturated fat; palmitic acid. However, It has a high O6 to O3 ratio of 13:1 and should be used sparingly. The total amount of Omega 6 is relatively small and should not therefore represent a health risk if used in moderation. There are endless objections by the meat, dairy, agriculture and pharmaceutical industries and politicians regarding the health benefits of animal products and why people should continue eating these products and that there is no link between dietary cholesterol and blood cholesterol. However, note there is zero cholesterol in plant foods compared with that in animal foods (Table 12.3). Assuming nil trans fat in the diet, the levels of total fat, saturated fat and cholesterol should be minimised as far as possible, with the key emphasis on minimising saturated fat. Saturated fat has no double bonds.

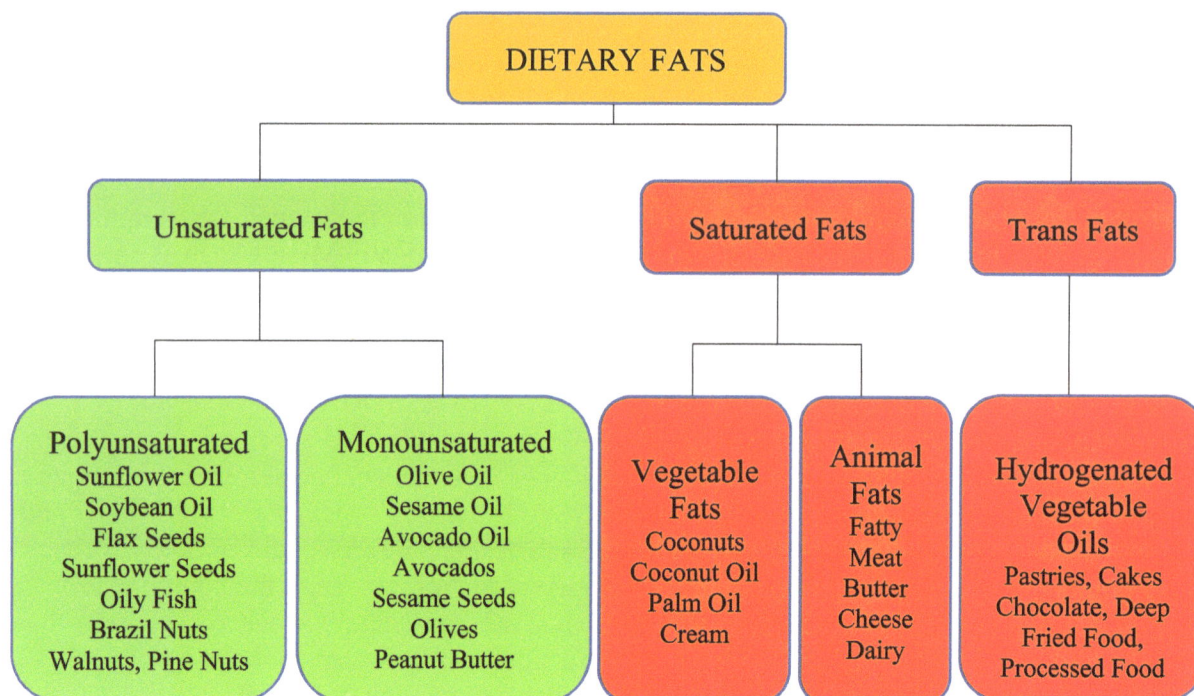

Figure 12.2 Source of various types of dietary fats from animal and plant sources. Green indicates healthier fats

When these molecules are oxidised for energy, more electron carriers are released into the mitochondria, causing Reactive Oxygen Species (ROS) and Superoxide damage. Polyunsaturated fats and monounsaturated fats have some double bonds and release less electron carriers on (β)-oxidation and are therefore, less unhealthy. Figure 12.2 provides a breakdown of dietary fats showing sources of dietary unsaturated, saturated and trans fats. The consensus is to eat less of the latter 2 groups of fats and marginal amounts of the healthier unsaturated fats. All fats, whether solid or liquid at room temperature (oils), should be consumed in moderation, especially when endeavouring to lose weight, principally restricting fats to moderate amounts if the least unhealthy fats, such as monounsaturated and polyunsaturated fats in plant foods.

If an individual is content with their weight, or has reached their weight loss goal, or indeed would like to gain some weight, then some healthy polyunsaturated and monounsaturated fats, such as nuts, seeds, avocado and a little olive oil or avocado oil can be reintroduced bearing in mind that small amounts of these foods have the highest caloric density. These fats can be used to fine tune weight management once weight loss goals have been achieved. In particular, remember that all oils including the highest quality extra virgin olive oil, is highly calorific, weight for weight. One tablespoon of olive oil, (just 14 grams), contains 119 kcal i.e., 5% of TEE (total daily calories), plus it contains a significant percentage of saturated fat. This is another illustration that calorie counting alone, is less relevant than caloric density and nutrition quality. The human stomach can either be filled to the brim with 500 kcal of nutritiously rich, low calorie salad and vegetables, or filled to just one tenth of its capacity with the same total 500 kcal but relatively low nutritional value, calorie dense foods such as butter, fries or cakes, thus contributing to overeating, weight gain and other health issues. As noted earlier, however, all fat is highly caloric, whether sourced from animal or plant sources and will lead to weight increase if not moderated. If people wish to eat fish, consumption of oily fish, up to twice per week, is recommended by many nutritionists and doctors.

12.29 UK Guidelines on Fats and Cholesterol

Table 12.4 UK guidelines on fat and cholesterol

Fat/Cholesterol	Guideline
Total Fat	Consume <35% Calories (TEE) from total fats
Saturated Fat	Women should consume <20 g/d and Men <30 g/d
Total Cholesterol	Total Blood Cholesterol <5 mmol/L. If trying to lower Cholesterol levels consume <200 mg/d

The author does not agree with some of the current UK guidelines (Table 12.4). The recommended total fat in the diet of approximately 30% calories is not necessarily conducive to healthy weight maintenance, nor optimal cardiovascular health. In addition, note that <20-30 g/d saturated fat is recommended and this seems high given the link between saturated fat and numerous health issues. The guidelines are not making much difference to increasing obesity levels. There remains no overall level on the level of dietary cholesterol unless blood cholesterol is measured from blood tests by doctors and considered to be too high (>5.5 mmol/L). Discussions on cholesterol are generally on limiting saturated fat to control cholesterol, rather than limiting dietary cholesterol unless the cholesterol in the blood is too high and the overweight individual is advised to reduce dietary cholesterol to <200 mg/d, i.e., equivalent to 1 egg per day and no other cholesterol. There is considerable debate on this topic as well as cholesterol types (LDL/HDL/VLDL and ApoB markers etc), including significant contributions from those with vested interest in the consumption of eggs and meat. There is 186 mg of cholesterol in just one egg and the total daily recommended amount is 300 mg/d for healthy people.

CHAPTER

13

Minerals Vitamins & Water

13.1 Minerals

Minerals are vitally important to health although only making up approximately 4% of body weight.

They have a multitude of functions, supplying the right conditions that are essential for cellular activity, influencing the condition of teeth and bones, helping organs control osmotic properties within the blood, serum, plasma, lymph and organ tissues, and they serve as essential metallic cofactors in enzymes which are themselves critical to all cells in the body.

They are generally categorised on amount and occurrence within the body, either as the major minerals (major minerals) or trace minerals (microminerals).

The major minerals of the human body traditionally include sodium, calcium, potassium chloride, phosphorus, magnesium and sulphur. Potassium and sodium chloride are key in maintaining electrolyte balance within body fluids. For microminerals or trace minerals, recommended dietary allowances (RDAs) have been established for six essential trace minerals; iron, zinc, copper, selenium, molybdenum and iodine. Three other trace minerals manganese, fluoride and chromium are important and form part of a healthy diet even though they are required in far smaller quantities.

Table 13.1 shows the principal functions and deficiency symptoms of 12 major and trace minerals required in the body and Table 13.2 shows plant sources of the minerals.

Table 13.1 Essential minerals and deficiency symptoms

Mineral	Function Health Benefits	Deficiency Symptoms
Calcium	For teeth & bones, blood clotting	Muscle spasm, cramping, osteoporosis
Magnesium	Muscles, nerves, bones, blood sugar levels	Low appetite, fatigue, weakness, nausea
Iron	Haemoglobin, cells, hair, nails, skin	Pale skin, fatigue, rapid heartbeat shortness breath
Potassium	Nerves, muscles, heart, cellular movement	Weakness, fatigue, cramps, irregular heart beat
Zinc	Immune system, metabolism, proteins	Diminished smell & taste, poor wound healing
Iodine	For thyroid hormone & metabolism	Severe fatigue, weight gain, puffy face, dry skin
Phosphorous	Filtering waste in kidney, bones & teeth	Anorexia, rickets, bone pain, muscle weakness
Selenium	Cell damage protection, DNA, thyroid	Impaired immune system, thyroid disfunction
Molybdenum	Proteins, genetic mater'l, toxin breakdown	Seizures, dislocated lenses, brain atrophy, lesions
Manganese	Connective tissue, blood clots, hormones	Poor growth, bone demineralisation skin rashes
Fluoride	Tooth enamel, stimulates new bones	Tooth cavities, weak bones
Chromium	Insulin & blood sugar, fatty acid synthesis	Confusion, weight loss, impaired coordination

Table 13.2 Essential minerals & healthy sources from plants

Mineral	Plant Sources
Calcium	Green vegetables kale, water cress, okra, broccoli, fruit, chickpeas, kidney beans, malted or wheatgerm bread, sesame seeds, tofu, tempeh, chia Seeds, almonds, figs, Brazil nuts (1 per week), soy milk, oat milk. (All improve absorption with vitamin D sunshine)
Magnesium	Spinach, malted or wheatgerm bread, quinoa, sesame seeds, sunflower seeds, almonds, Brazil nuts (1 per week) cashew nuts, hazel nuts, walnuts, whole grain foods; oats (steel cut), oat milk, lentils, soya beans,
Iron	Spinach, kale, broccoli, lentils, chickpeas, black-eyed peas, kidney beans soya & soya derived food quinoa, tofu & tempeh, cashew nuts, hazelnuts, almonds, peanuts, walnuts, pistachios, pumpkin seeds, hemp, apricots, sesame seeds (vitamin C for from citrus fruits such as lemons or peppers, Weetabix, shreddies, courgettes and onions improves iron absorption)
Potassium	Lentils, potatoes, kidney beans, soya beans, tempeh, lentils, avocado, bananas, spinach, butternut squash, apricots
Zinc	Lentils, potatoes, kidney beans, quinoa, soya beans, tempeh, avocado, bananas, spinach, butternut squash, whole meal spaghetti, pumpkin seeds, couscous, cashew nuts, whole g' rice
Iodine	Strawberries, cranberries, prunes, green beans, corn, potatoes, watercress, dried sea vegetables (dulse, kelp, nori), wakame, arame, kombu, iodized salt
Phosphorous	Whole grain foods; oats, (steel cut), rice, quinoa, pasta, lentils, tempeh, chickpeas, Brazil nuts (1/week) cashews, peanuts, hazelnuts, walnuts, pumpkin & sunflower seeds, corn on the cob
Selenium	Brazil nuts (1 per week), whole grain foods; oats, (steel cut), rice, tofu, quinoa, mushrooms, pasta, lentils, chickpeas, cashew nuts, peanuts, hazelnuts, walnuts, sunflower seeds
Manganese	Wholegrains, oats, pecan nuts, almonds, walnuts, lentils, chickpeas, green vegetables, peanut
Fluoride	Oatmeal, cooked spinach, boiled potatoes, cooked carrots
Chromium	Brazil Nuts (1 per week), wholegrains, bananas, whole wheat, broccoli, green beans, onions

13.2 Vitamins

Vitamins are chemicals or compounds of chemicals that are required, or are essential in the diet if the body is unable to make them from other substances. They have many different regulatory functions. The discovery of a vitamin was usually made when the results of dietary deficiencies were noted, such as beriberi (vitamin B_1) C, rickets (vitamin D or calcium) or scurvy (vitamin C) due to their absence in the diet. There are two main groups; water soluble and fat soluble. The water soluble vitamins are all the B vitamins and vitamin C and the fat soluble vitamins are vitamins A, D, E and K. The water soluble vitamins are managed differently in the body, compared with the fat-soluble vitamins and are absorbed into the hepatic portal vein blood bound for the liver. It is hypothesised that vitamin B12 (cobalamin) is transported by intrinsic factor (IF), haptocorrin (HC) in the stomach and transcobalamin (TC) or further active transport proteins. Retention time in the body for water-soluble vitamins is relatively short compared with the fat soluble vitamins which are retained over much longer periods. Storage of vitamins is achieved by linking with enzymes and specialist transport proteins. Water-soluble vitamins are excreted in the urine whenever the body detects that plasma levels are higher than levels in the kidney. Bile from the gall bladder produced by the liver significantly assists with transport of the fat soluble vitamins. Many of the B vitamins are essential in cell respiration, integral to the proteins and cofactors within the ETC to drive cellular energy (Chapter 15). Table 13.3 shows a list of the 13 vitamins, main functions and deficiency symptoms and Table 13.4 lists plant sources of vitamins.

Table 13.3 Vitamins and deficiency symptoms

Vit.	Name	Function Health Benefits	Deficiency Symptoms
A	Retinol	Vision, skin bone tooth growth, immune system	Night blindness, infertility
C	Ascorbic Acid	Antioxidant, protein metabolism, immune, iron abs'	Fatigue, scurvy bleed. gums
D	Calcitriol [a]	Absorption of calcium	Muscle/bone pain, tingling
E	Tocopherol	Antioxidant protects cell walls	Arms/legs loss of feeling
K	Phytonadione	Blood clotting efficiency	Bruising Wound Bleeding
B1	Thiamine	Energy metabolism, nerve function	Fatigue, irritability, memory
B2	Riboflavin	Energy metabolism, vision, skin	Fatigue, slow growth, sores
B3	Niacin	Energy metabolism, nervous system, skin, digestion	Dermatitis dementia diarrhe'
B5	Pantothenic Acid	Energy metabolism	Fatigue, insomnia, irritable
B6	Pyridoxine	Protein metabolism, red blood cells	Seizures, anaemia, rashes
B7	Biotin	Metabolism fats, carbs' & protein, nerves, hair, skin	Hair Loss, skin rash brit'nails
B9	Folic Acid	Making DNA & new red blood cells & other cells	Pale skin low appetite, irritab
B12	Cobalamin	New cells, nerve function	Fatigue, feel breathless, faint

[a] Vitamin D is essential for Calcium ion absorption & there is approximately 4 Months Storage in the Liver
UVB through skin epidermis converts 7-cholesterol to cholecalciferol which is carried to the kidneys
Parathyroid Hormone PTH stimulates the kidney to convert cholecalciferol to calcitriol i.e., the active form of Vitamin D

Table 13.4 Vitamins and healthy sources from plants

Vitamin	Name	Plant Sources
A	Retinol	Carrots, kale, watercress, red & green peppers, strawberries, raspberries, blackcurrants, blueberries, Brussel sprouts, broccoli, spinach, cabbage, oranges, kiwi, sweet potatoes, pineapple, mango, melon, lemons, limes, butternut squash
C	Ascorbic Acid	Kale, red & green peppers, strawberries, raspberries, blackcurrants, blueberries, Brussel's sprouts, broccoli, cabbage, oranges, kiwi, tomatoes, potatoes, peas, cauliflower, pineapple, papaya, melon, lemons, limes, mangoes, cranberry juice
D	Calcitriol [a]	Mushrooms (wild exposed to UV), broccoli, cabbage, soya, soy milk, oat milk, bananas, Fortified foods & cereals
E	Tocopherol	Hazelnuts, almonds, pine nuts, sunflower seeds, peanuts, avocado, butternut squash, canned tomatoes, mango, broccoli, spinach, kale, chickpeas, red peppers, kiwi, wholegrains
K	Phytonadione	Brussel sprouts, cabbage, kale, broccoli, spinach, spring onions, watercress, kiwi, Green beans, lettuce, parsley, pumpkin, grapes, blueberries, tomatoes
B1	Thiamine	Watercress, wholemeal spaghetti, oats, green beans, peas, pumpkin, sunflower seeds, wholegrain rice, corn on the cob, pecan nuts, oranges, avocado, Brazil nuts (1 per week), lentils, hazel nuts, soya milk. soya beans
B2	Riboflavin	Quinoa, soya milk, avocado, almonds, wholegrain rice, mushrooms, mange-tout
B3	Niacin	Mushrooms, quinoa, peanuts, wholegrain spaghetti, wholegrain rice, corn' cob
B5	Pantothenic Acid	Avocado, baked potato, sweet potato, legumes, mushrooms, peanuts, corn on the cob, mange-tout peas, oats, chestnuts, oranges
B6	Pyridoxine	Potato, banana, avocado, pistachio nuts, quinoa, wholegrains, navy beans, corn on the cob, sunflower seeds, Brussel's sprouts, spring greens, chestnuts, oranges, sesame seeds, tomatoes, walnuts
B7	Biotin	Tempeh, hazel nuts, oats, mushrooms, avocado, pecan nuts, walnuts, pistachio nuts, sunflower seeds, sesame seeds
B9	Folic Acid [b]	Green leafy vegetables, kale, rocket/arugula, Brussel's sprouts, citrus fruits, tempeh, lettuce, beetroot, legumes, asparagus, oranges, cherry tomatoes, tofu
B12	Cobalamin	Seaweed, supplement, best absorbed via spray

[b] Eating folate-rich foods is a good way to improve methylation. vegetables (especially dark leafy greens such as Broccoli, Spinach & Kale), fruit, nuts, peas, Whole grains, Asparagus, Avocado, Legumes, Sunflower Seeds

13.3 Vitamin B12

For reasonably healthy people, with no specific underlying health issues and not requiring other specific supplementation, vitamin B12 is probably the most important supplement of all, especially for those that significantly change their diet, for example eating less red meat. It is possible that people who abstain from eating meat may have lower B12 levels. Deficiencies are not uncommon in the general population, even amongst those eating large amounts of meat. Vegans who supplement often have an improved B12 status even compared with meat eaters.

The easiest available and most commonly ingested source is from animal meat. However, eating meat for protein or B12 reasons, must be considered along with all other components of eating meat, for example, saturated fat, lack of fibre, IGF-1 [6] and Trimethyl Amine oxides (TMAO), plus high levels of haem iron etc. [57]

Vitamin B12 is sourced is from bacteria in the soil and the origin of the availability to humans in the past may have been derived from foraging for foods in soils and that drained into water supplies. B12 was also present on potatoes and beet crops but no longer, due to agricultural washing and cleaning practice. In today's sanitized world, the water supply is commonly chlorinated to disinfect the bacteria thus, as for cholera, humans no longer obtain B12 from the water.

Vitamin B12 is essential for red blood cell synthesis and activation and protection of the Central Nervous System (CNS) and deficiency leads to fatigue and other symptoms listed in table 13.3. Many herbivorous animals e.g., cattle and sheep absorb B12 produced by bacteria in their own gut. Many animals accumulate B12 during their lives and livestock are often supplemented B12 in their feed. The animals are often exposed to manure in their living conditions where they derive B12 and some are even fed manure and some cows are fed poultry defecation waste.

The United States FDA has reported that most meats are contaminated with faecal bacteria. Plant foods contain much less B12 in modern times, due to antibiotics, pesticides, cleaning and sanitation.

Humans and animals do not synthesize B12 and must obtain B12 from bacteria from their diet. Some B12 is made in the human lower gut but absorption is fairly low.

Certain mushrooms and seaweed still contain some B12 but probably not sufficient amounts as part of a normal vegetarian diet. Some foods, such as breakfast cereals and nutritional yeast, are fortified with B12. This crystalline form of B12 is preferable to the protein bound form of B12 in meat, as it is easier to absorb but still likely to be of insufficient quantity day to day.

The crystalline non-protein bound form is also provided in supplements. The best supplement is probably in the spray format (300 ug/spray). Most of this spray format is absorbed directly unlike tablets. Seaweed is an alternative source but some people may find its distinctive taste unpalatable. For those starting out on a vegetarian diet and with adequate stores, deficiency may take years to develop. A few sprays per week should be more than adequate providing all the vitamin B12 needs for humans.

13.4   Benefits of Vitamin B12

- Essential for red blood cell synthesis (works closely with folate B9)
- Maintains development and function of healthy nervous system
- Essential for the synthesis of DNA and RNA
- Acts as a co-factor and helps release energy from food
- Essential in amino acid and fatty acid synthesis
- Prevents high levels of homocysteine causing increased cardiovascular disease risk

13.5   Deficiency Symptoms of Vitamin B12

- Fatigue
- Weakness, tingling sensation
- Light-headedness, blurred vision
- Loss of appetite
- Shortness of breath and palpitations
- Numbness
- Loss of memory
- Impaired DNA synthesis
- Impaired red blood cell synthesis
- Megaloblastic anaemia
- Neuropsychiatric symptoms
- Infertility
- Depression, confusion

13.6   Water

Humans can only survive for around 3 minutes without air. Water is the next most important and essential of all ingested nutrients for survival.

Humans can only survive an absolute maximum of 5 days (3 days on average) without water but up to 60 days (depending on stored fat levels) without ingesting food.

The human body is comprised of approximately 60% of water by weight. It forms the make-up of blood, digestive juices, urine, perspiration, muscle, glycogen, bones and all cells (cytoplasm). It is a vital part of metabolism and also formed from combining protons and oxygen to yield water as product of cellular respiration during the OXPHOS and the ETC process discussed in Chapter 15.

It is also vital for kidney function in maintaining the correct osmotic and sodium and potassium electrolyte balance. Water keeps the bloodstream flowing, enabling blood cells travel through capillaries. Water carries nutrients, oxygen and waste products to and from all cells, tissues and organs.

Water serves to maintain body temperature in equilibrium with fluctuating external temperatures. Water is essential in moistening mucous membranes such as the lungs and the mouth. It cushions joints and protects tissues and organs from shock or damage.

Water reduces the risk of urinary tract infections by keeping the bladder clear of bacteria. Water aids digestion throughout the digestive tract from mouth to waste and prevents constipation. Good hydration increases energy, gives improved mental clarity, leads to healthier skin, improved digestion and much healthier organs.

The majority of food eaten contains water. Approximately 20% of the body's daily requirement of water can be obtained from eating food. Processing and digesting this food produces approximately 10% more water as a by-product of cellular respiration and heat generation. The remaining 70% must be derived from drinking fluids.

The amount of water required depends on age, gender, weight, the weather and lifestyle and if a woman is pregnant or lactating and breast feeding. This daily amount ranges from 0.7 litres/d for infants, to 2.6 litres/d for lactating women. The average quantity of total fluid intake required for most adults is in the order of 2.5 litres/d for men (around 8 cups) and 2.1 litres/d for women. (around 6 cups).

This includes water, tea, coffee, low-fat milk, sugar-free drinks but excludes all alcoholic drinks.

Another guide is based on weight at 35ml/kg body weight. Thus, a 57kg adult female should consume approximately (57 x 0.035) i.e., 2.0 litres/d and an 80 kg adult man should consume approximately (80 x 0.035) i.e., 2.8 litres/d (these levels inclusive of water included in food).

Another most useful guide to the state of hydration is the colour of the urine. Adequate hydration is indicated by a very pale yellow or straw colour. If the colour of urine is darker, or deep yellow, hydration should be increased. If this is almost colourless, hydration should be decreased. It is equally important not to over-hydrate by drinking too much fluid. Excessive hydration increases the risk of losing essential salts such as sodium and potassium in the urine. This causes an imbalance in the body's electrolytes, where increased fluid is drawn into the cells causing hyponatremia, with several other risks, including kidney stones, raised blood pressure (hypertension), low heart rate (bradycardia), impacted brain cells, confusion, drowsiness or headaches and even coma.

Finally, when exercising such as jogging, walking, cycling, swimming or working out in the gym, an individual will be sufficiently hydrated by drinking a glass of water 20-30 minutes prior to most normal and moderate levels of exercise. Unless engaged in very high levels of endurance, such as half or full marathon, or 200 lengths in the pool, or the weather is very hot, or if in unfamiliar territory, it is not necessary to top up with water and not necessary to carry a 'heavy' (inconvenient) bottle of water. It is much better to top up prior to exercise rather than during exercise, therefore avoiding water sloshing around in the stomach. Moreover, it is certainly not a good idea to exercise outdoors in high temperatures, (>25 degrees C) unless this is absolutely necessary, or unavoidable and the correct risk safe and well-being assessment has been undertaken.

13.7 Alcohol

Alcohol, or more accurately, ethanol, is certainly not considered an essential nutrient, nor a nutrient at all. Alcohol does however, contribute significantly to weight gain (See the list below).

In essence, if possible, it is much better to avoid alcohol altogether for health reasons. If individuals consider that complete abstinence is not possible, then moderation on quantity, strength (units) and the number of days per week it is consumed should be followed and adhered to as much as possible.

Any reports, papers or anecdotes, current or historic, extolling the virtues of alcohol, for example, that a little red wine is beneficial, is incorrect. For example, resveratrol. An individual would need to drink many bottles of red wine to see any benefit from resveratrol levels but at what 'cost'. Grapes are of course far, far healthier. These studies have now been found to be flawed due to lack of scientific rigor in the sample set and fundamentally, the inclusion of subjects who were former heavy drinkers in the control group, i.e. those that do not drink. Therefore the overall health of the non-drinker group was diminished compared with those that drink alcohol (predominantly red wine) moderately and hence the conclusion that a little wine provides cardiovascular benefits for females over 50.

In short, alcohol, however dressed in flavours, mixers, additvies and sugars, is a harmful toxin, hence the need for current government advice regarding quantity and 14 units per week. Toxic ethanol is 2.77 to 5.55 mmol/L.

Alcohol molecules significantly damage many cells and tissues throughout the body, including the liver, brain, heart, pancreas and nerve cells. It can often be a significant proportion of an individual's daily caloric energy intake (EI).

It is a very highly addictive molecule and drug and a very strong toxin and a poison to the body. The addictive properties of alcohol make it very difficult to resist if consumed regularly, i.e., chronically (1 or 2 drinks per day). Many people look forward to it as 'an essential pleasure' in life each day.

Ethanol most closely resembles a carbohydrate in structure (CH_3CH_2OH - functional group hydroxyl -OH), however its' metabolism most closely resembles a fatty acid. Catabolically, each gram of ethanol yields 7 kcal and may be up to 20% of total energy intake (EI), or even more, especially when drinking is becoming the daily and chronic norm for a significant proportion of western culture. It is highly lipid soluble and therefore very easily absorbed in the body. Females generally, have a lower level of alcohol dehydrogenase enzyme leading to a significantly lower level of alcohol tolerance compared with males and there are many individuals living in, or originating from several Asian countries, who are intolerant to alcohol as they lack the dehydrogenase enzyme.

However, notwithstanding the above, alcohol is included in this book, not because of other health related issues but in terms of weight loss and weight management.

There is limited capacity to store alcohol in the blood. The body preserves glucose in the liver and body fat until alcohol is burned off. With alcohol in the blood, the body preserves glucose as a fuel, instead, oxidatively prioritising alcohol as the first burnt energy substrate at 7 kcal/g, until all the alcohol is removed. Therefore, relevant to weight loss, alcohol:

a) Is first on the list of oxidised priority of all the 4 fuels burned
b) Can shut down the burning of carbohydrate, fat and even protein for 12-36 hours
c) If imbibed regularly, will delay and reduce fat burning leading to weight gain
d) Has a higher calorific content than proteins or carbohydrate at 7kcal/g
e) Can be a large proportion of total daily caloric intake and lead to weight gain
f) Can slow metabolism leading to less fat burning leading to weight gain
g) Is very likely to lead to increased intake of highly processed calorie rich meals
h) Is also very likely to lead to increased intake of highly processed snacks, crisps and nuts
i) May lead to eating late in the evening reducing the time for cell repair and autophagy
j) Often leads to a reduction in exercise leading to weight gain
k) Causes the liver to make more cholesterol and fat leading to weight gain
l) Is particularly associated with weight gain around the belly

The liver oxidises alcohol to acetaldehyde, via alcohol dehydrogenase, then it is converted to acetate via acetaldehyde dehydrogenase, then to acetyl-CoA and burned off as energy, water and CO_2 in the TCA cycle, or used for fatty acid synthesis. Although short lived, acetaldehyde is the principal toxin, causing reactions, headaches, reactive oxygen species (ROS) cell damage and problems in the body. Table 13.5 shows the caloric content of different alcoholic drinks, 3 pints of beer can easily add up to 750 kcal or as much as 30% of total daily calories (TEE) and therefore, highly likely to lead to weight gain. It is also associated with increased likelihood of simultaneously eating more processed, low quality, high calorie food and snacks and also later in the evening, therefore extending the eating window, as judgement becomes impaired.

Table 13.5 Calories contained in typical alcoholic drinks

Beverage	Volume Alcohol (l)	ABV %	No. Units	kcal
Pint Real Ale	0.568	5.0%	2.84	250
Pint Estrella Lager	0.568	6.5%	3.69	225
Pint Cider	0.568	4.5%	2.55	210
Glass Red Wine	0.250	13%	3.25	190
Glass White Wine	0.250	12%	3.00	200
Single Gin & Tonic *	0.044	40%	1.76	170
One Whiskey	0.044	40%	1.76	105
One Cocktail	0.044	40%	1.76	186

* assumes normal tonic water

In conclusion, if not completely avoided, alcohol is a toxin and should at least be consumed in moderation. The addictive strength of ethanol is very high indeed and can lead to a very difficult, (sometimes daily) cycle to break. However, breaking the cycle and allowing the body (especially the remarkable and ever toiling liver) time to rest and recover for a few days, will certainly bring significant well-being benefits and certainly aid with weight loss targets whilst also improving overall health.

PART 2

NUTRITIONAL METABOLISM

&

BIOCHEMISTRY

CHAPTER

Metabolic Processes & Weight Management 14

On changing diet and reducing weight, with a concomitant reduction in energy intake in the short term, the body reacts and defends its stored energy position by reducing metabolic rate, whilst promoting an increase in calorie intake, through the release of hormones and neurotransmitters, invoking changes in appetite, hunger and various psychological behaviours. This often leads to a regain in weight over time for most people. Approximately 1 million calories (2,500 x 365 = 912,500,kcal/yr. for lean men), or 20 tonnes of food, are ingested over a year and compared with this order of intake, any changes in weight are actually fairly small. This suggests that many integrated, metabolic mechanisms are in simultaneous operation. Hence, just concentrating on reducing total calories alone is likely to fail. The body's metabolism, regulation and control of weight loss and weight gain in a changing environment are complex and scientists are still researching several key aspects of this. Table 14.18 provides a useful reference and overview summary of the various metabolic processes and pathways that will be discussed in Chapters 14 and 15.

14.1 Metabolic Changes on Weight Change

There is a persistent metabolic, hormonal and long term adaptation that occurs over time in the body's response to weight loss. [24] This is why the SMART, step by step, short term goals (Section 2.7) and gradual weight loss approach in this book is strongly recommended. The body adapts and reacts to preserve the status quo, using many tools and processes. The metabolic processes regulating body weight and the physiological and environmental conditions that drive them, are explored in this Chapter. It is true that two people can have the same body composition and overall size but a different metabolic rate.

Scientists do not yet know the precise reasons for this but it may well be down to how each body's cells have been 'trained' since birth, including environment, genes, changing body weight, exercise and diet. It is also the case that metabolism slows with age. Humans burn more calories at aged 30 than 60 and this gradual decline probably starts around the age of 20. After this age, on average across populations, BMR declines by approximately 1-2% per decade, mostly (there are variations between individuals), but not exclusively, due to the loss of fat free (largely muscle) mass.

There is also the body's individual set point, i.e., the theory where the body always tries to revert to the set point in weight, or a minimum of this set point but not lower (Section 1.9). This is one way to explain how the body fights to restore body weight after weight loss. Genes may well have a further role and there is some evidence that certain individuals do find it harder than others to lose weight. The *'personal fat threshold'* (Sections 1.7, 2.9, 14.1 and 14.5) [3] is an example. Once weight loss kicks in, cellular metabolism is slowed down, often referred to as metabolic compensation, metabolic adaptation or adaptive thermogenesis.

When this occurs, the body senses potential deprivation, in effect, adapting to survival, promoting anabolic or a preservation status. This is especially the case with crash diets or extreme diets, where a large amount of weight is lost over a short period of time, starving the body of a sufficient number of nutritious calories. It is also possible that metabolism slows down even further than that measured by the reduction in calory energy intake (EI) alone. Again, all set in motion for the organism's long term survival. By way of example of this metabolic slow down, a 10% reduction in non-lean body mass (fat mass) will cause a reduction of around 500 kcal to 600 kcal in burning energy metabolism. This leads to a plateau, where further weight loss then becomes more difficult.

To counter this and restart or kickstart the weight loss process, it is necessary to essentially fool the body, by increasing overall levels of activity and exercise. Say to an additional 600 kcal/d of exercise activity thermogenesis (EAT) and non-exercise activity thermogenesis (NEAT). The sensory 'memory' of fat cells or adipocytes can continue for many years after the weight loss, defending the original set point (Section 1.12), *'The fat cell is a harsh master with a long memory'* [2]. As a consequence, reaching a target new set point, or plateau, can take long periods, even months, where energy balance of expenditure, TEE against intake EI, for all organs and tissues eventually becomes the new and settled default status quo.

At rest, muscle burns calories at a faster rate than fat (Tables 14.1 and 14.2). As the individual becomes leaner, following loss of fat mass, the relative mass of muscle increases. Adding additional exercise and resistance training (Chapter 8) will therefore, contribute a little to higher calorie burning but more importantly, it builds and strengthens muscle. Hence metabolism is increased, serving to redress the slow-down of metabolism due to the reduction in weight and fat. Research shows that the increase in metabolism due to higher activity will not fully match the body's' fat preservation defence but it will certainly help drive energy processes. Key point 1.

This helps achieve the new set point. Once the new reduced weight is reached, the body requires fewer calories, i.e., a lower EI (Section 2.1 and 2.7). This reduction needs to be sustained and some find this stage difficult, especially reducing the calorically dense foods and 'appealing', high in fat salt and sugar. Recall from Chapter 5 that many nutritious, nutrient dense foods are lower in calorie density and higher in bulk, water and satiation and these foods will certainly help achieve that lower EI. In addition to all these factors, other psychological factors, for example cravings, have a very important and sometimes equal impact in returning to the original over-weight position. Key point 2.

The brain senses the comparative 'deprivation' and reduces the satiety and appetite control messaging, causing more likelihood of additional eating beyond satiety and hence the return of weight gain. Some people who have reduced weight, constantly crave food and their senses, smell in particular, and also imagination, are significantly enhanced, leading to unhealthy feeding relapses and weight restoration. Psychology kicks in to restore 'the old days'. Key point 3.

Once weight loss targets have been reached, many people feel great, satisfied and healthier and thinking they have achieved their goal. Once this is achieved, they start to repay the brains reward centre, by very gradually reintroducing the less healthy foods, such as more treats, after all they can now 'afford it'. The body, naturally responds to these messages by regaining the weight. The body strives to restore weight to improve survival capability. Key point 4.

14.2 Metabolic Rate and Body Composition

The body is essentially composed of water, fat, protein and minerals. For the purposes of weight management, the body can be seen as a two-component model of body composition comprising a fat component and a fat-free (lean) component. The major organs, adipose tissue and skeletal muscle comprise the bulk of the weight of the body. Table 14.1 shows an example of the mass specific BMR and mass of various body tissues and organs, where the total weight (mass) of the female is 58.02 kg and 69.44 kg for the male. The last 2 columns show the contributing percentage of total body (basal) BMR for each specific organ and tissue. It is likely that the gut will account for the majority of the residual (rest of the body) use, as the gut is highly metabolically active and requires considerable energy. [69,70, 71]

Table 14.1 Lean Adult mass specific metabolic rate and organ and body mass Proportion of Metabolic Rate

Organ	Organ Metabolic Rate (kcal/kg/d) [1]	Tissue/Organ Weight (kg)		Body Weight (% of Total)		Basal Metabolic Rate (% of Total BMR)	
Mass specific Metabolic Rate		Female	Male	Female	Male	Female	Male
Heart	440	0.24	0.33	0.41	0.47	8	9
Kidneys	440	0.28	0.31	0.47	0.44	9	9
Brain	240	1.20	1.40	2.07	2.00	22	21
Liver	200	1.40	1.20	2.41	2.57	21	15
Muscle	13.13	17.00	28.00	29.30	40.00	17	24
Adipose	4.50	19.00	15.00	32.74	21.40	6	4
Residual	12.00	18.90	23.20	32.60	33.12	17	18
Total		58.02	69.44	100.00	100.00	100	100

[1] Mass-specific Metabolic Rate or the Ki value % of Total BMR figures rounded

There are 78 organs in the human body, each with its own mass balance, metabolic rate, fat and muscle content and selection of glucose or fat (or other substrates) for energy, which is one of the reasons why weight management will likely never be a precise science. Body leanness, rather than weighing on the scales is a better long term objective and refers to the muscle, bone, and fat composition of body weight. Note how BMR depends on body composition, with additional muscle leading to a higher BMR in males compared with women and vice versa with the relative higher proportion of adipose tissue in females.

The rate of energy expenditure is more closely correlated with the total fat free or lean mass in all individuals. Once obese people reach their heavier weight and their higher, now stable, energy balance, both energy intake and expenditure tend to remain high. It is important to recognise that adipose tissue is not the only store of fat. Both EI and TEE have increased.

This fat is also present as visceral fat and ectopic fat in several organs, such as the liver, pancreas and other visceral organs. Therefore, both weight gain and weight loss include fat adjustment within other organs and tissues as well as the adipose tissue itself.

In a lean individual male weighing 69.44 kg at rest, the adipose (fat store) tissue weighs around 21% of total body mass but only accounts for around 5% of total BMR. Note from Tables 14.1

and 14.2 that this subcutaneous, abdominal adipose (fat) tissue, has the lowest tissue mass specific metabolic rate, of approximately 4.49 kcal/kg/d (18.8 kJ/kg/d) at rest therefore, using only 68 kcal/d.

By contrast it has been calculated that muscle, weighing around 40% of total body weight, has a significantly higher tissue mass specific metabolic rate of approximately 13.00 kcal/kg/d (54.4 kJ/kg/d) and uses approximately 367 kcal/d, or 24% of total BMR, far more than fat. Overall therefore, muscle tissue burns more than 5 times the energy compared with adipose tissue. The metabolic rate of adipose tissue varies depending on the location of the body fat, with the highest rate within the intra-abdominal and the lowest around the gluteal (backside) area.

Muscle is more metabolically active than fat meaning it requires more energy to maintain compared with fat. Muscle itself does not burn fat directly (any fatty acids that the muscle does burn for energy are provided in the blood stream by the liver) but having more muscle mass in the body leads to the body burning more total calories overall compared with the same total body weight with less muscle by proportion.

Therefore, total body tissue calorie burning certainly helps burn more fat. From Table 14.1 it can be seen that from 17% to 24% of total BMR (females and males respectively) is used up by skeletal muscle compared with just 4% to 6% for adipose tissue and yes, adding muscle will help increase metabolic rate overall and lead to some fat burning.

Table 14.2 calculates the total organ specific metabolic rate (total energy burned per day) from the mass specific metabolic rate and the weight of each organ. Note that the brain requires and burns the most energy of all organs and tissues in females at 288 kcal/d, compared with skeletal muscle of 221 kcal/d, even though the brain weighs just 2% of total body weight. Note, the last 2 columns of Table 14.2 showing the total metabolic rate, kcal/d, for each organ and tissue and these numbers calculate back to the percentage of total BMR shown in Table 14.1. [69,70,71]

Table 14.2 Lean Adult mass specific metabolic rate and organ body mass and total Metabolic Rate

Organ	Organ Metabolic Rate (kcal/kg/d) [1] Mass specific Metabolic Rate	Tissue/Organ Weight (kg) Female	Male	Body Weight (% of Total) Female	Male	Basal Metabolic Rate (kcal/d) Female	Male
Heart	440	0.24	0.33	0.41	0.47	105.60	145.2
Kidneys	440	0.28	0.31	0.47	0.44	123.20	136.4
Brain	240	1.20	1.40	2.07	2.00	288.00	336.00
Liver	200	1.40	1.20	2.41	2.57	280.00	240.00
Muscle	13.13	17.00	28.00	29.30	40.00	223.21	367.64
Adipose	4.50	19.00	15.00	32.74	21.40	85.50	67.50
Residual	12.00	18.90	23.20	32.60	33.12	226.80	278.40
Total		58.02	69.44	100.00	100.00	1332.31	1571.14

[1] Mass-specific Metabolic Rate or the K_i value

For example, in the female, the kidneys use a total of 123.2 kcal/d, which is 9.2% of the total BMR of 1332 kcal/d, whereas the brain uses 336 kcal/d, or 22 % of the total BMR of 1571 kcal/d in the male. Skeletal muscle mass burns the most total energy in males, at 367 kcal/d, compared with 336 kcal/d for the brain because the muscle mass is considerably higher by proportion, at 40% of the total body weight. This may be one of the reasons why some males work out in the gym so much, in order to keep building this mass on skeletal muscle as well as keeping trim.

From Table 14.2, comparing total body weight with total metabolic rate for these individuals yields a very similar average total body metabolic rate per unit weight, of 22.92 kcal/kg/d for females (1332.31/58.02) and 22.58 kcal/kg/d for males (1571.14/69.44) respectively.
The brain averages around 22% of total BMR, the liver at 21% females (15% in males), with the heart and kidney at around 8% each. In females, this leaves around 17% for skeletal muscle, 6% for adipose tissue and the remaining 17% for all other body tissues. The equivalent is 24%, 4% and 18% in males respectively.

However, even though the combined weight of the heart, kidneys, brain and liver, only represents approximately 5% of body mass, the total energy expenditure of these 4 major organs, is very considerable, at around 54% of BMR, significantly higher than the energy expenditure of combined muscle and fat at around 28% of BMR but muscle plus fat weighing 62% of total body mass.

The brain is probably the most metabolically active organ in the body per unit weight, as it represents in the order of just 2% of total body weight and yet uses 22% of total energy in the resting state. Small wonder, as it is estimated that there are approximately 400 miles (644 km) of capillary blood vessels within the adult human brain alone and a total of 60,000 miles (96,000 km) in the whole adult body, hence the need for rapidly available glucose. This increases significantly at up to 65% of total energy expenditure in young children. The brain uses proportionately higher energy in infants, at around 50%, due to relative proportion to body size at that stage of life.

However, total adult skeletal muscle mass represents in the order of 29% to 40% of total body weight (females and males respectively) and as its energy rate of activity is broadly similar to the brain (and liver), it is ultimately the skeletal muscles that have the major impact on overall BMR and nutrient requirements of the body. Muscle does not weigh more than fat; a kilogram is a kilogram for any tissue. However, muscle weighs more than fat by volume due to different morphology. Muscle is a much denser tissue and hence looks more taught, sculpted and toned and fat takes up more space.

A handful of muscle will weigh more than a handful of fat. For example,1 kg of fat will look different than 1 kg of muscle. An individual is more likely to notice a gain in fat than a gain in muscle. This is particularly important when recording weight each week. An individual's weight may change in composition (more lean and less non lean mass) if not in overall body mass.

A person will best be able to detect this with waist line measurement and above all observing overall body tone. Therefore, people should not be concerned if weight loss does not necessarily happen each week, if instead, fat mass is being exchanged with muscle i.e., fat free mass. This is a good thing and will help sustain the fat weight loss considerably. Increase in muscle mass should always be accompanied by a balanced wholesome calorie sufficient diet.

However, in terms of body building endeavour to build and grow additional muscle (hypertrophy), in order to increase metabolic rate, even a very large gain in muscle of 4.5 kg (10 pounds) will probably make a relatively small difference overall in metabolic rate, of around 60 kcal/d, or just 4% of total BMR and TEE. As described above, building more muscle per se can influence metabolism a little, as muscle burns at a higher rate than fat, however, it is possible that people will then, consciously or subconsciously, eat more (or be tempted to eat more) to match the increased metabolism. The psychology factor.

Overall, the relevance of managing non-lean body weight and higher metabolism due to increased muscle mass is more probably due to higher overall activity in the individuals that train and work out more, compared with those with less muscle mass and more fat mass.

Maintaining increased muscle mass is difficult and not necessarily always practical and sustainable for most people. This can lead to problems over time, such that without constantly exercising the enlarged muscle mass through regular training, year after year, this may lead to protein atrophy in the skeletal muscle and consequently surplus fat. Building muscle also increases the demand and adaptive responses on bone tissue and also heart mass and structure and there is an increase in post-exercise oxygen consumption relating to protein synthesis and adaptation and recovery.

Exercise and resistance training is however, all good news for fitness and health overall (Chapter 8). Increasing skeletal muscle mass may also result in an increase in total body mass which will increase the total expenditure TEE. More body weight will mean an individual burns more energy calories just by moving around, never mind the additional EAT in the gym, hence, the calculated 13.13 kcal/kg/d mass specific metabolic rate of muscle and the potential increased 60 kcal/d or 4% increase in TEE.

Ultimately, however, it may take in the order of 12 months and longer to add 5 kg of muscle mass to the body to increase calory burning by 60 kcal/d, assuming no additional dietary intake over this period. In conclusion, if the objective is to contribute to weight loss, it is probably not worth the considerable effort of adding this lean mass to increase TEE by 4%, especially when compared with managing weight loss, by eating improved nutrient quality calories.

This is not surprising, as the maximum rate of muscle growth from exercise and work is around 5 g/d.
The total body basal rate of metabolism (BMR) can-not be increased very significantly by exercise, (nor significantly increased by certain often quoted foods such as hot chilli's, coffee, cinnamon and spices etc, popular currently).

A further challenging and perhaps unexpected and confounding factor is that, on average, overweight and obese people tend to have higher metabolic rates. For example, an obese person may have a metabolic rate of 2,400 kcal/d compared with a lean individual whose BMR is in the order of 20% lower at 2,000 kcal/d.

This is probably because fat free mass also increases in obese people, including the significant quantity of non-fat mass in and around the adipose tissue and other connective and vascular tissue, that has developed to support the additional fat mass.

This is a very important factor and another example of the difficulty faced by people of significantly higher baseline weight when endeavouring to lose weight, compared to other individuals starting at much lower weight.

The foregoing supports the case that obesity is more closely related to excess intake over time, i.e., over nutrition, rather than reduced energy expenditure. Adding in the factor of reduced energy expenditure, then certainly serves to exacerbate the weight gain.

14.3 Hormonal Regulation of Nutrition Metabolism

There is a vast amount of knowledge and research based on all the incredible work and achievements of endocrinologists and biochemistry experts around the world, on how hormones play a pivotal role in nutrition, weight management, appetite, digestion and many other biological factors beyond the scope of this book, including the considerable number of digestive enzymes and other hormones, not included in this first edition. However, the following discussion provides an overview on the principal hormones involved in the regulation of glucose and fat burning and the storage and mobilisation of fat within and from adipose tissue, via lipogenesis and lipolysis respectively. In essence, hormones are the orchestrators of metabolic pathways and have a considerable role to play in energy oxidation, metabolism and weight management. Energy intake is regulated by complex interactions between nutrients, hormones, neuropeptides and several different parts of the brain. The interplay is highly elaborate and carefully regulated by the brain and other organs, including adipose tissue, which is itself an endocrine organ. These hormones are stimulated by lifestyle, energy maintenance and diet and the actions and consequences of each depend on energy conditions and blood chemistry at the time of stimulation and release. Simply separating these hormones into either fat storage (anabolic) or fat burning (catabolic) roles as listed below is not necessarily always applicable, as the regulatory processes are more finite and intertwined, involving positive and negative feedback mechanisms, for example, the full functions and mechanisms of Somatostatin are still being unravelled.

However, the following discussion is provided for information, to show 20 of these hormones at work; their key functions and the fundamental importance of the endocrine system in hunger, satiety, energy intake, expenditure, metabolism and weight management. The first 6 hormones (from Insulin to Dopamine) discussed, are generally considered to be fat storage promoting anabolic hormones that can, depending on prevailing conditions, work to promote fat storage. The next 12 hormones (from Glucagon to Enterostatin) are generally considered as catabolic hormones that can, depending on prevailing conditions, reduce feeding and promote fat oxidation i.e., fat burning. Most of these hormones work by suppressing appetite. The final 2 hormones (Thyroid Hormone and Somatostatin) act to carefully balance both catabolic, fat burning and anabolic fat storage.

Insulin
Insulin is the principal fat storage anabolic hormone and hence, this section on hormones will therefore concentrate mostly on this key hormone. It signals abundance of exogenous energy, promoting the storage of fat and uptake of glucose for oxidation in the cells, the storage of glucose as glycogen via glycogenesis in the liver and also the uptake of amino acids. Contrary to

popular rhetoric, apart from those people with diabetes or prediabetes, who need to carefully monitor blood glucose levels, insulin is a normal, vital, natural, healthy and essential hormone, carrying out functions in blood glucose regulation and metabolism. When insulin is high, glucose oxidation dominates over fatty acid (β)-oxidation and when insulin is low, fatty acid (β)-oxidation increases as the dominant substrate oxidised. A rise in insulin levels is a correct and perfectly normal response to blood glucose levels and should not be feared nor avoided at all costs, for example, the current trend in some perfectly healthy individuals becoming obsessed with glucose and insulin levels, including the regular use of continuous glucose monitors (CGM). Best to concentrate on a healthy diet promoting healthy and natural hormone levels. Insulin has a key role in both types of diabetes and is also linked with the Randle cycle (glucose-fatty acid cycle) (Section 14.13) and the selection of either glucose or fatty acid beta (β)-oxidation, dependent on conditions. Insulin is predominantly important in regulating the Randle cycle substrate selection system when neither fat, nor carbohydrate are dominant and where both are similar in proportion and both are in excess of the body's requirements. Insulin is the principle anabolic hormone that responds to periods of high available energy. Insulin therefore preserves fat, both by inhibiting lipolysis (fat mobilisation and breakdown) and inhibiting fatty acid (β)-oxidation (fat burning for energy). At the same time insulin promotes the glycolysis and oxidation of glucose (glucose burning for energy) and stimulates and enables the cell GLUT-4 transporters, facilitating more blood glucose to enter skeletal muscle and adipose tissue cells for oxidation. In contrast the brain, liver and kidney medulla do not require insulin for glucose transport. Insulin powerfully influences blood glucose levels or fat mobilisation and beta (β)-oxidation. In diabetes, the body is in a serious catabolic state, where there is a break-down of fuel stores and tissues. As insulin significantly affects glucose and fatty acid metabolism and weight gain and loss, some discussion of Type 1 and Type 2 diabetes mellitus is relevant here.

Type 1 diabetes is an autoimmune disease that usually starts in childhood. Most people with Type 1 diabetes tend to be thin and underweight. Insulin is not produced because the body has destroyed the pancreatic beta cells. Insulin must be prescribed and as it is destroyed by digestive enzymes it must be injected intravenously. If it is left untreated, the absence of insulin in Type 1 diabetics leads to very high and continuous levels of both fatty acid (β)-oxidation, leading to high ketone levels and diabetic ketoacidosis (DKA), together with very high blood glucose. In these conditions, glucose can-not enter cells and hyperglycemia (high glucose sugar levels) rises. The body tissues that are able to use fatty acids for energy (heart and skeletal muscles) continue to utilise fatty acids and body fat becomes depleted over time. In this disease, both blood fatty acid levels and blood sugar levels are high. Normal resting blood glucose level is around 5 mmol/L and with diabetic hyperglycemia, this can increase to 20 or even 50 mmol/L. The normal level of non-esterified fatty acids in the blood, is in the range of 0.2-1.0 mmol/L but in untreated diabetes mellitus, this may reach 3-4 mmol/L. Further, many key tissues, for example, the brain and CNS, are unable to use fatty acids for energy. The liver perceives that glucose is not available, nor can-not be used and that these key tissues can-not use fats for energy. With the ever-increasing rate of fatty acid mobilsation and oxidation, the liver switches the (β)-oxidation of fatty acids to the production of ketones that can be utilized by these tissues. The normal level of ketones in the blood is in the range of 0.2-1.0 mmol/L. In untreated diabetes, however, the combination of ketone bodies in the blood may reach 10-20 mmol/L increasing the blood acidity to a pH of around 7.1 (normally 7.4), leading to the very dangerous development of DKA, which is a life-threatening problem that affects people with diabetes. DKA more commonly occurs in patients with Type1 diabetes, though it can also occur in people with Type 2 diabetes.

People with Type 2 diabetes can also develop DKA, but it is much less common and less severe. For people with Type 2 diabetes, DKA is usually triggered by prolonged uncontrolled blood sugar, missing doses of medicines, or a severe illness or infection. The correct action of normal levels of insulin is to decrease hepatic glucose production by inhibiting glycogenolysis and gluconeogenesis i.e., the production of more glucose. Glucose uptake by skeletal muscle and adipose tissue is increased by insulin. Both of these mechanisms result in the reduction of blood sugar.

In DKA, however, insulin deficiency and increase of glucagon can lead to increased gluconeogenesis, accelerated glycogenolysis, and impaired glucose utilization. This will ultimately cause increased hyperglycemia. In addition, triacylglycerol levels increase causing hypertriglyceridemia. As sufficient glucose intake from the glucose already present in the blood is impaired, levels of glucagon and stress hormones increase. This, together with the lack of insulin, drives and increase of gluconeogenesis in the liver to increase hepatic synthesis of glucose. A further consequence is a net breakdown of tissue protein from amino acid substrate contribution causing a break-down in skeletal muscle. Without treatment, the significant reduction of body fat and structural protein leads to severe problems of emaciation and the individual losing too much weight.

Prior to diagnosis, a person may consider that they are unable to gain weight, no matter how much food, or what type of food is eaten. They are used to being lean, and are perhaps unaware of the highly damaging hyperglycemia, hypertriglyceridemia and ketoacidosis. On receiving treatment of insulin injections for Type 1 diabetes, weight is then gained as glucose can now be used and fat is stored rather than burned. However, with the developing signs of weight gain, some people then choose to misuse the treatment by reducing the insulin dose, which allows them to remain lean and eat what they want as before the diagnosis. This perpetuates a worsening situation for the body tissues.

Type 2 diabetes is far more common than Type 1 diabetes, mostly developing in middle age and the reasons for its development are more complex but principally due to lifestyle. Most people (but not all) with Type 2 diabetes tend to be overweight or obese. Obesity is not a consequence of diabetes but may occur many years or even decades before the onset of diabetes Type 2 and obesity is highly (not all the time but mostly) likely to lead to Type 2 diabetes. There are two common physiological reasons for the development of Type 2 diabetes; either, reduced secretion of insulin from the pancreas, or a failure of insulin, at relatively normal concentrations, to exert its normal effects and the body, where the body has become resistant to insulin i.e., insulin resistant, or a combination of both. Insulin resistance is prominent in obesity and Type 2 diabetes. There is now a general consensus that Insulin resistance and Type 2 diabetes is caused by cells being overloaded with fat. [35]

Free fatty acids (FFA) induce insulin resistance in muscle at the level of insulin-stimulated glucose transport or phosphorylation, by impairing the insulin-signaling pathway. It is also postulated that the accumulation of metabolites, including diacylglycerol (DAG) is the specific, direct cause of insulin resistance and not solely due to the accumulation of fat in muscle cells.

The breakdown of fats i.e., triacylglycerol (TAG) during lipolysis, is initiated by TAG lipases that generate DAGs and fatty acids. High dietary fat (and specifically DAG) causes insulin resistance over time, as endogenous fat builds up in adipose tissue and other key organs and tissues of the body including the liver, pancreas and skeletal muscle. Free sugars such as fructose worsen this fat accumulation as less fat is oxidised, then regular hypercaloric energy intake (EI), over nutrition, including all 3 macronutrients all exacerbate this situation. Insulin is a strong anabolic hormone. In the same way as for Type 1 diabetes, most people with Type 2 diabetes gain weight when placed on insulin treatment. Normally, the insulin released by the beta cells of the pancreas, would promote and facilitate glucose uptake, promote glucose oxidation and reduce or prevent the mobilization and (β)-oxidation of fatty acids. With insulin resistance, however, the fat cells are altered in their response to insulin.

This leads to a continued output of insulin from the pancreas, with a diminishing insulin impact and a rise in both blood sugar and fatty acids. Recall that carbohydrates (even healthy ones) spare the fat, that is, when there are plenty of carbohydrates in the system, in the absorptive state, less fat will be burnt overall, as the body's first choice is to use glucose for energy. The key is to burn both.

'Carbohydrate spares the fat' [14]

The metabolic changes due to diabetes resemble those of starvation. The liver makes excess glucose via gluconeogenesis instead of using it for energy. Muscle and adipose tissues fail to absorb glucose from the blood. Insulin deficiency and increased hormones such as glucagon, also lead to the release of free fatty acids into circulation from adipose tissue by lipolysis. The fatty acids levels continue to rise and as a consequence, undergo hepatic oxidation to ketone bodies (beta-hydroxybutyrate and acetoacetate), resulting in ketonemia and metabolic acidosis (DKA). During starvation, the liver and adipose tissue serve to maintain glucose levels by producing more and more glucose, fatty acids and ketone bodies. The excessive supply of fatty acids, reduces glucose break-down in the liver and skeletal muscle.

As a high fat diet increases insulin resistance, dietary treatment of Type 2 diabetes is total caloric restriction and not simply the replacement of carbohydrates by fat. Endogenous fat built up over the years is more problematic than the impact of recent and new dietary or exogenous fat. After an extended period, the beta cells of the pancreas become worn out and gradually become less able to produce insulin, resulting in clinically evident diabetes. Many of these symptoms can be reversed by a reduction in weight, requiring no further insulin or treatment. However, the insulin might be operating correctly within muscle tissue (which is not insulin resistant), correctly instructing the muscle to burn glucose not to use fat.

This can also occur in some other tissues that do not require insulin to activate GLUT-4 transporters and can use glucose without insulin. In contrast, the adipocytes and other tissues that are insulin resistant (i.e., perceiving that there is low insulin) must select the fatty acids as the substrate to be oxidised, rather than glucose. Lipolysis is stimulated to mobilise fats. Thus, the combination of tissues that are not insulin resistant and those that are insulin resistant, can lead to both high blood sugar and high levels of fatty acids in the blood. The Type 2 diabetic is generally insulin resistant and is therefore constantly using glucose and can-not shift out of glucose burning, or move to fat burning, even in the post absorptive, fasted state. It can be seen therefore, that the constant high insulin impacts on and alters the Randle cycle. There are areas of the hypothalamus (in the brain) that can become insulin resistant. This affects satiety and hunger.

In certain conditions, the free fatty acids-induce insulin resistance, preserving carbohydrate for use by these vital tissues. This is the case during starvation and naturally, during the second half of pregnancy, when the insulin resistance of the mother preserves glucose for the growing fetus. It is likely that all animals have developed an evolutionary mechanism during starvation, to trigger insulin resistance in hepatic and skeletal muscle, as a protective and survival measure, to ensure these tissues prioritise the oxidation of fat for energy, such that precious glucose can be saved for and utilised by the brain, CNS and RBCs.

Glucagon

Glucagon is produced in the pancreatic alpha cells and it mobilises fat for energy. It increases thermogenesis and energy expenditure. Glucagon increases blood sugar levels, in contrast to insulin which decreases blood sugar levels. Glucagon helps blood glucose levels rise in multiple ways, including glycogenolysis, signalling to the liver to convert stored glucose (glycogen) into glucose and then release it into the bloodstream. Glucagon can also prevent the liver from taking in and storing glucose, so that more glucose remains in the blood.

Glucagon helps the body synthesise glucose from other sources, such as lactate, glycerol and amino acids. It acts to reduce body weight and adiposity by suppressing appetite and increasing lipid metabolism by modulation. Glucagon promotes fat break down lipolysis by stimulating the activity of hormone-sensitive lipase. This is the key enzyme that mobilises the stored fats and triacylglycerol (TAG) hydrolysis (to DAG and MAG) in adipocytes. It also stimulates growth hormone, cortisol and adrenaline.

Ghrelin

Produced in the stomach. Ghrelin signals hunger to the hypothalamus in the brain, leading to an increase in appetite and further eating. Ghrelin also slows metabolism and decreases the ability to burn fat. Weight loss causes ghrelin levels to rise causing an increase in appetite.

Oestrogen

Produced in the ovaries and adipocytes - both high and low levels of oestrogen stimulate fat deposits and weight gain.

Gastric Inhibitory Peptide (GIP)

Produced in the duodenum and jejunum - GIP increases fat deposition.

Serotonin

Serotonin is a neurotransmitter and also acts as a hormone. It is produced in the peripheral organs and gut. Serotonin is mainly stimulated by ingested fat and protein. It blocks insulin secretion and reduces fat breakdown i.e., lipolysis.

Dopamine

Dopamine is produced in the (hypothalamus). It activates brain pleasure centres and rewards eating.

Leptin

Leptin is produced in the adipose tissue. It is a fullness hormone and signals to the hypothalamus in the brain that the body is full. Adipose-tissue-sourced leptin is secreted in proportion to adipose mass, and it conveys status information to the CNS about long-term energy stores. It helps regulate the long-term balance between the body's food intake and energy expenditure and helps to inhibit hunger and regulate energy balance, so that the individual does not eat when energy (calories) is not required.

The level of leptin, produced by adipocytes, reduces as the fat cells are diminished and hence the lower levels promote the sense, or feeling or craving, in the brain, that additional feeding is required.

Scientists have found that leptin diminishes to very low levels, in those people who lose a significant amount of weight, causing increases in feelings or perception of hunger. Even after several years, leptin levels do not return to the original level prior to the weight loss. This sends signals to the brain that promote additional eating and weight gain.

As leptin (causing satiety) levels do not return to original levels, people on crash diets, losing a large amount of weight (10 kg) over short periods of time (12 weeks) generally have lower leptin and lower satiety. As a result, there is often a tendency to want to eat more, with ensuing regain in weight.

This is not to say that defects in leptin levels is the reason people become obese. People remain obese even with high levels of leptin. However, these individuals often develop leptin resistance. A further school of thought is that leptin functions as a starvation warning and as leptin drops the urge to eat grows. With the modernity of 24/7 constant availability of food, the efficiency of the leptin system, which evolved during times of more frequent potential famine or food shortages, can become impaired and diminished.

Cortisol

Cortisol is synthesised from cholesterol and produced in the zona fasciculata of the suprarenal cortex. Cortisol is the principal stress response glucocorticoid, steroid hormone. It is regulated by the hypothalamus-pituitary-adrenal axis. Its main function is to regulate metabolism. It increases the availability of glucose to the brain.

When at high levels, it promotes gluconeogenesis and reduces glycogenesis (both preserving and maintaining glucose levels). Cortisol increases lipolysis, the break-down of fatty acids in adipose tissue. It acts on the pancreas to decrease insulin and increase glucagon and enhances the activity of adrenaline and other catecholamines.

Glucagon-like Peptide (GLP-1)

GLP-1 is a powerful 30 amino acid peptide hormone produced in the ileum. In response to feeding and particularly dietary glucose, GLP-1 inhibits glucagon, via stimulation of somatostatin which stimulates insulin, lowers blood glucose and suppresses appetite. It contributes to postprandial glucose regulation and augments meal-related insulin secretion from the pancreas. GLP-1 reduces food intake, slows gastric emptying and reduces hunger, to promote fullness and satiety. One of the latest, innovative weight loss drugs (Ozempic to be rolled out to over 1 million people in the UK in 2025 is based on this hormone. The equivalent drug Wegovy is used in the USA.

Growth Hormone (GH)

Growth Hormone is produced in the pituitary gland. Although it is highly anabolic, as it stimulates glucose uptake and promotes protein synthesis, it also stimulates lipolysis; the break-down of fat for energy.

Insulin like Growth Factor (IGF-1)

Insulin like Growth Factor is normally highly active in babies and children and decreases with age. It is produced in the liver and skeletal muscle. IGF-1 is triggered by GH and helps mobilises fat from adipose tissue. Too much IGF-1 in adulthood can negatively impact health causing serious diseases in humans. IGF-1 levels are elevated when eating meat and dairy products and eggs high in protein. This can cause the liver to release more IGF-1, stimulating cells to grow prolifically. Proteins in animal foods have a similar amino acid profile to humans and hence cause this a reaction to grow cells. In addition, farmers use a hormone to feed cows in order to speed up milk production. This hormone is called synthetic recombinant bovine growth hormone (rBGH) and will increase levels of human IGF-1 when consumed as milk and dairy. Avoiding meat and dairy and for example, changing to plant-based milk, will reduce levels of IGF-1, as will increased physical exercise.

Epinephrine (Adrenalin)

Epinephrine is produced in the suprarenal medulla. It mobilises fat ready for energy provision.

Testosterone

Testosterone is produced in the testes. At normal levels it reduces body fat, although a drop in levels increases abdominal fat and an increase in the level of oestrogen. A drop in testosterone may cause demotivation for exercise, therefore further reducing testosterone levels.

Cholecystokinin (CCK)

Cholecystokinin is produced in the duodenum and jejunum. CCK is an appetite suppressor that signals for a reduction in food intake.

Amylin

Amylin is produced in the pancreatic beta cells. It is released with insulin and is an appetite suppressor that inhibits food intake. Amylin is secreted with insulin to inhibit the secretion of glucagon. Several hours after feeding, insulin levels fall, leading to a reduction in the secretion of amylin, then glucagon levels begin to rise, stimulating the release of adipose tissue stored fatty acids into the bloodstream.

PYY (Peptide YY) & Enterostatin

PYY is produced in the intestine. It stimulates satiety and another appetite suppressor, Enterostatin. Enterostatin is produced in the ileum and is a fat intake suppressor.

Thyroid Hormone (T4 and T3)

Thyroid Hormone is produced in the thyroid gland and it controls and regulate the speed of metabolism. High thyroid hormone levels raise BMR and low levels decrease BMR.

Somatostatin (SST)

Somatostatin is produced in the stomach and neurons. It inhibits both glucagon and insulin secretion. It is both an appetite suppressor and appetite stimulant. It is a regulator of homeostasis and appetite and has a role in satiety and obesity. It is stimulated by increased GLP-1 levels.

14.4 Metabolic Energy Storage in Humans

Table 14.3 shows the total potential amount of energy stored in the body and the figures shown represent the average amount stored for lean individuals (70 kg male). These figures represent potential storage quantity and not how much is actually available that can be oxidised for energy.

Table 14.3 Macronutrient Metabolic energy storage in humans (male)

Fuel Substrate	Energy (kcal)	Energy (kJ)	Body Store (kg) [1]	Body Store (%)	Location within the body	Estimated Storage Period
Glucose	15	62	0.004	0.02 %	Blood/Tissues (4-5g/1tsp)	0.75 hours
Glycogen	1640	6,888	0.400	2.19 %	Stored in the Muscles	
Glycogen	410	1,722	0.100	0.55 %	Stored in the Liver	20 hours
Protein[3]	11,825	49,476	2.750	15.07 %	Mostly Structural Muscle	5 days Nominal
TAG (Fat)	132,600	554,798	15.000 [1]	82.17 %	Adipose, Muscle & Organs	Up to 53 Days [2]
Total	146,490	612,946	18.254	100.00 %		

Energy Density kcal/g: Glucose 3.7, Glycogen = 4.1, TAG = 8.84 Protein = 4.3
[1] For Lean individuals – More stored in overweight and obese individuals
[2] For Lean individuals – The record is a 32 stone man at 382 days who lost 20 stones during medically supervised starvation (including K+ Na+ & yeast/multivitamins and black coffee) [3] Around 25% of protein or 2.75 kg is potentially available for energy in fasting and starvation conditions

Under normal conditions, on average, fat storage accounts for approximately 92% of stored energy, with carbohydrate at just 3% of energy storage capacity and protein providing approximately 5%. Under starvation conditions, assuming all glycogen is depleted, fat storage is in the order of 90% with protein at 10%. Stored TAGs, (triacylglycerols) therefore represent the largest energy storage reserve in the body, almost 70 times greater than the amount of energy stored as carbohydrate glycogen.

The adult male human body contains approximately 11 kg of protein by mass and around 50% of this or 5.5 kg is located within skeletal muscle listed in Table 14.3. However, it is estimated that the body can only tolerate a maximum total loss of approximately half of this, i.e., 25% of body protein or 2.75 kg can, if necessary, be mobilised and utilised as a potential energy source in both day to day fasting and most in extreme starvation conditions. The remaining 75% of body protein, 8.25 kg is not metabolically active nor available as an energy source.

Under normal day to day activity including gluconeogenic contribution, protein can provide a range in the order of 5-15% of energy, with the remaining 85-95% of energy shared by carbohydrate, fat, glycerol and lactate.

If the body's protein was depleted by oxidising the more than 2.75 kg of protein for energy during starvation, it is likely that irreparable damage would set in, including weakened respiratory muscles, chest infections, pneumonia, impaired immune function and organ failure. In keeping with Starvation Priority 2, therefore, dietary protein contribution to gluconeogenesis prevents the excessive breakdown of skeletal muscle and other vital proteins and this is determined by glucose levels and tightly regulated by the liver.
During prolonged exercise, where glucose and fatty acids are fully available and dominate as fuel substrate, protein lacks the rate of energy provision needed to fuel intensive aerobic exercise in these conditions. This is because the metabolic tools for using the amino acids are not designed for speedy energy production. For example, the regulatory enzymes required to catalyse the required reactions in respect of the amino acids, threonine, glycine and serine and the oxidation of amino acids via gluconeogenesis or pyruvate, during intense exercise, have relatively long half-lives. Under these conditions, amino acid oxidation still probably accounts for no more than a maximum and quantity of 10% of the energy requirements at that time.

Under normal conditions, the carbon skeletons of amino acids, contribute in the order of 5-15% of total generated oxidative energy metabolism ATP, either directly, via the TCA cycle, or as precursors for fatty acid synthesis and gluconeogenesis. Some researchers quote that amino acids usually account for 10-20% of ATP generated (of total energy oxidised) during the fed state. Others report much smaller contributions from 2-5% in the fed state rising to around 15% of total energy used and higher during prolonged starvation.

The results of research and studies into gluconeogenesis contributions vary significantly based on the isotope tracers used, the test conditions, and the methods of calculation. Given the fairly wide range of reported data, Section 14.18 concludes an overall range from 5-14% of total energy production from protein and most (non-glycogen sourced energy) is derived from glycerol and lactate.

The role of amino acids to produce energy is however, less important than their role in helping to maintain glucose levels via gluconeogenesis (Section 14.16). Excess dietary amino acids not required by the body can-not be stored and are excreted in the urine. The source of this continuous energy is exogenous, or dietary protein, which is metabolised as energy, or lost in urea. Unlike carbohydrates, stored as glycogen, or fats, stored as TAGs in adipose tissue, amino acids are not stored for energy production. Protein is, however, used to build, maintain, and repair body tissues, as well as synthesise enzymes and hormones. Remember, excess calories, (EI), of all macronutrients, compared to energy expenditure (TEE), including protein, will lead to fat gain. The average intake of dietary protein and digested amino acids is in the order of 70-100 g/d and unless the body is growing (up to the age of 18-20), all of this must be utilised or disposed of each day. This comprises (a) synthesis of proteins, (b) fuel oxidation, (c) break down to urea via transamination in the liver and (d) the conversion to fats via DNL, and accounts for all of the ingested dietary protein. Growth stops after the age of 20. Working out in the gym does not 'grow' new muscle fibres but instead adapts and utilises more of the existing fibres.

The actual proportional rate of protein oxidation against total metabolic energy generation, depends on several factors, including the type of diet (low carbohydrate or high protein etc), the length of starvation period, or if the individual is at the end of endurance exercise, when glycogen levels are depleted and there is insufficient bio-availability of glucose and fatty acids.

Under these endurance conditions, when carbohydrate reserves are exhausted, protein amino acids could supply up to a theoretical 15% of the body's total energy needs and skeletal muscle can be broken down to be used as fuel helping to preserve glucose for the brain, RBCs, and other glucose reliant and dependent tissues and cells.

14.5 The Storage of Fat

Eating 'high calorie' fat and processed, simple sugar foods and not allowing time to burn off this energy, prior to refeeding, inevitably leads to fat storage. The desire to re-feed too soon is caused by the high blood sugar, fat, salt and insulin rush and then crash, leaving cyclical, perceived 'hunger'. This is all part of the complex nature of hunger and the evolutionary relationship with 'out of sync' natural hormones, fooled by modern 24/7 food available society and the sweet spot (fat, salt, sugar) of attractive and addictive taste, purposefully designed and perpetuated by the vested interest food industry, striving to maximise profit on human desire for these foods. We have seen how fat is metabolised and the body's natural choice for long term energy storage. Fat is the most energy-dense and most stable macronutrient for storage of energy, both during food shortage and long term everyday use.

Fat is far easier to store than glucose. As the body does not need to produce much insulin to store it, fat is last in line to be burned for energy due to the fuel substrate oxidative hierarchy. Therefore, when the intake is iso-caloric (sufficient), or hypercaloric (over-nutrition), the body does not burn fat, the body stores fat. This is what occurs when eating high quantities of fat and free sugar based carbohydrates together i.e., junk food. It costs only 2-3% of ingested energy from fats to store this fat. The fatty acid chemical profile of stored fat closely resembles the fatty acid profile of the fat in the diet and not that of de novo lipogenesis (DNL).

If an individual eats 100 calories from complex carbohydrates, 25 of those calories are oxidised and used up in the glycogen storage process. However, comparing this with 100 calories from fat, only 2 calories are oxidised in the process of storing this fat in adipose tissue.

DNL conversion of glucose to fat is highly endergonic, requiring considerable metabolic energy i.e., approximately 25-30% of the dietary energy consumed, hence, whilst it has sufficient intake and energy needs, the body chooses to store exogenous, dietary, ingested fat. Section 14.40 discusses DNL in more detail. Only 1-4% of added fat tissue to adipose stores is derived from dietary carbohydrates via DNL. [38] The other 95- 99% of stored fat is derived from exogenous, dietary fat. The body has a tendency to gravitate towards 'autumn' storage mode, as it gears up to storing energy for the potentially long winter of possible famine ahead. This is effectively a dopamine response, causing a person to eat more than is necessary for fat storage or hibernation.

This is also one of the reasons why some people are attracted, or can-not resist the odour and or taste of highly caloric, high energy fat, high spice, high cholesterol and high sugar based food, especially as autumn and winter approaches along with colder weather. Add a little low quality protein (burgers/pizza) and the result is a super 'tasty', additive cocktail, ideal for fat storage. This results in a perpetual evolutionary storage preparation autumnal season.

Armed with constant high fat, high blood sugar and cholesterol, the body prioritises oxidation of the more volatiles fuel first. For those people wanting to lose weight, it is therefore, very important to understand that in order to oxidise and burn body fat, it is firstly necessary to deplete glucose and fat in the blood. Stop eating high caloric rich, fatty and high free sugar foods. Dietary fatty acids in triacylglycerols are not used for energy and they are taken up by the adipocytes and stored as fat. Have a rest from eating for a few hours. Take advantage of the overnight fast (Chapter 5).

Furthermore, if an individual eats a considerable amount of carbohydrates, whether these carbohydrates are simple processed sugars (junk food, pastries, chocolate) or healthy, wholegrain, complex carbohydrates, then the body will not need to burn up its fat reserves, it has plenty of fuel for the time being. If the individual continues to eat healthy carbohydrates, without a sufficient break, or fast, the body will be perfectly content to use this glycogen and glucose for energy first and will, therefore, reduce the amount of fat burned overall. [38] A break in eating, for example, TRE (Chapter 5), allows more fat mobilisation and fatty acid beta (β)-oxidation. In addition, if fat stores are full, added carbohydrates impacts more on blood glucose levels and increases the propensity for fat storage even more significantly.

Any fat eaten with these carbohydrates will go straight to fat storage and net lipogenesis will be positive, hence the term, *Carbohydrates spare the fat* [14], i.e., glucose has a fat storing, sometimes referred to as a 'fat sparing', effect. If energy caloric excess continues, adipocytes are enlarged but the total number of adipocytes is relatively unchanged. The total number of adipocytes increases at its fastest rate in late puberty. Thus, the weight loss aim should be to reduce total body stored energy.

Overall, eating nutritious food, sufficient healthy carbohydrates, limiting total fats (especially saturated fats), eating slightly fewer total calories (EI) and ensuring sufficient breaks between refeeding and exercise, is the best recipe for losing excess weight. Studies on low carbohydrate or low fat diets have shown that more net fat loss is likely to arise from a low fat diet, rather than a low carbohydrate diet.

Data from studies done by Kevin Hall et al, demonstrate that:

'calorie for calorie, restriction of dietary fat led to greater body fat loss than restriction of dietary carbohydrate in adults with obesity'. 'Cutting carbohydrates increased net fat oxidation, but cutting fat by equal calories had no effect'. 'Cutting fat resulted in more body fat loss as measured by metabolic balance'. 'In contrast to previous claims about a metabolic advantage of carbohydrate restriction for enhancing body fat loss, our data and model simulations support the opposite conclusion when comparing the Restricted Fat and Restricted Carbohydrate diets. Furthermore, we can definitively reject the claim that carbohydrate restriction is required for body fat loss'. [24]

Obese people have more and larger adipocytes compared with lean people and if body fat is lost, the number of adipocyte cells does not decrease, hence the future propensity to regain weight if dietary habits and conditions are repeated. Weight loss from fat is essentially a reduction in the volume of the fat and therefore the volume of the cells. The body retains these cells as the storage reservoir 'with a long memory', to be refilled once sufficient food becomes available. The adipose cells can expand and shrink. Fat storage capacity varies in lean/overweight people, which is, to some extent, also influenced by genetics before fat overflows into the bloodstream. Other people's fat cells have a smaller capacity to store fat before the overflow into the bloodstream occurs. These people may often have a higher propensity than others to go on to develop diabetes. This difference is called the 'Personal Fat Threshold' (Sections 1.7 and 2.9). [3]

In summary, the body must deplete upstream more volatile fuels before excess fat is mobilised and oxidised for energy rather than retained in storage.

14.6   Lipolysis - The Breakdown and Mobilisation of Fats

Before we move on to the (β)-oxidation of fats for energy, we will firstly pause to see how fats are broken down and made ready for this energy production. This is achieved via the process of lipolysis. Lipolysis, literally fat splitting, or fat breakdown, is a catabolic pathway that occurs mainly in the in the cytoplasm of adipose tissue. It promotes mobilisation of metabolic fuel from adipose tissue to peripheral tissues, in response to appropriate energy demands, for example during fasting and or exercise.

Table 14.5 A Summary of Lipolysis

What is it?	The Breakdown of fats by hydrolysis to release fatty acids & glycerol
What is its Principal Role?	To maintain energy levels during low glucose conditions
Why does it occur	To mobilise and break down fats energy for the body
When does it occur?	During fasting or exercise or between meals when glucose is low
Where does it occur?	Predominantly adipose (fat tissue) cytoplasm
How is it stimulated?	A fall in insulin levels, a rise in glucagon, adrenaline, GH, TSH & T3
How is it inhibited?	A rise in insulin levels, a fall in glucagon
What is needed to action it	Triacylglycerol lipase enzyme
What are the start reactants?	Triacyl glyceride (TAG) Fats
What is the end product?	Fatty acids & glycerol (these burnt for energy or re-esterified to TAG

The major determinant of total free fatty acid release is the amount of total fat. To obtain energy from fat, triacylglycerols (TAGs) must first be broken down by hydrolysis into their two principal components, fatty acids and glycerol. In Section 14.3, we noted that insulin is the most important overall anabolic regulatory hormone, where lipolysis can only be invoked when insulin levels fall. When insulin levels rise and the body has sufficient energy, lipolysis is inhibited. Lipogenesis and fat re-esterification is stimulated and fat remains in storage.

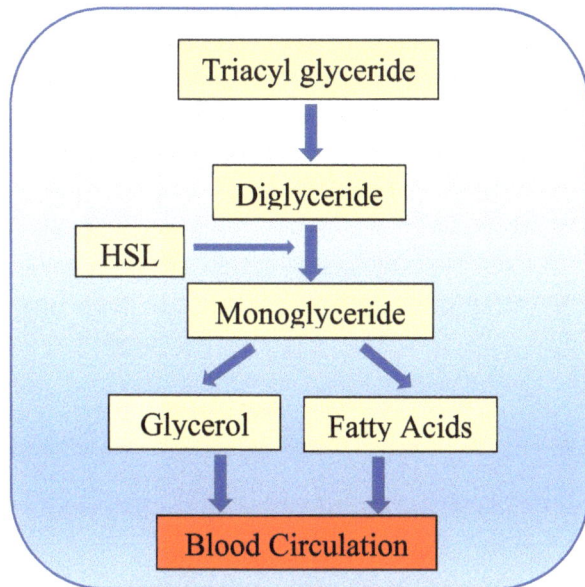

Figure 14.1 Overview of the Pathway of Lipolysis: the mobilisation and breakdown of fats

The key specific enzyme for the mobilisation of fat is hormone sensitive triacylglycerol lipase (HSL) found in adipose tissue. Adrenaline (epinephrine), norepinephrine (noradrenaline), adrenocorticotropic hormone (TSH), glucagon, growth hormone (GH), and thyroxine (T4 and T3) all act as catabolic hormones and stimulate lipolysis.

Therefore, insulin, the principle anabolic hormone, acts antagonistically to these hormones. The resulting fatty acids from lipolysis are then oxidized by (β)-oxidation into acetyl-CoA, which runs through the TCA cycle for the production of energy via OXPHOS. Therefore, when glucose levels are low, triacylglycerols can be converted into acctyl-CoA molecules and used to generate ATP (the energy carrier) through aerobic respiration.

The glycerol that is released from triacylglycerols after lipolysis, directly enters the glycolysis pathway in the liver or kidney, via phosphorylation by glycerol-3 phosphate (G3P) to dihydroxyacetone phosphate (DHAP). From here glycerol can be oxidised via glycolysis, or go through gluconeogenesis via glycerol-3-phosphate (G3P) to produce glucose. It is also possible that some peripheral organs and tissues, including skeletal muscle, can convert glycerol to lactate.

As one triglyceride molecule yields three fatty acid molecules and each triglyceride contains as much as 16 or more carbons, fat molecules yield far more energy than carbohydrates and are an important source of energy for the human body both at rest and during higher metabolic work. Each step of hydrolysis leads to the removal of one fatty acid.

Brown adipose tissue (brown fat-high levels in human babies) contains mitochondria whose structure differs from mitochondria found in white adipose tissue (white fat). Brown adipose tissue generates more heat. Section 14.8 shows that triacylglycerols generated by lipolysis, yield significantly more energy per unit (x 6.75) mass when compared to carbohydrates (and proteins). This is due to the fact that:

- Fats have significantly more ATP yield at 106 ATP fats and 30 ATP from carbohydrate
- Fats have much higher energy density due to high number of acyl groups & long chains
- Fats are non-polar, low water content compared to glycogen, hence more fat is stored
- Fats are stored in a very efficient storage form
- Fats are relatively unreactive with other substances
- Fats have a lower oxygen content and higher ratio of hydrogens and are far more reduced
- Fats rich in double bonds from long carbon chains high numbers of energy rich electrons
- Fats produce more reduced NADH & $FADH_2$ due to different metabolic pathways
- Fats have less prominent functional molecular synthesis roles than proteins (enzymes etc)

14.7   Glycogenesis - The Synthesis of Glycogen

Glycogenesis is the process of anabolic synthesis of glycogen from glucose molecules in the liver and skeletal muscle and it occurs when there is excess energy and blood glucose levels are high, for example, after feeding in the fed or absorptive state. As many as 6 enzymes are required for this process.

Table 14.6   A Summary of Glycogenesis

What is it?	The Process of synthesising glycogen chains from glucose molecules
What is its Principal Role?	To store excess energy for later use or when the body requires energy
Why does it occur	To enable the storage of excess glucose
When does it occur?	When there is excess energy after feeding, and during rest periods
Where does it occur?	Predominantly in the liver and muscle tissues
How is it stimulated?	A rise in insulin levels, a rise in blood glucose and a fall in glucagon
How is it inhibited?	A rise in glucagon levels, a fall in blood glucose and fasting (AMP)
What is needed to action it	Glycogen synthase & hexokinase (HK) in muscle & Glucokinase liver)
What are the start reactants?	Glucose
What is the end product?	Glycogen

The glucose molecules, linked by glycosidic bonds are phosphorylated and added to large numbers of long glycogen chains. This storage process is usually activated during rest and fasting periods and primarily stimulated by insulin in response to high blood glucose levels.

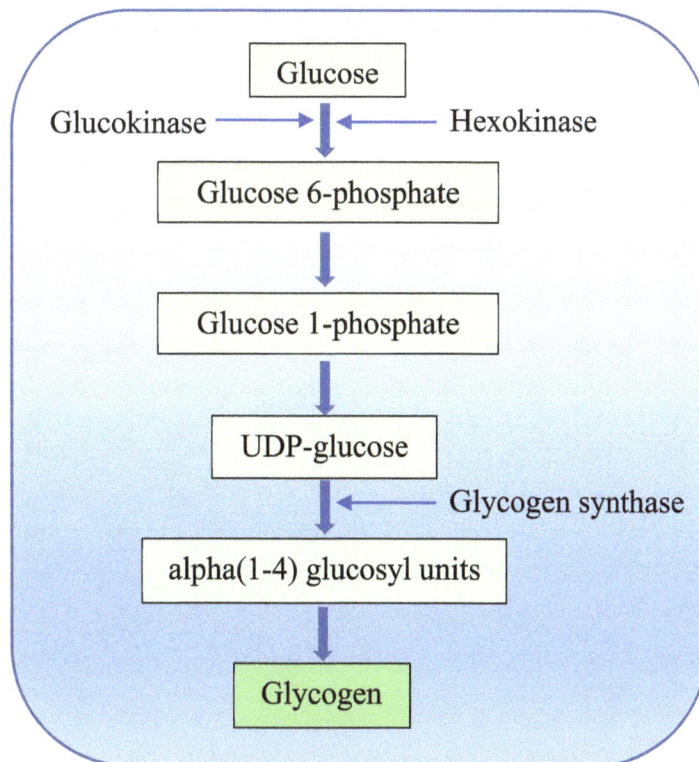

Figure 14.2 Overview of pathway of Glycogenesis from glucose in the liver and muscles

Insulin, glucagon and cortisol regulate the relationship between glycogenesis (formation and build-up of glycogen when blood glucose is plentiful) and gluconeogenesis (synthesis of glucose for energy or and glycogen for storage, from non-carbohydrate precursors when blood glucose is insufficient). When glycogenesis occurs, glycogenolysis is inhibited and vice versa.
For a 70 kg man at rest, the maximum rate of glycogenesis is 118 g/h, significantly higher than, or more than 7 times the maximum rate of glucose oxidation at 16 g/hr.

14.8   The Storage of Glycogen

Of total carbohydrate stores in the body, approximately 80% of carbohydrate is stored in skeletal muscle and 14% is stored in the liver. The remaining 6% of carbohydrate circulates in the blood in the form of glucose. Although most glycogen is stored in skeletal muscle, liver glycogen is the most important site for glucose level in terms of homeostatic regulation for the rest of the body. Glycogen is also present in most other tissues in small quantities, even adipose tissue. Muscle glycogen is used locally by muscle tissue during physical exertion. Glucagon and adrenaline stimulate glycogenolysis (glycogen break-down). Storing energy as glucose polymers is common to all forms of life (plants and animals).

Total glycogen stored in the body is around 500g. As total muscle mass in the body is overall far greater than the mass of the liver, muscle provides far more total mass of glycogen i.e., approximately 400g stored (75%) as a proportion of the whole body's glycogen needs than the liver i.e., approximately 100g stored (25%). In very highly trained individuals, total glycogen stores can rise from 500 grams to as much as 700 grams. However, it is the liver that controls and orchestrates the rest of the body's essential requirements for glucose, supplied from the liver's glycogen stores. The process of glycogenesis is the synthesis of glycogen from glucose molecules. Glycogen is also far less reduced (less hydrogen-carbon bonds) than fatty acids and thus not as energy rich. Further detail of the energy yield difference between glucose and fatty acids is provided in Sections 15.10 to 15.13.

Carbohydrate has two important advantages, over fat as a metabolic fuel; firstly, it is the only fuel that can produce ATP anaerobically, in the absence of oxygen, and secondly, 2.5 times more ATP is produced per amount of oxygen consumed when glucose is oxidized, compared with ATP produced per oxygen consumed when fat is oxidized (Section 15.13). Another reason why glycogen is the selected form of carbohydrate storage, is that high concentrations of glucose disrupt the osmotic balance of the cell and may cause Chronic hyperglycaemia. Therefore, it is stored as a non-osmotically active polymer like glycogen. Also, glycogen is less chemically reduced (more oxidised) compared to fats and is thus less energy rich overall and hence, the body has evolved to use fats for long term energy reserve. During rest or moderate activity, fatty acids are the major proportional volume source of total energy, at around 62% of total energy used as muscle fuel (38% glucose). At 65% VO_2 max, the body is using a larger amount of total energy, albeit at a lower fat oxidation proportion, at around 50% of fats and 50% glucose as muscle fuel.

The reason that all excess fuel is not stored as fat and instead some is stored as glycogen, is that the body needs to sequentially control the release of glucose from glycogen carefully, in a controlled manor, to maintain the correct and finite level of blood glucose between feeding. This is not possible using fats. It is also energetically far easier and more rapid to obtain energy from glycogen and glucose than fat and adipose tissue. As discussed in Section 14.2, muscle cells burn approximately 13 kcal/kg/d compared with only 4.5 kcal/kg/d for fat cells in adipose tissue – hence, weight for weight, lean mass to non-lean mass i.e., increased muscle mass in a body, compared with fat mass, will burn more energy in total and lead to a higher metabolism. Each mole of glycogen contains around 3 moles of water (often leading to initial quick weight loss experienced during the first 2 weeks on many diets).

Triacylglycerol fats are non-polar and therefore stored in a virtually anhydrous (very low water content), highly reduced form, whereas carbohydrates are polar and thus much more hydrated. ('Carbon hydrated').

If the amount of energy stored as fat in adipose tissue was instead, stored as glycogen, it would add approximately 64 kg (c140 pounds) to human body mass, adding a further 90% of weight to a typical lean weight male and hence would certainly not, from an evolution perspective, contribute to humans' ability to cope with physical exertion for flight, fight or chase.

Glycogen is approximately 77% water by weight. As one triacylglycerol (TAG) molecule yields three fatty acid molecules, with as many as 16 or more carbons in each one, fat molecules yield much more total energy than carbohydrates.

In addition, triacylglycerol fatty acids yield more than twice the energy per unit mass (circa 9 kcal/g) when compared to carbohydrates and proteins (circa 4 kcal/g).

The predominance of fat as a long term storage form of total energy calories, is because when water is factored in, fat is about 6.75 times more calorically dense than glycogen. That is, one gram of triacylglycerol fat stores contains approximately 6.75 times the total energy of 1 gram of glycogen. Therefore, fatty acids can hold more than 6.75 times the amount of energy per unit of stored mass.

1 gram glycogen	= 0.33 gram glycogen & 0.77 gram water
1 gram triacylglycerol(TAG)	= 1gram triacylglycerol & 0.00 gram water
1 gram triacylglycerol(TAG)	= 3 times the mass 1 gram glycogen
TAG energy yield	= 9 kcal/g
Glucose energy yield	= 4 kcal/g
Glycogen water content	= 0.33g/g
Delta caloric energy density TAG vs glycogen	= 6.75

Glycogen is necessary to fuel very active muscle and is around 7% of liver fresh weight. Small amounts of glycogen are also found in the kidneys, male reproductive tract, adipose tissue, heart, and erythrocytes (RBCs). The circulating blood keeps the brain and CNS supplied with glucose.

The readily mobilised glucose from glycogen is a good source of energy for sudden and also very strenuous activity. Unlike fatty acids, the released glucose can be metabolised in the absence of oxygen and can thus supply fuel for anaerobic energy, especially essential for the RBCs.

The overall efficiency of storing glycogen to be used for rapidly mobilised energy is 97% (30 mole ATP costing only 1 ATP of glucose 6-phosphate for storage).

However, ingested carbohydrate is oxidized to glucose relatively quickly and is thus more readily available to burn as energy and is the preferred metabolic fuel of the absorptive state, rather than being extensively stored. The brain only has very small glycogen stores as it has very little water per unit mass.

14.9   Glycogenolysis - The Breakdown of Glycogen

Glycogenolysis is the process of glycogen degradation to produce glucose for energy into the bloodstream when required. For the purposes of balancing blood sugar, i.e., glucose for the rest of the body, glycogenolysis occurs in the liver and kidney.

It also occurs in the skeletal muscle directly, when muscle tissue requires additional glucose locally for muscle contraction work load. Muscle contraction, primed by adrenaline, is the main stimulant for glycogenolysis in muscle tissue locally and unlike the liver and kidney, the skeletal muscle does not have any role in glucose homeostasis for the rest of the body.

Glycogenolysis in the liver is stimulated by glucagon from the alpha cells of the pancreas and the catecholamine hormones adrenaline (epinephrine) and noradrenalin (norepinephrine) from the suprarenal glands.

Table 14.7 A Summary of Glycogenolysis

What is it?	The breakdown of glycogen to generate glucose for the body
What is its Principal Role?	Maintenance of normal blood glucose levels preventing hypoglycaemia
Why does it occur	To restore blood glucose levels for energy when energy levels fall
When does it occur?	Between meals and during fasting and hypoglycaemia, low ATP in cells
Where does it occur?	Primarily in the liver and muscles and some in Adipose tissue
How is it stimulated?	A rise in glucagon levels in the liver and adrenaline in the muscles
How is it inhibited?	A rise in glucagon levels, a fall in blood glucose and fasting (AMP)
What is needed to action it	Glycogen phosphorylase, phosphorylase kinase, phosphoglucomutase
What are the start reactants?	Glycogen
What is the end product?	Glucose

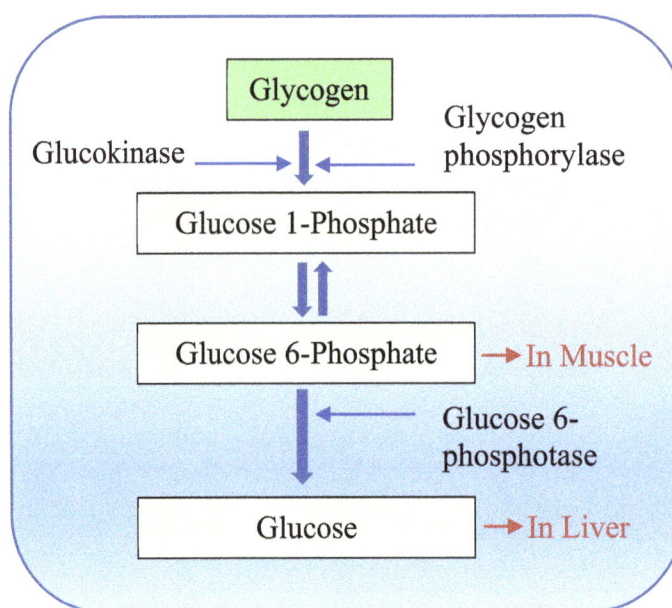

Figure 14.3 Overview of the pathway of Glycogenolysis from glycogen to glucose in the liver and muscles

There is also a second mechanism via elevation of raised $[Ca^{2+}]$ in the cytoplasm. Hypoglycaemia (fall in blood glucose) is a major reason for the stimulation of glycogenolysis, leading to rapid restoration of blood glucose levels, as long as glycogen levels are adequate. Increased levels of cyclic adenosine triphosphate (cAMP) also stimulate glycogenolysis, signalling that more energy is required via release and oxidation of more glucose. By responding to noradrenaline via this cAMP mechanism, glycogenolysis maintains stability during hypoglycaemia. Hepatic (from the liver) glucose production derived from glycogenolysis and gluconeogenesis ranges from 70% down to 16% depending on time from feeding, through fasting and the level of exercise.

14.10 Energy Substrate Utilisation and Metabolism

This section provides an overview on the body's use of the various energy substrates to generate energy and synthesise glucose for energy in various tissues and organs during different conditions. Sections 14.27 to 14.37 review these substrates in further detail.

The body's starvation priorities and metabolic pathways are, of course, not evolved, nor in place to be utilised as deliberate mechanisms to satisfy diets, such as 'low carb' carbohydrate deprivation or 'high protein', 'high fat', 'keto', 'paleo' or 'carnivore' diets etc. Unless medically necessary, due to specific requirements of certain patients undergoing medically supervised treatment, it is inappropriate to promote attempts to manipulate highly complex metabolic systems, many of which are still not fully understood even by the world's leading scientists and researchers. However, some internet bloggers and even people who are medically qualified, continue to recommend some of these approaches to reduce weight or grow muscle.

In addition to the main two energy substrates of glucose and fatty acids and the fatty acid derived ketones, there are several other key substrates that are used in times of fuel shortage via the process of gluconeogenesis and these are discussed in Section 14.16. The liver and to a lesser extent the kidneys can utilise these substrates to synthesise glucose, when necessary, both to maintain blood glucose levels for the body's energy requirements (Starvation Priority 1) and also preserve structural protein for the body's integrity (Starvation Priority 2).

These substrates are either converted to glycerol-3-phosphate or pyruvate, or enter the TCA cycle at a later stage, as intermediates, such as oxaloacetate or dihydroxyacetone phosphate. The major non-carbohydrate precursor substrate contributors are glycerol, lactate and amino acids. The body principally burns both carbohydrates (glucose) and fats (fatty acids) and to a lesser extent, oxidises ketones and the carbon backbones of amino acids at all times, 24 hours a day in different proportions, depending on metabolic and environmental circumstances at the time. It is important to note that each cell within different tissues will be burning either glucose or fatty acids at any one time but not both simultaneously. However, the body as a whole does burn both glucose and fatty acids continually and simultaneously in varying proportions. We have noted the body's two metabolic priorities for survival during starvation. In addition to the genuine, emergency, longer term starvation conditions, these priorities are also fundamental to the body's requirements to maintain health in general overnight fasting each day. Glucose or fatty acids both enter the cytosol of the cell and convert into the common metabolic molecule of acetyl-CoA, which is finally oxidized to ATP for energy production.

Fatty acids are derived by β-oxidation, whereas glucose is the substrate mainly for glycolysis and the pentose phosphate pathway (PPP). Oxidation of both substrates takes place in the cell mitochondria. The relative contribution of glucose or fatty acids for energy depends on several factors.

Substrate utilisation during rest is highly dependent on carbohydrate availability. The higher the glycogen and glucose availability, the lower the fat (β)-oxidation and vice versa. Moderate levels of exercise and lower carbohydrate intake will increase fat beta (β)-oxidation. Higher intensity exercise will decrease the proportion of fatty acid (β)-oxidation and increase the proportion of glucose oxidation, providing there are sufficient glycogen reserves.

Per unit time, higher intensity exercise will, however, burn more total (both fat and carbohydrate) energy overall than total energy burned at lower levels of exercise intensity.

At rest, the body is using carbohydrates and predominantly fat (with very small amounts of lactate, glycerol, amino acids and ketones) to maintain energy supply and glucose levels for day to day cellular and physiological requirements and BMR. For most of the time, these other fuel substrate molecules, are not used significantly, until conditions change, for example on exercise intensity, fasting, carbohydrate deprivation and starvation.

A further substrate, creatinine, is used (only) in muscle tissue, over very short period (6 seconds) on first burst intensive energy demand, for example athletic sprinting and is a relatively small overall contributor to energy demand and hence, will not be discussed further here.

The respiratory quotient (RQ) is the ratio of CO_2 production against O_2 consumption and is used to assess the amount and proportion of fat and carbohydrate utilisation for energy. An RQ of 1.0 indicates a predominance of carbohydrate oxidation and an RQ of 0.71 represents a predominance of fat beta (β)-oxidation. From many experiments, it has been determined that the average resting RQ is approximately 0.82 and this value therefore indicates a mixture of substrates. This calculates at approximately 62% of resting energy provided by fats with approximately 38% from carbohydrates (glucose and glycogen) and a small amount from other substrates.

The RQ of 0.82 indicates that other substrates including amino acids contribute small amounts to resting energy generation. Therefore, it can be further estimated that 33% of total macronutrient substrate oxidation is derived from carbohydrate and a maximum of 5% from other substrate degradation via gluconeogenesis during normal resting energy requirements, either as TCA cycle intermediates, then oxaloacetate or into pyruvate.

The precise rate of oxidation and proportional contributions of the macronutrient substrates and also, lactate, glycerol and ketone bodies, against total energy requirements vary and are dependent on the individual, the diet type, whether in the fed (absorptive) state or fasted (post absorptive) state, starvation status and exercise intensity and endurance conditions. The actual numbers at any one time and over a period can only be determined by measurement in the laboratory during test conditions.

14.11 Glucose is Essential For Rapid Energy

Recall that glucose is an essential energy source for virtually all forms of life. Glucose is the preferred, most flexible, soluble and easiest to digest and oxidise molecule fuel substrate for the body, fully aerobically, or with less available oxygen. It is the only form of molecule which directly satisfies all the energy requirements of the human or for that matter, all mammals. Other metabolic fuel substrates provide the back-up contingency to preserve glucose.

Unlike fats, glucose is a universal fuel for most, if not all cells in the body and carbohydrate is both the cheapest (energy efficient) source of calories and also the major source of dietary fibre. Glucose is the preferred metabolic fuel for the brain and healthy, nutritious carbohydrate should form the majority of macronutrient calories in a healthy diet.

Glucose is also at the source of the pentose-phosphate-pathway, needed to replenish the electron carrier NADPH, which supplies reducing power for biosynthesis, free radical defence and cell detoxification.

When humans eat carbohydrate, the carbohydrates are broken down into 6 carbon glucose (C6), which can then be used as an energy source for all tissues, most especially the brain, CNS, RBCs, kidney medulla and retina.

Glycogen is stored in the liver and muscle, ready for release of glucose to the body and muscle tissue respectively as and when required via the process of glycogenolysis. Unlike the relatively long process of the digestion of fats and also withdrawal of fatty acid currency from 'the bank of adipose', the 'banks of glycogen' in the liver and muscle' are much better prepared to respond rapidly to additional glucose needs of the body for rapid oxidation and hence, in the fed state, it is an order of magnitude easier to facilitate the withdrawal of glucose currency from 'the bank of glycogen' at any time using glycogenolysis than burning additional fat.

Tissues such as the heart and kidney cortex utilise fatty acids constantly for their needs as well as glucose and other fuels and this helps glucose homeostasis especially if required for skeletal muscle rapid activity or exercise.

Glucose is converted into two 3 carbon molecules of pyruvate (3C), via glycolysis and, each of these molecules then enters the tricarboxylic acid (TCA) within the mitochondria and are further broken down into 2 carbon molecules of acetyl-CoA (2C). Acetyl-CoA is the common intermediate in the oxidation of carbohydrates, fatty acids, and certain amino acids. The acetyl-CoA molecules then enter the TCA cycle and combine with the 4-carbon molecule, oxaloacetate (OAA), to form the 6-carbon molecule citrate and the cycle continues to produce NADH and energy via OXPHOS and the ETC to produce energy via the energy carrier ATP.

A total of 30 molecules of ATP are produced from one molecule of 6 carbon glucose. Hence, glucose provides a much lower total amount of energy compared with energy released from fatty acid terminal (β)-oxidation (106 ATP) but much more readily available per unit time. Fats will still continue to be oxidised at the normal rate, unless there is a glut of dietary carbohydrate and fats together. In this case, carbohydrates will be always burned preferentially.

As available carbohydrates are diminished and the length of time between the absorptive state increases, this preferential oxidation will shift and fats will then become the dominant substrate.

One of the key advantages in using carbohydrate and glucose for readily dispensable fuel, is that more ATP energy can be generated per unit time, compared with the (β)-oxidation of fat. During exercise, fat cannot synthesize ATP fast enough for the contractile demands of skeletal muscle fibers at higher intensities which is why fast twitch muscle fibers must use glucose.

This is also highly relevant to high performance, high intensity and high endurance athletes, where approximately 90 g/hr of mixed carbohydrates is recommended for training or events of more than 4 hours duration. In addition to this advantage, in order to the yield the same amount of ATP, (β)-oxidation of fatty acids requires more oxygen and thus more metabolic energy than the oxidation of carbohydrate.

The complete oxidation of one molecule of glucose requires 6 molecules of oxygen, whereas the complete oxidation of a typical fatty acid such as palmitate requires 23 molecules of oxygen (Section 15.13).

The body needs a minimum of 160 g carbohydrate glucose/d and 75% of that i.e.,120 g/d is required by the brain. Hence, the recommended daily allowance (RDA) of carbohydrate is 130 g/d, calculated and based on the needs of the brain and to ensure sufficient levels of blood glucose can be maintained.

This minimal amount of dietary carbohydrate calculates at around 26% and 21% of daily EI for adult human females and males respectively. If dietary protein is maintained at around 15%, this would mean around 60% and 65% of EI respectively is the proportion of daily dietary fat which, in the opinion of the author, governments, nutrition experts and most of the scientific community is far too high and may well lead to health consequences.

The 3 primary sources of blood glucose are exogenous from the diet, glucose released from the hepatic breakdown of glycogen via glycogenolysis and finally the hepatic synthesised glucose from non-carbohydrate precursors via gluconeogenesis.

Glucose is the only fuel that the brain uses under non-starvation conditions and the only fuel that RBCs use under any conditions. The body constantly works to preserve glucose to be used as the prominent, preferred monosaccharide fuel because:-

(1) It was probably originally available to primitive biochemical systems, as this monosaccharide was formed from formaldehyde under prebiotic conditions and was, as such, available as an energy source for primitive biochemical systems.

(2) Glucose has a strong tendency to exist in the hexose ring conformation and consequentially, relatively little tendency to modify proteins and therefore, has a low tendency, relative to other monosaccharides, to nonenzymatically glycosylate unstable proteins due to its ring structure.

In their open-chain forms, monosaccharides contain carbonyl groups that can react with the amino groups of proteins to form Schiff bases, which rearrange to from more stable amino-ketone linkage. Such non-specifically modified proteins often do not function effectively and would be highly disadvantageous to the body.

(3) All the hydroxyl groups in the hexose glucose ring conformation of beta-glucose are equatorial, contributing to the natural sugar's high relative stability.

The following calculation shows the amount of total glucose and glycogen stores available in the human body. Two molecules of ATP are generated in the conversion of glucose into 2 moles of pyruvate.

Glucose + 2P + 2ADP + 2NAD$^+$ = 2 NADH + 2 H$^+$ + 2 H$_2$0 + 2 Pyruvate + 2 ATP		
Energy released approximately	=	⁻90kJ mol ⁻1 (⁻22 kcal mol^{-1})
Daily glucose needs of the brain	=	120 g/d (75% of the body's requirement)
Total body requirement	=	160 g/d
Approximate glucose in body fluids	=	20 g/d
Available glycogen from direct reserves	=	190 g/d
Total Period of direct reserves of glycogen	=	1 day

To further preserve precious glucose, under normal day to day activity, the heart uses from 60% to 90% of its total energy from triacylglycerol fatty acids and the kidney cortex also uses fat predominantly. The skeletal muscle also uses both fat and glucose at rest and more glucose and less fat as intensity increases. (Table 6.1). Sufficient glucose must always be available to the brain and other tissues (such as RBCs) that are absolutely dependent on this fuel. There is a very small capacity to store glucose in the blood, liver and muscle, a different order of magnitude to the energy held in fat.

Just one teaspoon, or 4 grams of glucose (16 kcal) normally circulates in the blood stream. Despite this very small amount, a large number of cells rely on this glucose and are sensitive to its presence. The liver works to keep blood glucose within a narrow range at 4.4 to 6.7 mmol/L for (euglycemia) to be utilised as the preferred substrate by the RBCs brain CNS, kidney medulla and retina. Unlike fats, glucose can be used (anaerobic conditions) in the absence of oxygen. This glucose level range is maintained by liver production, absorption from the gut and uptake by peripheral tissue.

The blood-glucose concentration is kept at or above 4.4 mmol/L by three processes; firstly, the mobilisation of glycogen and the release of glucose by the liver, secondly, the release of fatty acids by adipose tissue and thirdly, the shift in the fuel used from glucose to fatty acids by muscle and the liver.

When the liver glycogen stores are depleted, gluconeogenesis from the amino acid alanine continues but this process just replaces the glucose that had already been converted into lactate and alanine by muscle tissue and RBCs. The brain oxidises glucose completely to CO$_2$ and H$_2$O. Another form of carbon is therefore required for the net new synthesis of 'new glucose' and some of these carbon atoms are derived from glycerol, released by lipolysis of triacylglycerol from adipose tissue, with the remaining carbon atoms supplied from hydrolysis of muscle proteins to amino acids via gluconeogenesis. Blood sugar (glucose) is regulated by a highly elaborate and integrated and coordinated suite of neuroendocrine responses, including both positive and negative feedback components. These regulatory processes are now reviewed.

The body's responses to glucose availability during exercise involves the sympathetic nervous system (SNS) and its capabilities via post-ganglionic nerve. This releases epinephrine (adrenalin) directly from the adrenal medulla for direct and endocrine signalling and also norepinephrine (noradrenalin), either directly, from the adrenal medulla or the nuclei of the locus coeruleus in the brain. A further response yields a neurotransmitter from the ganglia near the spinal cord and abdomen. These stimulants together, initiate cardiopulmonary, cardiovascular, and other fight and flight autonomic responses. The cardiovascular system controls and directs the limited and precious glucose supply to where it is needed (heart, brain, kidney cortex, RBCs, retina and working skeletal muscle), and therefore works to shunt blood flow and glucose delivery away from other tissues.

If levels drop below 4.4 mmol/L the hormonal response is triggered by:

- Glucagon: Promoting hepatic glycogenolysis and gluconeogenesis
- Epinephrine (adrenaline): Inhibiting glucose by muscle, increasing muscle glycogenolysis
- Cortisol and Growth Hormone: Delaying their responses
- Insulin: Acts antagonistically reducing serum glucose levels

Typical low glucose symptoms (hypoglycaemic) may be androgenic i.e., feeling faint, general weakness, muscle weakness (legs feel like jelly), sweating, tachycardia and tremors. Insulin induced hypoglycaemia usually means a very low blood glucose level of 3.3 mmol/L and some patients can be even lower at 2.7 mmol/L and some even asymptomatic at that point. In the organs within the abdominal cavity (the splanchnic bed), the pancreas secretes insulin (from the β-cells), glucagon (from the α-cells), and somatostatin (from the δ-cells). In addition, gastric secretion of 3 hormones, ghrelin, glucagon-like peptide-1 (GLP-1), and pancreatic peptide YY (PYY) affect appetite and eating behaviour. This provides major support of glycemia (sufficient blood sugar levels). Feedback control includes secretion of myokines (e.g., interleukin 6 (IL-6)) and lactate from working muscle, and also changes in blood (glucose), as well as adrenalin (epinephrine), from the adrenal (gland) medulla, that inhibits pancreatic secretions. In adipose fat cells, lipolysis is also inhibited by lactate via cyclic adenosine mono phosphate (c-AMP) and cyclic AMP response element binding protein (CREB), as well as transforming growth factor beta 2 (TGF-β2) released from the liver which is under the influence of circulating lactate.

It is equally important to ensure blood glucose levels are not raised too high, as prolonged hyperglycaemia can cause damage to many key cells in the body, for example, in the eye, this is crucial in preventing the development of retinopathy.

Glucose can-not be stored because high concentrations disturb the osmotic balance of the cell which is why it is stored as the non-osmotically active glycogen. High glucose levels can damage the endothelial cells of capillaries that nourish the retina and would negatively impact vision. Stable glucose levels therefore, help preserve the vision and prevent retina capillary damage.

Glucose toxicity can lead to chronic hyperglycaemia that damages a number of cell types and is strongly correlated with the myriad of Type 2 diabetes mellitus related issues. Most vulnerable to the effects of prolonged elevated plasma glucose levels include pancreatic β cells and vascular endothelial cells. β-cell dysfunction promotes decreased insulin synthesis and secretion, further worsening perpetuating hyperglycaemia. Chronic hyperglycaemia is strongly correlated with retinopathy, nephropathy, and neuropathy.

14.12 Fatty Acids Provide Substantial Energy

Fat is by far the best naturally evolved, long term energy storage molecule to safely keep in the deep secure vaults of the adipose tissue energy bank ('the bank of adipose') in humans and all mammals. However, that very responsibility comes with a higher bioenergetic cost price when withdrawing additional fatty acids currency from the bank compared with glucose.

Both fats and carbohydrates are burned all day as part of normal cellular processes to produce acetyl-CoA the universal molecule used to process release the break-down of carbon-to-carbon bonds to drive the release of energy from food.

Different cells in different tissues burn one or the other at any one time (The Randle Cycle Section 14.34), but the whole body is burning both substrates simultaneously. The heart and skeletal muscle use fatty acids for energy preferentially and constantly. The rate of mobilisation and (β)-oxidation of fats depends on the amount and proportions of carbohydrate and fat ingested and if there is a caloric surplus or deficit or the intensity of exercise. The triacylglycerols are mobilized by lipolysis and broken down to glycerol and fatty acids.

Fatty acid (β)-oxidation is a significant source of metabolic energy between feeding and also in high energy demand states, such as moderate exercise in the fasted state or during long endurance exercise.

These metabolic conditions stimulate the secretion of epinephrine and glucagon which increase the rate of lipolysis and the release of fatty acids from adipose tissue. From day to day, fatty acids provide a large portion of the normal energy requirement of heart muscle, skeletal muscle, the cortex of the kidneys, adipose tissue and the liver, especially when glycogen and gluconeogenic precursors are scarce, providing high quantities of metabolic energy, while also sparing muscles from catabolic breakdown.

At rest, apart from the RBCs the brain, kidney, medulla and retina, which must utilise available supply of glucose constantly, those other body tissues constantly using energy, such as the heart and kidney cortex, are able to apply for, sequester, metabolise and burn fats as a major source of energy for most of their energy needs and especially during sleep or fasting.

As these tissues undertake this basal process all the time they have no immediate need for a more rapid supply of energy from glucose, unless physical activity is increased significantly and the demand for ATP rises and they will then require higher levels of glucose to supplement to fat for energy needs.

The reason why dietary, exogenous fats are more difficult to utilise as immediate energy from the blood are multifold; firstly, unlike glucose, they are insoluble in water, secondly, they form very large particles that require special agents such as bile, for emulsification, before digestive enzymes can start to degrade them. Thirdly, they can-not be absorbed into the blood because their molecules would damage the gut, which means that the digestive products i.e., free fatty acids then need to be put through a convoluted system of processes to be reassembled and re-esterified into triacylglycerols and chylomicron particles and fourthly, they need to travel to the lymphatic system before finally being released into the blood supply.

Most of these are stored as fat in the 'bank of adipose' tissue. Unlike dietary carbohydrates, especially free sugars, the amount of fat oxidised from dietary fat is insignificant. The vast majority of fat oxidised for energy comes from free fatty acids (FFA) also referred to as non-esterified fatty acids, via the lipolysis of fat stored in adipose tissue or in muscle (intramuscular triacylglycerol) and glycerol.

When conditions are right to utilise the energy from fats, the body's reserves of fat (i.e., triacylglycerols TAGs) are burned for energy oxidation. For example, when insulin is low and for heart tissue pumping activity, skeletal muscles or kidney cortex at rest, fasting overnight, or before breakfast via moderate walking, or for the majority of steady state, normal energy processing.

In non-ketogenic tissues, such as muscle adipose tissue, the end product of fatty acid (β)-oxidation is acetyl-CoA, which is used directly as an energy source within the TCA cycle and OXPHOS to produce high quantities of energy for the cell.

Each fatty acid has in the order of 12 to 18 carbon atoms and during fatty acid breakdown, 2 carbon atoms are cleaved off at a time, ultimately producing the 2 carbon molecules of acetyl-CoA (the same acetyl-CoA as produced from the pyruvate molecules during carbohydrate oxidation) via the process of beta (β)-oxidation.

Fatty acid degradation prior to acetyl-CoA oxidation in the TCA cycle and OXPHOS, is a multi-cyclical, oxidative process that converts a fatty acid into a set of activated acetyl units (acetyl-CoA). These can then be processed by the TCA cycle. This first stage process of degradation follows 4 stages:

1) Oxidation: of acyl-CoA by acyl-CoA dehydrogenase to yield a trans alkene enoyl-CoA
2) Hydration: of the trans alkene by enoyl-CoA hydratase to yield hydroxyacyl-CoA (=OH)
3) Oxidation: of hydroxyacyl group by NAD^+ & dehydrogenase to a carbonyl ketoacyl-CoA
4) Cleavage: acetyl-CoA cleaved by thiolase yield an acyl-CoA 2 carbon atoms shorter

Stages 1 through 4 are repeated to yield acetyl-CoA. The cleaved acetyl-CoA can now enter the TCA cycle and ETC.

Hence a relatively large number of 106 molecules of ATP are produced from the full terminal oxidation (via stage (β)-oxidation and stage 2 TCA cycle and OXPHOS) one molecule of 16 carbon palmitate fatty acid and therefore, a far richer source of energy than glucose oxidation (30 moles ATP) but not as rapid to free up for quick bursts of energy.

The chemistry reasons why fatty acids produce so much more energy per mole compared with glucose is discussed in further detail in Section 15.10.

Glycolysis is anaerobic and yields 2 pyruvate, 2 NADH, 2 ATP per glucose molecule. Glucose oxidation is aerobic and yields pyruvate, acetyl-CoA and normally, 30 ATP per molecule of glucose. Fatty acids are far more reduced than glucose. Palmitic acid only contains two oxygens per sixteen carbons, whereas glucose has six oxygen atoms per six carbons.

Consequently, when palmitic acid is fully oxidized, it generates more ATP per carbon (106/16) than glucose (30/6). Fatty acids require several rounds (example 7 or 8 rounds) of (β)-oxidation and therefore produce far more ATP or around 3 and a half times the mass of ATP compared with glucose. However, fat (β)-oxidation is slower because approximately 4 times the amount of oxygen is required, less CO_2 is produced although both have similar 32% efficiency overall. See section 15.13 for further detail on this metabolic cost and yield comparison.

Fat within adipose (fat) cells in the liver can be converted to triacylglycerol and phospholipids, or transported to mitochondria for oxidation to release energy.

The fatty acids are released into the blood to the tissues that require this energy, entering the cells' mitochondria, via carnitine palmitoyl transferase (CPT-1) and are then broken down into acetyl-CoA via beta (β)-oxidation. This enables entry to the TCA cycle in those tissues. Therefore, acetyl-CoA, derived from fatty acid beta (β)-oxidation, enters the TCA cycle and is processed for ATP carrier energy in the same way as for glucose oxidation.

In the absorptive state, therefore, for the vast majority of people eating sufficient carbohydrates, the processes of glycolysis and the TCA cycle is producing adequate quantities of acetyl-CoA and oxaloacetate from both glucose, via pyruvate and fatty acids, via beta (β)-oxidation.

However, as we will see later in Section 14.26, during prolonged fasting, or carbohydrate deprivation diets, or under starvation conditions, the carbohydrate stores of glycogen in the liver and skeletal muscles become more and more depleted, especially after 24-48 hours. A level of glycogen does remain however, for emergency activity.

This often produces some rapid and temporary weight loss, initially mainly due to the loss of water in glycogen and subsequently over longer periods, due to overall caloric restriction. We know that the body prefers to use glucose as its primary source of rapidly available energy (especially those key tissues mentioned above). With continuing depleted glucose levels, however, the body i.e., the liver, turns to synthesising glucose from non-carbohydrate sources, via gluconeogenesis, including glycerol, lactate and amino acids.

Oxaloacetate (OAA) can also be made from pyruvate via glycolysis, malate and amino acids aspartate and asparagine (Table 14.13).

Figure 14.4 shows the 'replacement' inputs (anaplerotic) and outputs (cataplerotic) of the TCA cycle, including the source of acetyl-CoA and TCA cycle intermediates (particularly OAA) and the process where TCA intermediates can supply into other metabolic (catabolic) pathways. Recall that the TCA cycle is responsible for directing molecules to both catabolism, including processing metabolic food substrates as energy, and anabolism, for the biosynthesis of other molecules. However, in the liver's role to preserve glucose throughout the body, the final products of fatty acid (β)-oxidation in the liver are ketone bodies, that can be used by tissues such as brain, which has very little capacity to undertake (β)-oxidation capacity.

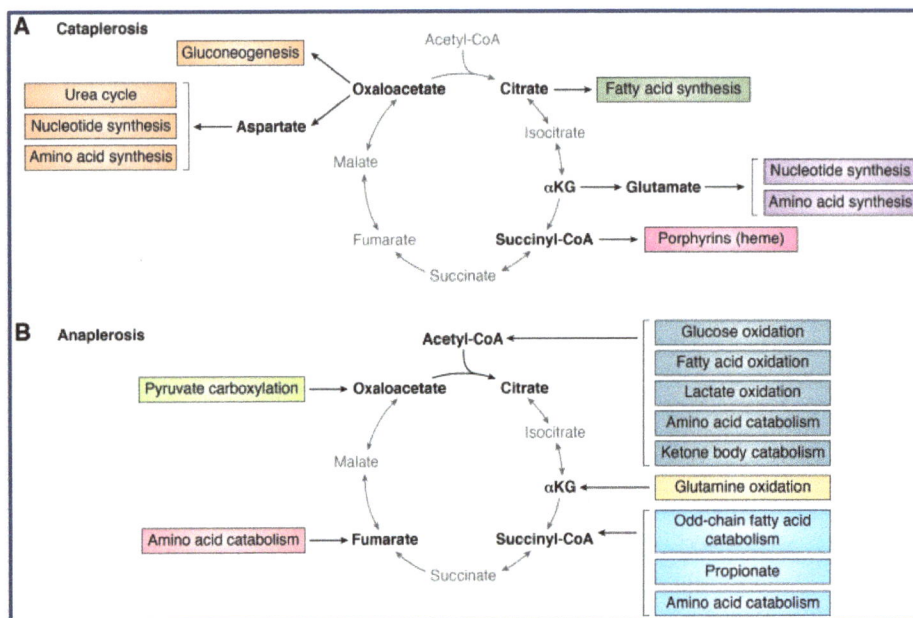

Figure 14.4 Outputs and inputs of the tricarboxylic acid (TCS) cycle.

Source: Regulation of the mammalian tricarboxylic acid cycle. Arnold PK, Finley LWS. Regulation and function of the mammalian tricarboxylic acid cycle. J Biol Chem. 2023 Feb;299(2):102838. doi: 10.1016/j.jbc.2022.102838. Epub 2022   Dec 26. PMID: 36581208; PMCID: PMC9871338.open access article under the CC BY license (http://creativecommons.org/licenses/by/4.0/).

A: Cataplerosis removes intermediates from the TCA cycle that are no longer needed, for example amino acids for gluconeogenesis or further amino acid synthesis.

B: Anaplerosis is the process of replenishing the TCA cycle's catalytic intermediates, that have been used for biosynthesis. Note, the input of glucose, fatty acids, lactate and ketones for oxidation, all feeding through acetyl-CoA and amino acids feeding through acetyl-CoA and other TCA intermediates. Both processes regulate the rate of biosynthesis and degradation. These two processes are involved in the ultimate disposal of all metabolic intermediates. Triacylglycerols are therefore, highly concentrated stores of metabolic energy. They are reduced and anhydrous and thus able to store more total energy compared with glycogen.

Fat energy yield	$= 38$ kJ/g^{-1} (9 kcal/g)
Carbohydrate energy yield	$= 17$ kJ/g^{-1} (4 kcal/g)
Protein energy yield	$= 17$ kJ/g^{-1} (4 kcal/g)

Fatty acids are not soluble in aqueous solutions and hence to reach tissues for energy burning, the released fatty acids bind to albumin, a blood protein. Albumin delivers fatty acids to the tissues. Fatty acids can be converted to acetyl-CoA for oxidation and high energy yield;

however, it is key to note that animals are unable to convert fatty acids into glucose. This is because the conversion of pyruvate to acetyl-CoA is irreversible, and hence, animals (lacking the requisite enzymes) are unable to achieve the net synthesis of glucose from fatty acids. Acetyl-CoA can-not be converted to pyruvate or OAA. The 2 carbon atoms of the acetyl group of the acetyl-CoA enter the TCA cycle (Figure 15.2) but then, 2 carbon atoms leave the cycle (as CO_2) in the decarboxylation reactions catalysed by isocitrate dehydrogenase and alpha-ketoglutarate dehydrogenase. Therefore, OAA is regenerated but it is not newly formed when the acetyl unit of acetyl- CoA is oxidised by the TCA cycle.

Hence, 2 carbon atoms enter the TCA cycle as an acetyl group but 2 carbon atoms leave the cycle as CO_2 ,before OAA is generated. As a result of this, no net synthesis of OAA is possible and there is no route to pyruvate or glucose.

Plants do, however, possess the necessary enzymes to complete this process. Almost 50% of both exogenous (dietary) and endogenous fats are unsaturated, and provide a substantial proportion of energy to the body. These unsaturated fats are catabolized in the same way as saturated fats except that 1 fatty acyl-CoA dehydrogenase. As the only exception, odd chain fatty acids (OCFA) can in fact be converted to acetyl-CoA, then to propionyl CoA and succinyl CoA, then up the chain to glucose via the TCA cycle, although OCFA are not a significant contributor for energy generation. Glycerol can be converted to pyruvate or glucose in the liver (gluconeogenic). The glycerol backbone of the triacylglycerol molecule can itself provide another source of energy discussed in Section 14.20. Glycerol from the TAGs, formed by lipolysis, is absorbed by the liver and phosphorylated. It is then oxidised to dihydroxyacetone phosphate, which is isomerised by glyceraldehyde-3-phosphate dehydrogenase to glyceraldehyde-3-phosphate. This enzyme can catalyse reactions in both glycolytic and gluconeogenic pathways and glycerol can be used as a source of energy via gluconeogenesis. Muscles have their own store of fats – hence the term 'marble cuts of beef' Migratory birds use their triacylglycerol fat reserves for their migration across thousands of miles from Northern Europe to Africa.

14.13 The Randle Cycle: Fat & Glucose Burn Competition

There is a balance between carbohydrate and lipids that influence their utilization. This is referred to as the Randle cycle (also known as the Glucose Fatty Acid Cycle) and the process is reciprocal; either inhibiting the level of glucose oxidation, or free fatty acids (non-esterified fatty acids (NEFA) beta (β)-oxidation, depending on cell conditions and the level of insulin. [59, 64,]

The acetyl CoA formed in fatty acid oxidation enters the TCA cycle only if fat and carbohydrate degradation are appropriately balanced. Acetyl CoA must combine with oxaloacetate to gain entry to the TCA cycle and the availability of oxaloacetate principally depends on an adequate supply of carbohydrate. Hence burning fats efficiently relies on sufficient carbohydrate.

Table 14.8 A Summary of the Randle Cycle

What is it?	The reduction in glucose uptake & use when fat burning is intense
What is its Principal Role?	The efficient burning of either glucose or fats in each cell but not both
Why does it occur	To maintain fuel substrate homeostasis, fine tuning hormonal control
When does it occur?	Constant regulation
Where does it occur?	The liver and adipose tissue
How fat burn inhibited?	High Fed State or relatively high glucose intake increases glucose burn
How glucose burn inhibited?	Fasted Sate or relatively high fat intake & exercise
What is needed to action it	Hexokinase II & IV (glycolysis), phosphofructokinase-1 (PFK-1)
What are the start reactants?	Glucose and fatty acids (some ketones depending on conditions)
What is the end product?	Energy from ATP via glucose or fatty acid (or ketone oxidation)

Fatty acid oxidation for energy is inhibited by conditions favouring glycolysis and glucose oxidation, which are:- Stimulation of PDH, a rise in NAD+ and mitochondrial calcium and a fall in NADH and by an increase in malonyl-CoA and CPT-1 (preventing the entry of fatty acids into the mitochondria by inhibiting Carnitine palmitoyl transferase-1 (CPT-1)). Malonyl-CoA is used by fatty acid synthase to produce long chain fatty acids for storage as triacyl glycerides, whilst promoting glucose utilisation.

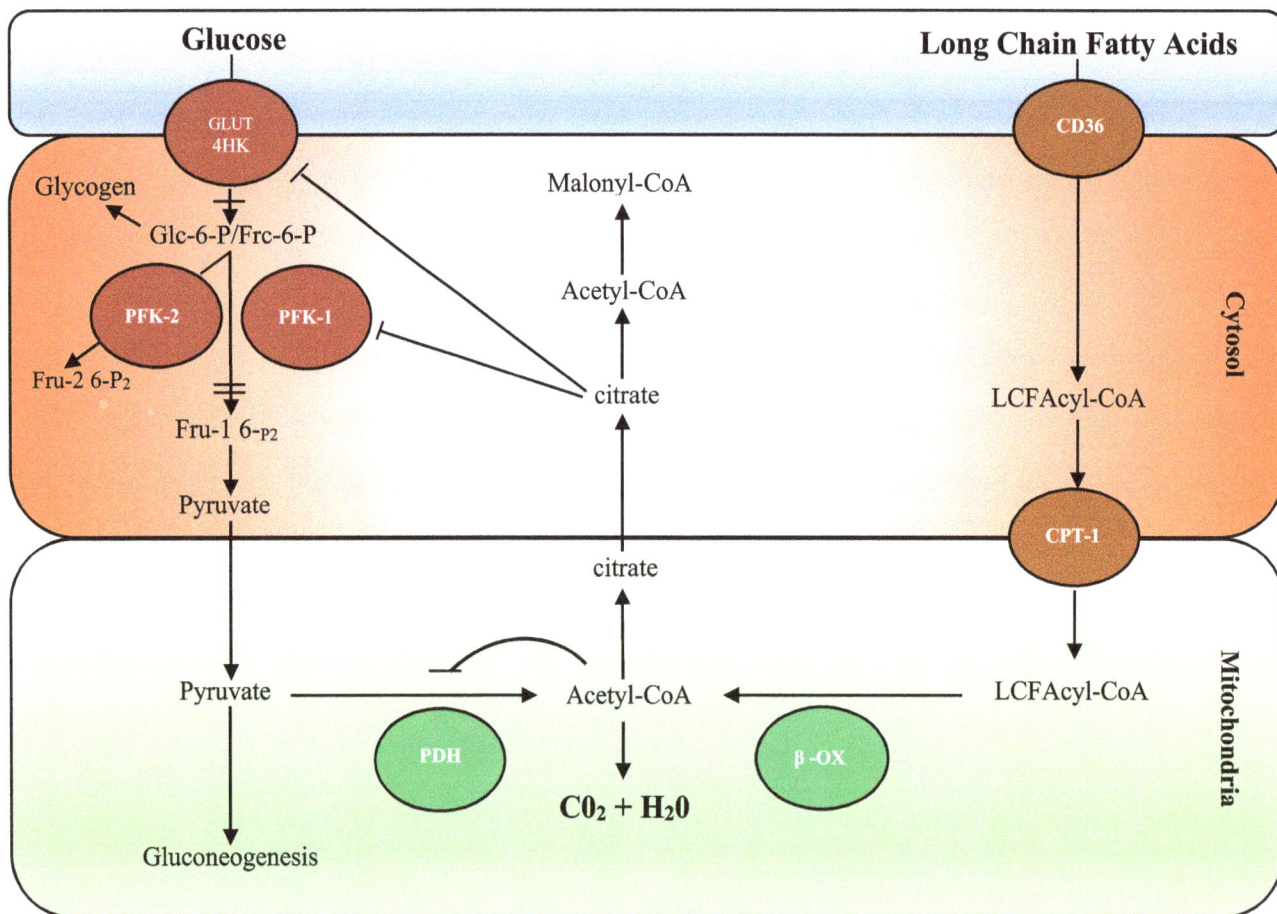

Figure 14.5 The Pathway of The Randle Cycle: Hue L, Taegtmeyer H. [59]

Glucose oxidation for energy is inhibited by conditions favouring fatty acid oxidation which are:- An increase and excess of acetyl-CoA and citrate in the mitochondria and a rise in NADH leading to the interruption of the actions of PDH and PFK-1 necessary for glycolysis.

The metabolic relationship between glucose and fatty acids is reciprocal and not dependent. [65] Provision of NEFA promotes fatty acid oxidation and storage in adipose tissue and inhibits glucose oxidation and may promote the storage of glucose as glycogen, if the glycogen stores are incomplete. [65] As the corollary, glucose promotes glucose oxidation and the storage of glucose as glycogen and fat in adipose tissue and inhibits fatty acid β)-oxidation.

Oxidation of glucose is decreased by the oxidation of NEFA and by high fat diets, starvation and exercise. NEFA β)-oxidation also stimulates glucose formation from lactate in the liver by gluconeogenesis.

The rate of glucose oxidation is dependent on the activity and activation of the pyruvate dehydrogenase complex and pyruvate dehydrogenase kinase within the mitochondria of each cell. This activation is affected by $[Ca^{2+}]$ ions, insulin and end products of fatty acid and ketone body oxidation. The inhibitory effects of NEFA oxidation on glucose disposal and oxidation has been proved on numerous occasions since 1963.

It is important to recall that at cellular level, each cell oxidises only one fuel at any time, for example, either glucose or fatty acids but not simultaneously. Two or more metabolic fuels are of course utilised simultaneously at whole organism, organ or tissue level. The Randle Cycle occurs where accelerated plasma free fatty acids (FFA) (β)-oxidation inhibits muscle glucose transport, glucose oxidation, and glycogen synthesis. It is the body's mechanism for the mutual inhibition of simultaneous fatty acid and glucose metabolism and prevents both processes from occurring at the same time in any given cell. The cycle can be seen as a fine tuning of substrate metabolism control based on available energy, nutrient type and availability to the cell and supplements hormonal control selecting the substrate dependent on the supply and demand of the cell.

This control system prevents simultaneous substrate oxidation in each cell, as this would create a futile metabolic cycle, in which any fatty acids synthesised from continual and significantly excess glucose, would immediately be oxidized. Therefore, whichever substrate is being eaten in excess determines what is oxidised. The Randle Cycle depresses glycolysis and this occurs during times of high energy intake, when there is a significant reduction in the uptake and utilization of glucose in muscle and when fatty acid (β)-oxidation is intense. This allows the blood glucose concentration to be maintained relatively constant, at the expense of the fatty acid concentration in the blood. The cycle also prevents hypoglycaemia during prolonged exercise. [59, 64, 65]

The release of free fatty acids for energy provision during long period endurance exercise will inhibit glucose utilisation in skeletal muscle tissue with the ultimate aim to preserve glucose or glycogenesis, for the key cells that are glucose dependent. This also occurs, for example, under various conditions, such as in stressed (insulin resistance and cortisol, adrenaline, glucagon, growth hormone) fasted, or starvation states, when carbohydrate is genuinely, or artificially kept low, such as keto or high fat diet, or carbohydrate deprived dieting and the body is awash with free fatty acids, predominantly burning fatty acids. Stressed states trigger glycolysis and gluconeogenesis. In addition, inhibition of glucose oxidation at the level of pyruvate dehydrogenase (PDH), and PFK-1 also preserves pyruvate and lactate, which can be used as precursors for gluconeogenesis, though producing less ATP than glucose in the TCA cycle. The Randle cycle can also occur in heart and liver tissue that can utilise lactate as fuel. In these organs, lactate inhibits the oxidation of both glucose and fatty acids. Excess pyruvate can be converted to oxaloacetate (OAA) by anaplerosis, a process that can replenish TCA cycle intermediates.

What happens when both substrates are elevated and equally high? In that event, hormones step in and regulate metabolism and in particular insulin. Insulin orchestrates and acts as referee in the final selection of fuels for burning. Reference to Section 14.3 showed how this mediation operates.

14.14 Ketones Preserve Glucose

Ketone bodies are normally present in small concentrations in the blood (0.1 to 0.5 mmol/L), providing small amounts of energy to the muscles and heart and a little (<5% energy) for the brain. In circumstances when the body is unable to produce sufficient glucose, from carbohydrate or fatty acid beta (β)-oxidation, or even from gluconeogenesis, the body's first and primary aim from the diet is to seek an alternative way to generate sufficient glucose.

As we have seen, when the liver does not have enough OAA from carbohydrate (glucose) oxidation, to process all of the acetyl-CoA produced from the breakdown of fats, for example, due to carbohydrate deprivation or starvation, the excess acetyl-CoA from fatty acid (β)-oxidation builds up and can-not enter the TCA cycle.

Under these metabolic stress conditions ketones are predominantly formed in the mitochondria of the hepatocytes from acetyl-CoA generated primarily from β)-oxidation of fatty acids. They are also synthesised in brain astrocytes, most of these in the hypothalamus.

Increased dietary fatty acid intake, decreased glucose availability and high levels of fatty acid β)-oxidation stimulate an increase in ketogenesis in the liver. When significant gluconeogenesis has been stimulated by low insulin and high glucagon concentrations in the blood glucose and OAA is used up for gluconeogenesis then sufficient OAA is not available to drive the TCA cycle and OAA is unavailable for condensation with acetyl-CoA.

Accumulated acetyl-CoA from the non-esterified fatty acids (NEFA), is instead converted into significantly higher levels of acetoacetate and then hydroxybutyrate and acetone. These ketone bodies are formed in the liver in higher quantities and diffuse out of the liver into the circulation as an emergency source of energy. Ketosis is associated with a 20% to 50% decrease in circulating glucose and insulin concentration.

Ketones provide approximately 20 to 22.5 moles ATP/mole of ketone (Table 15.12) rising to 60% and even up to 75% of the brain's total energy requirement and in times of prolonged starvation, ketones, glucose and lactate, along with glucogenic amino acids, sustain the brain and preserve essential levels glucose levels in the blood. In these circumstances, the brain can survive on only 22 to 28 g/d of glucose or approximately 4% of normal total dietary caloric energy intake (EI).

Therefore, ketone production preserves glucose for the brain and other glucose dependent tissues, especially the totally glucose obligate RBCs. By alleviating the need for gluconeogenesis, it also serves to reduce protein degradation and spare and preserve structural protein. The ketones are shipped to the tissues that need glucose the most. These tissues, principally the brain, skeletal muscles and heart, are equipped with special enzymes within the mitochondria to convert ketones to acetyl-CoA and to use this pathway to generate ATP. If this build-up of acetyl-CoA and ketosis continues, however, it can cause damage to the body including ketoacidosis. The body reacts by trying to remove the toxins, detectable on the breath and perspiration as a pungent odor and in the urine. Ketogenesis is discussed in further detail in Section 14.39.

14.15   Excessive Fatty Oxidation and Cell Damage

Even with the body's natural Randle cycle control process, discussed in Section 14.13, continued excessive fatty acid (β)-oxidation to the cost of glucose oxidation will lead to metabolic damage, due in part to reactive oxygen species (ROS) and also impaired insulin sensitivity. The biochemical mechanism of the build-up of ROS is described below. [59]

With a high, or sufficient amount of carbohydrate, oxidation of glucose in the TCA cycle yields a higher ratio of NADH to $FADH_2$ and a higher ratio of NAD^+ to NADH, compared with the ratio produced by the (β)-oxidation of fatty acids.

Higher NAD^+ means that there is more NAD^+ to be recycled, starting with the linchpin enzyme, complex, pyruvate dehydrogenase (PDH) or (PDC) and the other 3 TCA cycle reactions, requiring NAD^+ and thus prevents potential road blocks in the TCA cycle, for example the build-up of citrate. PDH is fundamentally key in converting glucose to pyruvate and it catalyses the irreversible oxidative decarboxylation of pyruvate from glucose into acetyl-CoA. Acetyl-Co A joins with oxaloacetate (OAA) to form citrate in the TCA cycle. Citrate then continues through the TCA cycle, driven by normal glucose or through normal fatty acid terminal oxidation.

Therefore, adequate carbohydrate and glucose oxidation inhibits excessive (β)-oxidation of fatty acids and effectively reduces lipolysis. PDH is key in maintaining lipid homeostasis.

Fatty acid (β)-oxidation inhibits various glycolytic steps, including blocking the main enzyme Pyruvate dehydrogenase (PDH) and 6 -Phosphofructokinase (PFK-1), reducing glycolysis hence the Randle cycle. Increased, excessive fatty acid beta (β)-oxidation, therefore, effectively serves to reduce glucose oxidation through the TCA cycle, via the inhibition of PDH activity and occurs when the ratio of either acetyl-CoA to CoA or NADH to NAD^+ increases.

The level of citrate builds up and leads to inhibition of PDH. When fatty acid (β)-oxidation (burning) is intense, the levels of acetyl-CoA, NADH and citrate build up and accumulate. With the inhibition of PDH and PFK-1, required by glycolysis this shuts down glucose oxidation.

The excess of NADH also causes a back log and build-up of acetyl-CoA and citrate in the cytoplasm, impairing the movement of glucose into the cell. This excess of fatty acid (β)-oxidation therefore also impairs insulin sensitivity.

Conversely, the reverse occurs when hyperglycemia occurs. The increase in glycolysis leads to the build-up pyruvate and malonyl CoA. The malonyl CoA enters the mitochondria and shuts down the enzyme carnitine palmitoyl-transferase (CPT-1). Malonyl-CoA regulates fatty acid (β)-oxidation by inhibiting the activity of CPT-I, which is the rate limiting enzyme of mitochondrial fatty acid uptake. It controls the rate of fatty acids entering the mitochondria for subsequent full-term oxidation and OXPHOS. This enzyme is critical to the entry of fatty acids into the mitochondria and in this case the fatty acids can-not be oxidised, hence, the reduction and shut down the beta (β)-oxidation of fatty acids. Oxidation of glucose delivers 5 oxidised molecules of NADH, for 1 molecule of $FADH_2$. For fatty acid oxidation, this ratio changes to only 2 molecules of NADH being oxidised for every molecule of $FADH_2$.

In the ETC, Complex 1 receives its electrons from NADH and Complex 2 normally receives its electrons from $FADH_2$. (For the ETC, refer to Figure 15.4). In carbohydrate oxidation, the 5:1 ratio of NADH to $FADH_2$ favours the transfer of electrons from Complex 1 to Coenzyme Q. [59]

However, in fatty acid beta (β)-oxidation, the reduced ratio, of NADH to $FADH_2$, means that Coenzyme Q has a closer association with Complex II. As a consequence, Complex 1 is therefore less able to receive its electrons from NADH and thus, less able to transfer electrons to Co-enzyme Q. This results in a build-up of NADH and electrons and hence an increase in the ratio of NADH to NAD^+. The build-up of electrons caused by this shift causes the increased formation of the superoxide hydrogen peroxide (H_2O_2) and other reactive oxygen species (ROS). ROS are normally produced inside the mitochondria during metabolism and act as regulators of cell multiplication and differentiation.

Normally, less than 5% of oxygen molecules in the mitochondria convert to ROS. Antioxidants (mostly ingested from plants in the diet) normally scavenge the ROS and maintain the ROS at healthy levels. But excessive ROS are inflammatory toxins and damage mitochondria and further elicit the body's defence action to release an uncoupling antioxidant protein to remove the ROS. However, this defensive action causes a knock-on effect, disrupting the proton gradient which decreases ATP production and leads to significantly more waste heat.

Excessive ROS irreversibly damages proteins, DNA, cell membranes and lipids with many adverse effects on cellular functions. Furthermore, increased ROS production can lead to excessive fat accumulation and oxidative stress. The reason why ROS are so toxic is that the single oxygen atom is highly reactive and unstable and therefore, each one tends to bind a twin atom, forming molecular oxygen.

However, the stability of this bond is compromised because only one pair of electrons is shared and two unpaired electrons remain; free radicals often generate. Examples of ROS are hydrogen peroxide, superoxide anion radical, singlet oxygen and hydroxyl radical. ROS are highly reactive molecules with unpaired molecules in their outer orbital electron shells. [59]

In summary therefore, excessively high, continual (β)-oxidation of fatty acids relative to glucose oxidation causes the reverse and shifting of the ratios of $FADH_2$ to NADH and a higher ratio of $NADH$ to NAD^+. This has a knock-on effect through to the ETC. Too much $FADH_2$ in ratio to NADH will cause a back log of NAD^+ in the ETC and eventually a build of reactive oxygen species (ROS) which is stressful and damages mitochondria and mitochondria DNA.

Consequently, a perpetual state of inadequate carbohydrate and excess fatty acid leads to metabolic inflexibility, cell stress and cell damage. The build-up of the ratio of NADH to NAD^+ causes a backlog of NADH, citric acid and acetyl CoA in the TCA cycle. This impairs the movement of glucose into the cell. In excess fatty acid oxidation, PDH is inhibited and glucose can-not therefore, be oxidised and ATP energy production is reduced also leading to a decrease in insulin sensitivity. This scenario is essentially similar to insulin resistance. This metabolic stress is known as the Randle effect. Oxidative stress is an important underlying factor that leads to development of human diseases including obesity and non-alcoholic fatty liver disease. Excessive and continual fatty acid (β)-oxidation compared with glucose oxidation, for example deliberately invoked by following low carbohydrate diets, leads to increased ROS which (on so many grounds) the author considers should be avoided.

14.16   Gluconeogenesis - The Maintenance of Blood Glucose

Gluconeogenesis is the process of synthesising glucose from non-carbohydrate sources to help maintain sufficient glycemia or euglycemia, i.e., correct blood glucose levels. These sources are derived from glycerol, lactate, two amino acids, plus small contributions from other amino acids and also the branched chain fatty acid propionate. Muscle and adipose (fat store) tissues lack the ability to form glucose from non-carbohydrate precursors because they lack the necessary enzyme, glucose-6-phosphatase and muscle and adipose tissue can-not therefore, contribute to the maintenance of blood glucose levels.

Table 14.9 A Summary of Gluconeogenesis

What is it?	The formation of new glucose from non-carbohydrate precursors
What is its Principal Role?	To maintain glucose homeostasis and build up glycogen reserves
Why does it occur	A response to detected depletion of glycogen stores and glucose levels
When does it occur?	Between meals & during fasting 8 hours after eating, starvation, Low Carb
Where does it occur?	Primarily in the liver, to lesser extent in the kidneys, muscles, gut, brain
How is it stimulated?	A rise in glucagon and cortisol, fasting and starvation
How is it inhibited?	A rise in insulin levels a decrease in glucagon and eating
What is needed to action it	Phosphoenolpyruvate carboxykinase & fructose-1-6 biphosphatase
What are the start reactants?	Glycerol, lactate, amino acids (alanine), pyruvate, OAA, propionate
What is the end product?	Glucose production (and glycogen if sufficient glucose is present)

Gluconeogenesis takes place during periods of fasting, when there is insufficient or limited dietary glucose consumption or glycogen store or blood glucose availability in the cells, for example, overnight while asleep and in more extreme carbohydrate deprivation diets or during prolonged fasting and starvation. Gluconeogenesis is under the control of hormones, cell nutrient intake, stress conditions, redox (oxidation and reduction) states, and energy substrate concentrations.

The body values glucose as such an important fuel that other metabolic products, such as glycerol and lactate and protein are salvaged to synthesise glucose in this process and gluconeogenesis, conducted by the liver and kidneys, is a vital process forming glucose from non-hexose (non-carbohydrate) precursors during times of glucose shortage. Although only a moderate amount of glucose is synthesised by gluconeogenesis in normal circumstances, this process is critical to the maintenance of blood glucose levels. Acute hypoglycaemia can damage the brain and kidneys and other tissues because of their dependence on glucose for fuel.

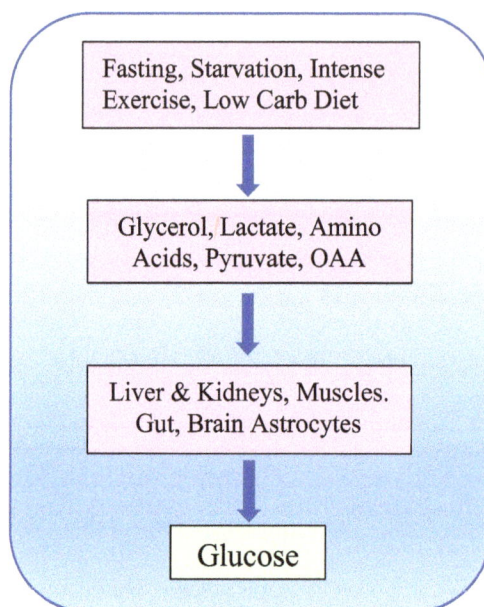

Figure 14.6 Overview of the pathway of Gluconeogenesis from non-carbohydrate precursors to form glucose

The renal medulla, RBCs, retina and testes are unable to survive for long periods without glucose. The gut, skeletal muscle and brain cells can conduct a measure of gluconeogenesis locally for their own tissue demands when conditions require this. Glutamine is the primary substrate in the intestinal gut cells and also the preferred amino acid for gluconeogenesis in the kidneys.

Gluconeogenesis mainly occurs in the liver, normally (70-90%), to a lesser extent (10-30%) in the renal cortex corticoid cells and a small amount in the intestine (intestinal gluconeogenesis or IGN), skeletal muscle cells and brain astrocyte cells. Some researchers report much higher renal contribution up to 40% of total systemic gluconeogenesis depending on the severity of conditions. This process maintains supply of glucose in the blood so that all other tissues, including the brain CNS, muscles and RBCs, are able to utilise their needs, taking sufficient glucose from the blood. Gluconeogenesis also delays the onset of ketosis by around 2 weeks. (Figure 14.10). The process must not be excessive because high rates of gluconeogenesis will lead to problems of high levels of ammonia (Section 14.16). Gluconeogenesis is also one of the body's main clearing mechanisms for the muscle and red blood cell erythrocyte principal metabolite, lactate. In the kidneys, ammonia (NH_3) is produced as a by-product, which normally serves as a major role in maintenance of the acid-base balance, serving as the counterion for excretion of ketoacid anions.Glycogenolysis (the break-down of glycogen to glucose) is principally used during shorter fasting periods, when blood sugar decreases between meals or after sleeping, and gluconeogenesis is mostly invoked during longer periods of fasting. Both processes, however, occur to some degree because glucose is required for energy production. Gluconeogenesis is typically active in catabolic states, e.g., post-prandial, low carbohydrate diets, high fat diets prolonged exercise or starvation. Technically, the true major non-carbohydrate precursors of 'new glucose' from new carbon skeletons are glycerol and amino acids. Lactate is formed by skeletal muscle when glycolysis needs oxidative metabolism and when the rate of glycolysis exceeds the rate of oxidative metabolism. Lactate is readily converted to pyruvate by lactate dehydrogenase. Thus, it is not truly gluconeogenic as it effectively recycles glucose from existing carbon skeletons. Although gluconeogenesis essentially resembles a reversal of glycolysis, where the latter converts glucose to pyruvate and gluconeogenesis converts pyruvate to lead to glucose, it is not simply a reversal because several reactions differ, requiring 4 unique reaction steps not in glycolysis (compare Figure 14.7 and Table 14.10). Glycolysis and gluconeogenesis are coordinated to ensure that all cells receive their required energy and are reciprocally regulated by allosteric enzyme stimulation and inhibition by insulin, glucagon, growth hormone, epinephrine and cortisol, so that both pathways do not occur simultaneously. For the glycolytic breakdown of glucose, the most important control is the enzyme phosphofructokinase (PFK), converting fructose-6-phosphate to fructose 1,6 biphosphate and this enzyme is allosterically controlled by the level of ATP such that high ATP inhibits the conversion and high AMP stimulates the enzyme. Therefore, glycolysis is stimulated as the energy charge of the cell falls. The critical regulatory enzyme of the last reaction stage glycolysis is pyruvate kinase (PK), converting phosphoenolpyruvate to pyruvate and the critical regulatory enzyme of the first stage gluconeogenesis is phosphoenolpyruvate carboxy kinase (PEPPK), converting oxaloacetate to phosphoenolpyruvate. Both pathways are reciprocally regulated. Table 14.10 shows the 11 stages of gluconeogenesis from pyruvate to glucose. Note that a total of 6 high transfer potential phosphoryl groups (4 ATP + 2 GTP) are hydrolysed to synthesis glucose from each of two 3 carbon pyruvate molecules (2 ATP and 1 GTP from each). Stages in bold text are the 4 unique and key gluconeogenic reactions necessary to by-pass the 3 irreversible steps of glycolysis. The other 7 reactions are common to both pathways. (Figure 14.7 curved arrows). The ATP and GTP is needed to power the 3 bioenergetically unfavourable reactions shown in Table 14.10.

Table 14.10 Key Reaction Steps in Gluconeogenesis from each mole of Pyruvate to Glucose & ATP/GTP cost

Stage	From	To	ATP/GTP
1	Pyruvate	**Oxaloacetate** (Pyruvate carboxylase)	1 ATP
2	Oxaloacetate	**Phosphoenolpyruvate** (Phosphoenolpyruvate carboxykinase)	1 GTP
3	Phosphoenolpyruvate	2-Phosphoglycerate	
4	2-Phosphoglycerate	3-Phosphoglycerate	
5	3-Phosphoglycerate	1,3-Biphosphoglycerate	1 ATP
6	1,3-Biphosphoglycerate	Glyceraldehyde 3-Phosphate	
7	Glyceraldehyde 3-Phosphate	Dihydroxyacetone phosphate (Glycerol enters at this stage)	
8	Glyceraldehyde 3-Phosphate	Dihydroxyacetone phosphate/Fructose 1.6-bisphosphate	
9	Fructose 1,6-bisphosphate	**Fructose 6-phosphate** (Fructose 1,6-bisphosphatase)	
10	Fructose 6-phosphate	Glucose 6-phosphate	
11	Glucose 6-phosphate	**Glucose** (Glucose 6-phosphatase) (in the endoplasmic reticulum)	

Stage 10. Glucose 6-phosphate can synthesise glycogen rather than glucose if conditions are right Glucose 6 phosphatase only in liver & kidneys

Note, for every mole of glucose, 2 moles of pyruvate are required, hence, the total energy required per mole of glucose is 4 ATP and 2 GTP. In summary, the anabolic process of gluconeogenesis is highly endergonic compared with the exergonic catabolic pathway for glycolysis:

Glycolysis yields:	$+84 \, kJ \, mol^{-1}$ ($+20 \, kcal \, mol^{-1}$) and 2 ATP
Gluconeogenesis uses:	$-38 \, kJ \, mol^{-1}$ ($-9 \, kcal \, mol^{-1}$) and 6 ATP (4 ATP + 2 GTP)

The mitochondrial pyruvate dehydrogenase complex (PDC) is one of the most important drivers of cell metabolism, linking two of the principal catabolic pathways of; 1) glycolysis, culminating in the irreversible step oxidising pyruvate to provide acetyl-CoA and releasing CO_2 and 2) the TCA cycle and OXPHOS for cyclical full oxidation of glucose and fatty acids. Pyruvate carboxylase (PC) is another important mitochondrial enzyme that also reacts with pyruvate but this time catalysing the conversion of pyruvate to oxaloacetate. This latter reaction is the critical reaction of gluconeogenesis, for the anaplerotic replenishment of TCA cycle intermediates and for numerous biosynthetic reactions.

The selection of the pathway to glycolysis or gluconeogenesis is regulated reciprocally to ensure energy is used as efficiently as possible, producing net ATP synthesis and avoiding futile and wasteful reactions. Glycolysis is turned off when the energy charge of the cell is high and no additional ATP is required and gluconeogenesis is then activated. Conversely, when the energy charge of the cell diminishes, the cells begin to produce more ATP via glycolysis and gluconeogenesis is turned off to conserve ATP. The following shows the major activation and inhibition factors for each of these two pathways.

PDH: Pyruvate to Acetyl CoA for the TCA Cycle and Oxidation of Glucose or Fatty Acids

Activated by: Acetyl CoA↓ ATP↓ ADP↑ NAD⁺↑ Cell Energy↓ Blood Glucose↑

Inhibited by: Cell Energy↑ Acetyl CoA↑ ATP↑ NADH↑ Fatty acid oxidation↑

PC: Pyruvate to Oxaloacetate for replacement of TCA cycle intermediates & Gluconeogenesis

Activated by: Fasting↑ Glucagon↑ Insulin↓ Blood Glucose↓ Cell Energy↓ ADP↓ ATP↑ Acetyl CoA↑

Inhibited by: Feeding↑ Insulin↑ Glucagon↓ ADP↑

The common substrate pathway for the non-carbohydrate gluconeogenic precursors is via pyruvate, consisting of the eleven enzyme-catalyzed reactions converting pyruvate to glucose Table 14.10). Gluconeogenesis commences with the formation of oxaloacetate (OAA) through the carboxylation of pyruvate but is inhibited in the presence of high levels of ADP.

This reaction is catalyzed by pyruvate carboxylase, which is stimulated by high levels of acetyl-CoA, i.e., when fatty acid oxidation is high in the liver. In the mitochondria, pyruvate is converted to OAA with pyruvate carboxylase, then OAA is converted to malate using up NADH. The malate can then be transferred into the cytosol via the Malate-aspartate shuttle and reconverted and oxidised back to OAA using NAD^+. In the cytosol, OAA is converted to phosphoenolpyruvate using the enzyme phosphoenolpyruvate carboxykinase (PEPCK).

Reversing the glycolytic process, the phosphoenolpyruvate is converted to fructose 1,6-bisphosphate, then to fructose-6-phosphate, then glucose-6-phosphate, finally generating glucose in the cell's endoplasmic reticulum.

For gluconeogenesis to operate, the body must be in the fasted state and the energy state of the tissue must be low to stimulate precursor supply, for example, glycerol, lactate, alanine. Initial stimulation is from cortisol and increased glucagon, leading to lipolysis, mobilising the oxidation of free fatty acids to providing the required ATP to power the process. Eighteen of the twenty amino acids can contribute to gluconeogenesis, with alanine and glutamine the predominant contributors. Amino acids (from the diet) are provided by the muscles during starvation. In the liver and kidneys, both gluconeogenesis and (β)-oxidation of fatty acids, increase the level of acetyl-CoA and decrease the levels of OAA. Table 14.11 shows the relative proportions of the gluconeogenic precursors as fasting progresses. Gluconeogenesis, is a 'stop gap' regulatory process that has evolved to get the body through low glucose levels overnight, between feeding or extended fasting and starvation times until carbohydrate food arrives. We have seen that the gluconeogenesis pathway is highly endergonic bioenergetically (requires and absorbs energy) and the energy equilibrium from the conversion of pyruvate to glucose is heavily weighted on the glycolytic side to form pyruvate.

The process requires ATP, GTP and NADH. Conversion of two molecules of pyruvate to one mole of glucose requires six high-energy phosphate bonds (4 ATP + 2 GTP, Table 14.10) and results in the oxidation of two NADH molecules. However, glycolytic conversion of one mole of glucose to two moles of pyruvate produces two high-energy phosphate bonds (2 ATP) and reduces two moles of NAD^+.

The requirement of 6 ATP equivalents to produce a single glucose molecule, are therefore bioenergetically expensive, compared with using glucose from dietary carbohydrate. As the very principle of gluconeogenesis is preserving glucose levels for energy, it is not appropriate to obtain this energy from glucose via glycolysis of the TCA cycle and therefore, the bioenergetic cost of gluconeogenesis is met by the (β)-oxidation of fatty acids.

The Alanine cycle (also known as the Glucose-Alanine cycle or the Cahill cycle) (Figure 14.9), is utilised in times of low blood glucose where skeletal muscle protein is broken down to create glucose for ATP and contraction and the resulting nitrogen is transaminated to pyruvate to form alanine. The alanine is transported to the liver and converted to pyruvate to synthesis glucose. Here the nitrogen enters the urea cycle via ammonia. The cost of excreting urea is a further 4 ATP. Therefore, the true cost of gluconeogenesis is a total net of -10 ATP to produce one mole of glucose. Assuming this glucose proceeds to oxidation via glycolysis, the TCA cycle and OXPHOS yielding 30 ATP, then the total net cost of gluconeogenesis from amino acids is approximately 20 ATP. Table 14.15. Notwithstanding this, the body cell's overall need for glucose is prioritised and outweighs the bioenergetic efficiency of ATP yield.

The day-to-day function of low levels of gluconeogenesis and in fact ketogenesis is to sustain the body's needs for glucose in times of fasting, food shortage and starvation. Figure 14.7 shows the 2 pathways of glycolysis and gluconeogenesis. Note the curved arrows illustrate the 4 unique steps of gluconeogenesis to by-pass the 3 irreversible reaction stages in glycolysis shown in table 14.10.

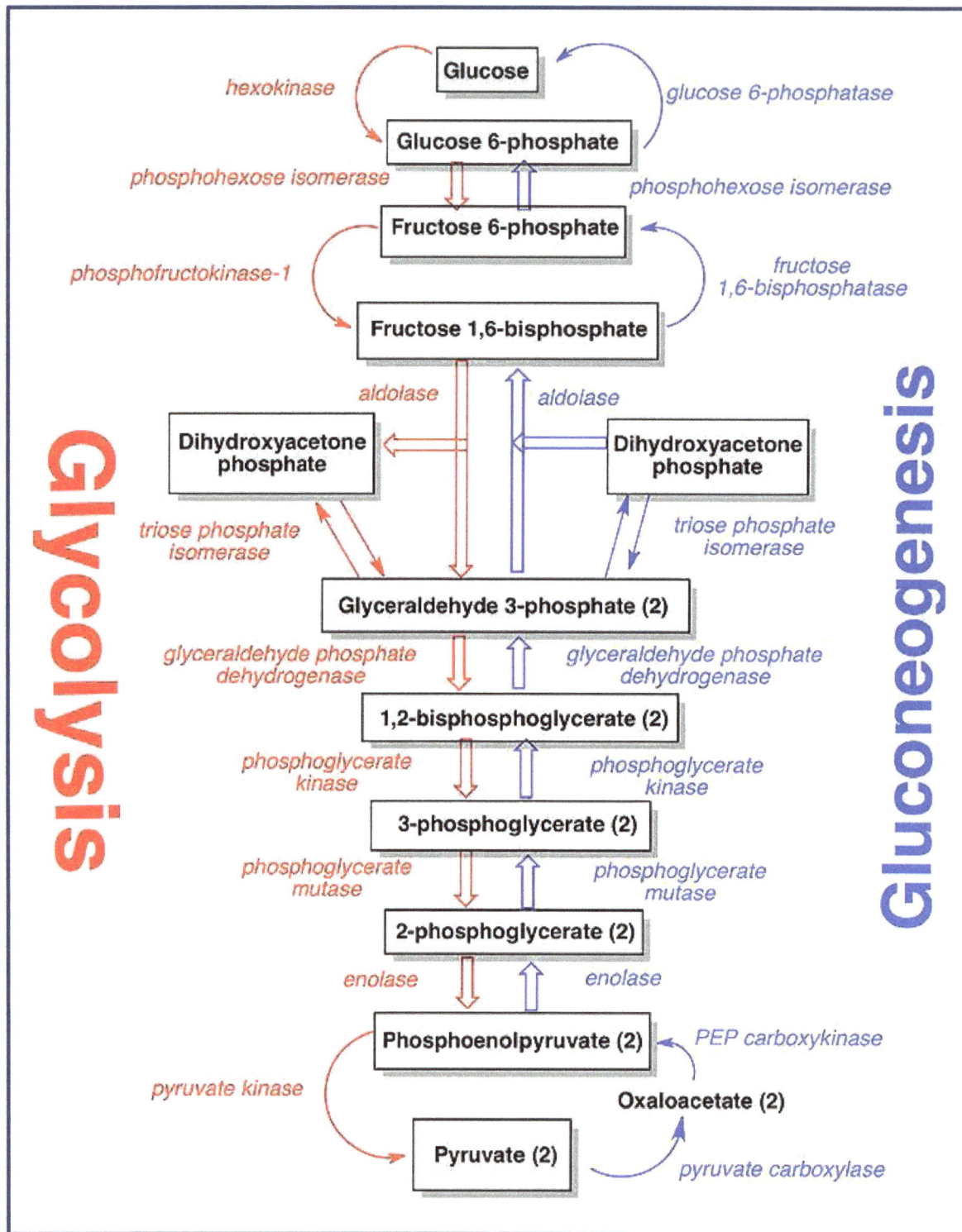

Figure 14.7 Glycolytic and gluconeogenic pathways. Dwong527 (wikimedia.org/wiki/File: Glycolytic and gluconeogenic pathways.jpg), CC BY-SA 3.0 (https://creativecommons.org/licenses/by-sa/3.0), via Wikimedia Common

Excessive and continual gluconeogenesis leads to several metabolic stress symptoms. Enhanced hepatic glucose production from gluconeogenesis leads to increased glucose released into the blood which is especially problematic for people with Type 2 diabetes as it can cause hyperglycemia. Insulin, which normally inhibits hepatic gluconeogenesis, works in close regulatory harmony with glucagon, which stimulates gluconeogenesis and these 2 hormones maintains healthy gluconeogenesis levels to maintain euglycemia. However, this balancing process is disrupted in those individuals with obesity, NAFLD, prediabetes, diabetes and or insulin resistance and for these individuals, with much higher available levels of plasma gluconeogenesis substrates, the gluconeogenesis contribution can double with the concomitant decrease in glycogenolysis leading to significant increase in hepatic glycogen content and no change in endogenous glucose production.

All this leads to far higher rates of gluconeogenesis to synthesise glucose compared to simply obtaining glucose derived from glycogen stores. This can lead to hyperglycemia even for those people that do not have diabetes. These processes are natural and essential mechanisms, normally activated and tightly regulated during fasting each day (overnight).

Ultimately and alongside ketogenesis, gluconeogenesis is utilised in starvation, to ensure critical minimum levels of glucose are maintained for the cells that rely on glucose. Abnormally high levels of gluconeogenesis or ketogenesis are not 'steady state' processes, designed for long term or continuous utilisation, in order to burn proteins, fats and other substrates, rather than using glucose from dietary carbohydrates.

Gluconeogenesis is governed by the liver. Overall, exogenous dietary consumption of gluconeogenic substrates within ingested food make only a small immediate contribution to endogenous glucose production and does not significantly increase the rate of gluconeogenesis. High insulin level is the key inhibitor of gluconeogenesis and falling insulin levels, during fasting and the availability of gluconeogenic substrates stimulates gluconeogenesis.

The build-up of acetyl-CoA may lead to the production of ketones and a reduction in full oxidation of fatty acids to yield the full complement of ATP via the TCA cycle and OXPHOS.

The reduction of OAA slows the TCA cycle and leads to a build-up citrate and further accumulation of acetyl-CoA because the reduced OAA can-not combine as efficiently with the increased amount of acetyl-CoA. However, unlike the depleted OAA levels in the liver, when the brain receives the ketones, the brain does have sufficient levels of OAA and although the brain has much lower levels of acetyl-CoA from glucose and pyruvate, the ketones take over this role and are converted into acetyl-CoA within the mitochondria of neurons. The acetyl CoA then enters the cell TCA cycle and OXPHOS to yield a full complement of ATP.

To reiterate, this process has not evolved to be unnaturally adopted long term, for example, deliberate carbohydrate deprivation. Figure 14.7 shows the pathways and 11 reactions of glycolysis from glucose to pyruvate (catabolic) and gluconeogenesis (anabolic). Before we look at some of the detailed mechanisms of gluconeogenesis, it is worth spending some tome describing the role that the liver and kidneys play in this process.

14.17 Gluconeogenesis - The Role of The Liver and Kidneys

The liver and kidney and to a much lesser extent (and their own local needs), the gut, the brain and muscles, are the only organs capable of making glucose via the process of gluconeogenesis because of the presence of two rate-limiting enzymes for glucose production. These enzymes are phosphoenolpyruvate carboxykinase (PEPCK) and glucose-6-phosphatase (G6Pase) and enable the process to convert glucose 6-phosphate into glucose or glycogen. The liver is the major site of gluconeogenesis but the kidneys are also an important contributor in prolonged fasting. The kidney is one of the largest total consumers of energy requiring 440 kcal/kg tissue/day (Table 14.1), the same mass specific metabolic rate as the heart and is the second most bioenergetically active organ. The kidney proximal cells require significant quantities of reliable energy sources to fulfil its roles of filtration, reabsorption and excretion. The liver and kidneys provide similar relative amounts of gluconeogenesis in the post absorptive state. Following an overnight fast, 72% to 80% of glucose is released into the circulation from the liver and 20% to 28% of glucose is released from the kidney. Many scientific studies have confirmed that the kidney renal cortex proximal tubules can in fact contribute up to 40–50% of endogenous gluconeogenesis under very high levels of stress, acidosis, fasting and obesity or in diabetics and insulin resistance.

Triacylglycerol (TAG fat) build up in the renal proximal tubules drives further renal gluconeogenesis through increased acetyl-CoA concentrations. This has the effect of reducing the role of insulin to suppress gluconeogenesis. Ectopic fat accumulation also leads to increased renal gluconeogenesis and eventually to kidney damage. As fasting duration increases, glycogen levels are depleted in the liver. After >42 hours of fasting, around 85% of glucose released is from gluconeogenesis and as fasting progresses further, the percentage of glucose from kidney gluconeogenesis also increases. (Table 14.11). The glucose synthesised in the kidney is used locally by the renal medulla which is heavily dependent on glucose. The maintenance of blood glucose levels via renal gluconeogenesis becomes more important during prolonged fasting and in circumstances of liver failure. The kidneys, unlike the liver, have no significant glycogen stores and therefore contribute to the maintenance of blood glucose homeostasis through gluconeogenesis only and not, like the liver, through the dual mechanisms of both the breakdown of glycogen via glycogenolysis and gluconeogenesis. Just as skeletal muscle glycogen serves to provide glucose energy locally for muscle contraction, likewise, the glucose produced in the kidneys is the substrate used to produce the energy consumed by the kidneys, producing ammonia, the urinary excretion that is essential to regulate metabolic acidosis.

The liver, kidneys and also the lungs act as metabolic sensors to regulate vascular blood pressure and homeostasis, cardiopulmonary pressure and chemoreceptor response using the renin-angiotensin-aldosterone system (RAAS). In the liver, glycerol and lactate are the predominate gluconeogenic precursors. However, in the kidneys, the main gluconeogenic precursors are lactate, glutamine and glycerol. Under conditions of metabolic acidosis, glutamine becomes the most significant precursor for gluconeogenesis in the kidneys. The liver's preference after glycerol and lactate is the amino acid alanine. Prevention of hypoglycaemia may be controlled by both the liver and kidney. This is known as hepatorenal glucose reciprocity. In the post absorptive state, after an overnight fast, the kidneys utilise approximately 10% of all the body's glucose. Section 14.18 provides an examination of the contribution of the various glucogenic precursors to glucose sparing and the homeostatic role and importance of glycerol, lactate, amino acids and gut bacteria fermented branch chain fatty acid, propionate from dietary fibre.

14.18   Gluconeogenesis - The Substrate Precursors

There are 5 principal precursors feeding liver and kidney metabolism of gluconeogenesis. These are glycerol, lactate, alanine, glutamine in the kidneys and propionate from the gut.

These probably account for 95% of all gluconeogenesis with the remainder from other amino acids: glycerol and lactate combined, provide approximately 80-95% of gluconeogenic substrates (excluding glycogenolysis), with around 5-15% provided by the amino acids, alanine and glutamine, from endogenous catabolism of skeletal muscle.

Much work has been done using isotope tracers to track the source and destination of carbon atoms from gluconeogenesis and the proportional contribution to glucose synthesis during fasting. There are various assessment of gluconeogenic proportional contributions in relation to the overall contribution of amino acids (alanine and glutamine) to the synthesis of glucose compared with the contributions from glycerol and lactate. The author estimated relative proportions of gluconeogenic precursors in synthesising glucose during the fasting state are shown in Table 14.11.

The precise levels will vary with each individual and the amount and type of food ingested prior to the fast, together with any activity or exercise during the fasting period. Therefore, the overall estimate average and the range values (in parentheses) are provided.

Table 14.11 Estimated Proportional Contribution of Blood Glucose from Glycogen Stores and Gluconeogenesis

Hrs of Fast	% Glycogen	% Lactate	% Glycerol	% Amino Acids	Combined
8 hours	70% (40-70%)	15% (7-18%)	10% (5-11%)	5% (3-5%)	30%
12 hours	45% (18-46%)	27% (13-32%)	18% (9-20%)	9% (5-9%)	55%
20 hours	35% (14-36%)	32% (15-38%)	21% (11-24%)	11% (6-11%)	65%
40 hours	15% (6-16%)	42% (20-50%)	28% (15-32%)	14% (8-14%)	85%

After 8 hours of fasting, glycogen from the liver and skeletal muscles is contributing 70% of glucose and the combined gluconeogenic precursors are contributing 30%. The author has then calculated the proportions for the 12, 24 and 40 hour fasting states from the combined proportions of the 8 hour fast as shown in parentheses. After 14 hours fasting, glycogen contribution falls to 45% of glucose and the combined gluconeogenic precursors are contributing 55%.

Many researchers estimate this as a more or less equal 50% share of glucose production. After 24 hours, the glycogen contribution again falls to 35% with 65% from gluconeogenic precursors, then after 40 hours the glycogen levels have depleted significantly contributing just 15% with gluconeogenesis providing as much as 85% of glucose. Amino acid contribution to glucose synthesis ranges from around 5 to 14% by 40 hours. This indicates that lactate provides the majority from 15-50% with significant amounts from glycerol from around 10-32% of glucose.

Other studies report that glycerol is the dominant precursor for glucose synthesis as the fast progresses, at nearer 50% after 20 hours with lactate at 20%. and that amino acids for gluconeogenesis is not significant ranging from less than 5% to 10% after overnight fasts. This makes sense and seems to correspond with the body's second Starvation Priority during starvation or food deprivation, i.e., to preserve protein, reduce proteolytic catabolism and retain protein for structural duties. This also aligns with the abundant availability of glycerol and lactate within the significant quantity of triacylglycerol in adipose tissue stores and lactate in the skeletal muscles and RBCs.

Some researchers have postulated that endogenous amino acid gluconeogenic contribution may be higher, particularly on extensive carbohydrate deprivation and this might well be the case but it also seems logical that the body will only call on significant amounts of the structural protein reserve as the principal contributor in the most extreme of emergency, such as prolonged starvation but in that instance, ketogenesis also steps up to supply the brain.

See the relative quantities of ketogenesis and gluconeogenesis after 2 weeks starvation in Figure 14.10. More research is needed. It is also necessary to consider both direct and indirect contribution to gluconeogenesis, where for example, lactate is recycled, enabling higher quantities of lactate and therefore higher contribution, whereas glycerol is not recycled.

Also, glycerol also contributes carbons to glucose directly in the liver and also by converting to lactate and it is probable that glycerol is ultimately overall the net highest true gluconeogenic contributor to gluconeogenesis.

Lactate is effectively recycling existing glucose carbon skeletons and glucose itself is a major contributor to many gluconeogenic precursors via the Cori cycle, providing large proportions of carbon atoms for lactate, alanine and glutamine. Alanine also contributes carbons to lactate via pyruvate. Glucose produced by gluconeogenic precursors is also stored in liver glycogen by glycogenesis for subsequent glycogenolysis release as glucose as required and estimates are in the order of 30% stored as glycogen and 70% more rapidly released as glucose into circulation. Refer to figure 14.11 to see how each process, substrate and hormone varies in the fasted state as the fasting period extended.

In summary, there is still much research ongoing and much to be discovered, regarding the precise contributions of the various precursors to gluconeogenesis, however, the general consensus on current knowledge and in keeping with how the body is set up metabolically, is that during each overnight fast, glycogenolysis is the principal contributor to glucose released from glycogen via glycogenolysis, ranging from around 15-70% of glucose from long term fasting to overnight fasting.

Then, under most conditions and most of the time, glycerol (from stored triacylglycerol in stored fat) and lactate (from pyruvate, RBCs and skeletal muscle) are the principle gluconeogenic contributors, combining to around 40-70% of net glucose synthesised.

Less significant, but important contribution of around 5-14% is provided from amino acids and mainly endogenous alanine and glutamine, from catabolised (transamination) protein breakdown, with gluconeogenesis from all substrate precursors increasing with extended starvation. These highly elaborate mechanisms ensure glucose levels are correctly maintained under the varying conditions encountered. We will now review the main gluconeogenic substrate precursors.

14.19 Lactate and the Cori Cycle

The Cori cycle is a most important system used by the many cells and tissues throughout the body and has various functions, enabling cells to obtain energy anaerobically, maintain appropriate levels of lactate in the blood and to resynthesise glucose. Lactate is formed from within active skeletal muscle, where the synthesis and release of lactate generates ATP for the muscle at most times, supplying approximately 50% of muscle energy needs, rising to 80% in strenuous demand, including some anaerobic degradation but also aerobic degradation.

Table 14.12   A Summary of The Cori Cycle

What is it?	Conversion of lactate (from anaerobic glycolysis in muscle) to glucose
What is its Principal Role?	To provide energy to muscles during exercise & the removal of lactate
Why does it occur	A response to the muscles requirement for energy
When does it occur?	Between meals and during fasting 8 hours after eating
Where does it occur?	Primarily in the liver, muscles & in the RBCs & lymph immune cells
How is it stimulated?	A drop in circulating blood glucose leading to a rise in adrenaline
How is it inhibited?	A rise in insulin levels a decrease in glucagon and eating
What is needed to action it	Hexokinase, phosphofructokinase, pyruvate kinase
What are the start reactants?	Pyruvate to lactate back to glucose
What is the end product?	Glucose to ATP, NADH

The Cori cycle (also known as the Glucose-Lactate cycle) was discovered by Gerti Cori and her husband Ferdinand Cori, who, along with Argentinean physiologist Bernardo Houssay, were awarded the Nobel Prize in science in 1947. The cycle, shown in Figure 14.8, is a pathway in skeletal muscles in which lactate, accumulated in the muscle when the rate of glycolysis exceeds the rate of oxidative metabolism, is transported to the liver to be converted to pyruvate and glucose by gluconeogenesis. Under these conditions, pyruvate is produced from the break-down of muscle glycogen via glycogenolysis and anaerobic glycolysis of glucose.

However, the Alanine cycle, using alanine for glucose synthesis, is more likely to be used in aerobic conditions. The Cori cycle, using lactate for glucose synthesis, is more prevalent during lower oxygen or more anaerobic conditions. In these conditions, anaerobic fermentation to lactate occurs via the enzyme lactate dehydrogenase, and NAD^+ is regenerated to enable further glycolysis. The lactate is then transferred via the blood to the liver.

The glucose produced in the liver is then returned to the muscle via the bloodstream and from there three pathways are available, firstly, to be recycled as lactate if low oxygen conditions continue, or secondly, to be fully oxidised when conditions improve via the TCA cycle and OXPHOS, or finally, to be stored as glycogen via glycogenesis, if muscle activity has ceased.

The skeletal muscle would not be able to sustain the net cost of 4 ATP and 2 GTP during sustained intensity and the process is instead managed by the liver. For many years, it was the general consensus that lactate is a waste product (lactic acid) only occurring in anaerobic conditions, causing for example, muscle cramps and fatigue.

This has recently been proved not to be the case and pyruvate and lactate are easily interchangeable, depending on metabolic need.

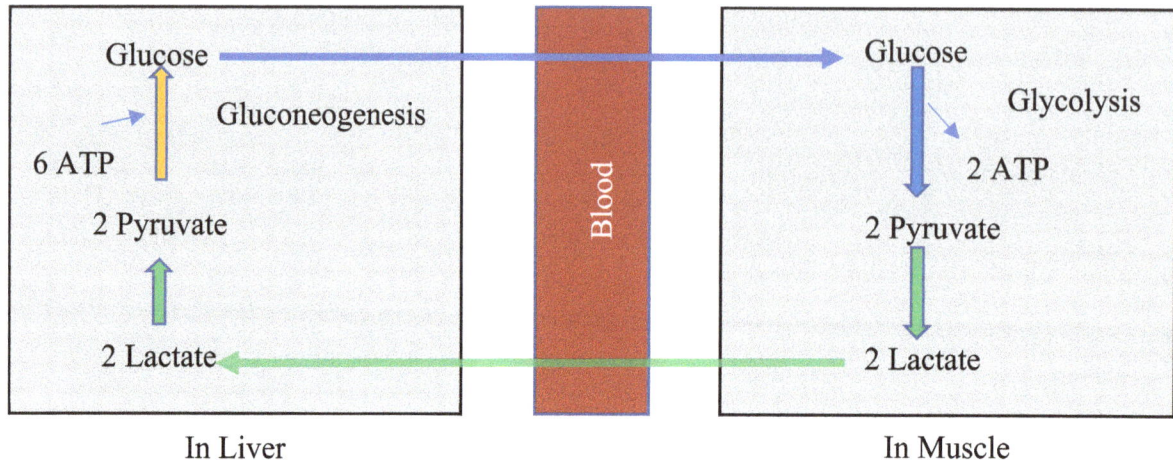

Figure 14.8 Overview of the pathway of the Cori cycle Anaerobic Production of Lactate

The lactate molecule is similar to the pyruvate molecule but with the addition of 2 hydrogen atoms and lactate is thus technically, a positively charged acid. It is currently hypothesised that lactate is in fact the dominant molecule of the two. Lactate also acts as a buffer for the body, holding on to potentially damaging protons.

Lactate is easily converted back to pyruvate in the presence of oxygen by adding water. A significant proportion of the glucose delivered to the skeletal muscles, adipose tissue and RBCs is recycled as lactate.

This lactate then returns to the liver as a substrate for gluconeogenesis. Lactate is also provided by those tissues using glucose fully anaerobically, such as the RBCs and kidney medulla. Therefore, as we have seen, lactate contributes to glucose recycling as it is derived from glucose. This is effectively recycling the existing glucose carbons, whereas other gluconeogenic substrates provide new carbon skeletons from fats or protein.

Lactate formed in contracting muscle and RBCs can also be used by other tissues, including the heart cardiac muscle and slow-twitch skeletal muscle. In these tissues, lactate reverts back to pyruvate and runs through the TCA cycle and OXPHOS to generate ATP.

Once again, we see that this use of lactate, rather than glucose, is yet another glucose sparing process available to the body, allowing more glucose to remain in circulation and a further example of how the body values glucose. Gluconeogenesis from one mole of lactate to glucose, then full oxidation of the glucose to CO_2, yields a net total of 24 ATP (Table 14.15).

14.20 Glycerol

Glycerol is one of the main precursors of gluconeogenic synthesis of glucose and along with the 3 fatty acids, is a component of each triacylglycerol (TAG) molecule, released from adipose tissue as a result of lipolysis and hydrolysis into the component parts. It is also used in endogenous lipogenesis to reassemble TAGs.

Glycerol can be used for energy by the liver and other tissues and is very useful to the body in providing a source of energy because it is relatively easily converted into glucose as it enters stage 7, high up in the gluconeogenic pathway at the dihydroxyacetone phosphate stage (Table 14.10 and Figure 14.7). Glycerol can be quickly incorporated into the glycolytic pathway and unlike fatty acids (not formed into ketones), it can also be used to supply the brain with energy. Glycerol is mainly metabolised and absorbed by the liver where it is phosphorylated to glycerol-3-phosphate.

It is then oxidised to dihydroxyacetone phosphate, which is isomerised by glyceraldehyde-3-phosphate dehydrogenase to glyceraldehyde-3-phosphate, then either oxidation, via the TCA cycle or to form glucose via gluconeogenesis. Glyceraldehyde-3-phosphate dehydrogenase can catalyse reactions in both glycolytic and gluconeogenic pathways.

Many tissues are capable of oxidising fatty acids by beta (β)-oxidation. Triacylglycerol hydrolysis in adipocytes is converted to glycerol and fatty acids. The pathway for glucose synthesis from glycerol is as follows:

> Glycerol to glycerol-3-phosphate to dihydroxyacetone phosphate to glyceraldehyde-3-phosphate to fructose 1,6 biphosphate to fructose-6-phosphate to glucose-6-phosphate to form glucose.

Gluconeogenesis from one mole of glycerol to glucose then full oxidation of the glucose to CO_2 yields a net total of 27 ATP (Table 14.15).

We have seen that glycerol based synthesis of glucose is estimated at approximately of 25% total glucose production overnight and this may increase to up to 44-50% during prolonged fasting.

14.21 Amino Acids and The Alanine Cycle

The amino acid alanine, formed from muscle, is often referred to as the major amino acid precursor of gluconeogenesis in the liver along with some glutamine from the kidneys when required. Eighteen of the twenty amino acids can in fact contribute to gluconeogenesis although alanine and glutamine are by far the most significant gluconeogenic contributors.

Alanine also constitutes a high percentage of the structure of all amino acids in most proteins, such that it is a relatively easy step to convert alanine to other amino acids and vice versa, for example, valine and leucine into alanine. In times of strenuous activity or low blood sugar in skeletal muscle cells, pyruvate from glycolysis is diverted away from forming acetyl-CoA for oxidation in the TCA cycle and instead, amino acids are broken down for energy and converted to alanine.

Figure 14.9 illustrates the pathways of the Alanine cycle where alanine is formed from other amino acids in the skeletal muscles, then transported in the blood to the liver and converted to glucose with the cleaving of the amine group and loss of the nitrogen via the urea cycle. The Cori cycle (Section 14.19) is also shown and both pathways occur at different times depending on muscle activity and energy status of the cells.

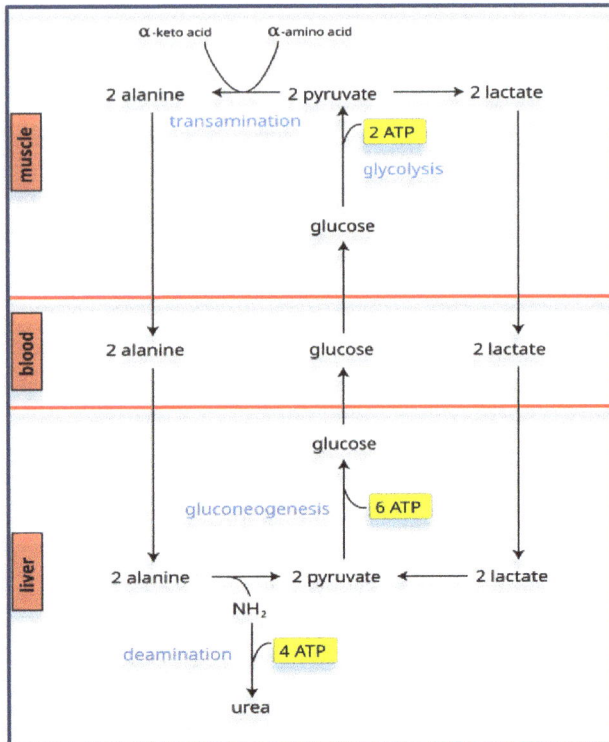

Figure 14.9 The Cori & Alanine Cycle: Source by kind permission of Arya Bima edited by wrekadora.svg 2011 Wikipedia Creative Commons Attribution-Share Alike 3.0 Unreported

This is achieved by transamination of nitrogen from glutamate to alpha ketoglutarate using the enzyme alanine transaminase. The alanine is then released from the muscle into the blood and taken up by the liver for conversion back to pyruvate, then to glucose by gluconeogenesis.

This liver glucose can then be reabsorbed by skeletal muscles for continuing energy. The nitrogen amine group removed from the alanine in the liver is then converted to urea via deamination, for excretion by the kidney i.e., the urea cycle.

The process of transferring alanine to the liver and glucose to the skeletal muscle is called the Glucose-Alanine cycle (also the Alanine cycle or the Cahill cycle). Figure 14.9 also shows the Cori cycle, (see also Figure 14.8) for lactate recycling to glucose.

The Alanine cycle, using alanine for glucose synthesis is more likely to be used in aerobic conditions, whereas the Cori cycle, using lactate for glucose synthesis is more prevalent during lower oxygen or more anaerobic conditions. The Alanine cycle performs a vital role of transporting waste nitrogen from tissues back to the liver for safe disposal by the urea cycle and excretion by the kidneys.

The total amount of ATP required in the Alanine cycle (6 ATP using alanine) is more than the Cori Cycle (4 ATP using lactate) because the Alanine cycle requires the removal of nitrogen then urea utilising a further 4 ATP for this additional process, meaning a total of 10 ATP are required for gluconeogenesis using alanine. In contrast with the Cori cycle, pyruvate is not reduced to lactate by NADH in the Alanine cycle conserving NADH and therefore more electrons can proceed to the TCA cycle and OXPHOS.

For amino acids that are catabolised in the TCA cycle as TCA cycle intermediates, thermogenesis ranges from 1 to 7 ATP per mol of amino acid and increased oxidation of metabolic substrates leading to increased thermogenic heat production.

We saw in table 14.10 and Figure 14.8 that for each glucose molecule synthesised from 2 moles of pyruvate, an additional 4 high-phosphoryl-transfer potential molecules plus 2 GTP are consumed i.e., 6 ATP.

For alanine as the substrate, an additional 4 ATP are required for the disposal of nitrogen from the urea cycle. Table 14.15 shows the estimated ATP generated from various fuel substrates.

The total net energy generated from amino acids, oxidised via the TCA cycle through to OXPHOS, (for example alanine), is approximately 20 ATP/mole, which is broadly similar to that available from ketones (c21.5 ATP/mole) and as expected, lower than glucose (30 ATP/mole) or fatty acids (106 ATP/mole). (Table 14.15 and Section 15.13).

Hence the relatively high thermic effect, or thermogenesis of protein. Gluconeogenesis is instead, heavily ATP dependent and there is a considerable bioenergetic cost of using amino acids from protein as a source of energy, compared with glucose, lactate, fatty acids, glycerol and glucose. Recall, its role is to help maintain blood glucose levels. The body stores approximately 10 kg of protein or 40,000 kcal (Table 14.3) but the vast majority of this is needed for cellular and structural requirements and a pool of around 5 kg is theoretically, potentially available but in fact only sparingly used as an energy source to synthesise glucose for the brain or in high intensity needs in muscles and other tissues reliant on glucose when carbohydrate levels fall, or during sleep or prolonged starvation.

As discussed in Section 14.18, the contribution of the oxidation of all available amino acids to total body energy requirements in these pathways is normally around of 5% of total energy expenditure but combined amino acid glucogenic contribution in more extreme conditions, such as carbohydrate deprivation, low glucose and depletion of glycogen reserves, can increase to around 14% or more of the total and in particular alanine, rising from 2 to 10% and glutamine, rising from 2 to 4% during prolonged fasting. In long-term fasting or starvation conditions, gluconeogenesis must therefore oxidise amino acids to form glucose via oxaloacetate and pyruvate intermediates, although as we have seen, this is bioenergetically expensive and a relatively high, thermogenic bioenergetic process compared with glucose or fatty acid beta (β)-oxidation via the TCA cycle and OXPHOS.

In times of carbohydrate, glycogen and glucose shortage and during the first few days of starvation, it is also possible to generate pyruvate, oxaloacetate and 5 other TCA cycle intermediates from increased use of glucogenic amino acids to synthesise i.e., preserve glucose via anaplerosis, i.e., filling in or 'repleting missing elements' in the TCA cycle. This vital gluconeogenic process, in tandem with glycerol and lactate, therefore ensures sufficient glucose is always available to the brain, CNS, RBCs and other dependent tissues.

We will now look at how amino acids are broken down to synthesise other molecules in transamination and cleave the nitrogen in deamination (Figures 14.4 and 14.9). Firstly, these amino acids are broken down in the liver where alpha amino group nitrogen is removed via the urea cycle, converting the remaining carbon skeletons into pyruvate and other TCA cycle intermediate compounds (shown in Table 14.13).

Cataplerotic reactions (or the process of cataplerosis) are the reactions that remove intermediate metabolites in the TCA cycle, for example utilising oxaloacetate (OAA) for gluconeogenesis or to synthesise aspartate, which means that there is less OAA to condense with acetyl-CoA. Another example is the use of ketogenic amino acids for the formation of ketones or fatty acids. Anaplerotic reactions (or the process of anaplerosis) are the opposite and are those reactions which replenish the TCA cycle intermediates that have been extracted for biosynthesis and in the first example provided, pyruvate carboxylase could convert pyruvate to OAA to replenish the OAA. Anaplerosis and cataplerosis are reciprocal reactions involved in the maintenance and function of the TCA cycle in the ultimate disposal of all metabolic intermediates.

There are a total of 9 intermediates (including cis-aconitate) in the TCA cycle. There are two co substrates of the TCA cycle, pyruvate and acetyl-CoA and also acetoacetyl CoA for the production of ketones and fatty acids in lipogenesis. The substantial proportion of acetyl-CoA comes from pyruvate, derived from carbohydrate and glucose. Although appropriate and sufficient levels of pyruvate from exogenous carbohydrate are key in driving the efficiency of the TCA cycle, pyruvate is not essential to turn the TCA cycle, as fatty acids and amino acids can also provide the acetyl-CoA. However, acetyl-CoA is essential for the TCA cycle, feeding in the 2 carbon atoms to condense OAA to citrate.

Table 14.13 provides a summary of the various fates of the 20 amino acids. Column 1 lists the amino acids. The first fourteen listed (blue font) are glucogenic and can produce glucose or glycogen. Five amino acids (black font), phenylalanine, tyrosine, isoleucine, threonine and tryptophan are both glucogenic and ketogenic and 2 amino acids (red font), leucine and lysine are solely ketogenic. These last 7 glucogenic/ketogenic and ketogenic amino acids (black and red font) can go on to produce either ketone bodies or fatty acids. Table 14.13 column 2 lists the 7 intermediate molecules resulting from the breakdown of 20 amino acids via protein catabolism. These comprise the 3 TCA cycle co substrates, pyruvate and acetyl-CoA and acetoacetyl CoA, plus 4 of the TCA cycle intermediates; oxaloacetate, succinyl CoA, fumarate, alpha-ketoglutarate. Column 4 'Energy Substrate Synthesised' shows production of glucose or glycogen and ketones or fatty acids. Three of the TCA intermediates, succinyl CoA, fumarate, alpha-ketoglutarate can be converted to OAA. The OAA generated from pyruvate or these other 3 TCA cycle intermediates can either be converted to glucose, via gluconeogenesis (via phosphoenolpyruvate and PEPK, Figure 14.7), or continue to contribute the OAA initiating role in the TCA cycle, or be used as a cataplerotic agent to synthesise other amino acids.

The products of amino acid degradation can therefore be summarized as:

Gluconeogenic	- Glucose and Glycogen
Ketogenic	- Ketone Bodies or Fatty acids
Cataplerotic	- Synthesis of other amino acids
Anaplerotic	- Replacing TCA cycle intermediates & oxidised

The selected route of these optional pathways depends on the energy, glucose and substrate level, controlled by hormonal influence and regulatory enzymes within each cell. For example, if the cell needs energy, insulin levels in the blood will be low, less OAA will be converted to glucose and more OAA will be available to the TCA cycle for energy production. If the body is in a fasting state and requires glucose, OAA moves towards gluconeogenesis. If there are low levels of glucose but high levels of acetyl-CoA from oxidation of fatty acids, this can trigger the need for more energy generation again, leading to more OAA for the TCA cycle and OXPHOS.

On the other hand, if there is plentiful supply of food substrates and insulin is high and energy and pyruvate and ATP are in abundance, the pyruvate can be converted to OAA via pyruvate carboxylase for gluconeogenesis. If there is sufficient glucose, this new synthsesised glucose can also be directed for glycogen synthesis in the liver and muscles and the synthesis of amino acids alanine, aspartate and glutamine. When in excess, amino acids can-not be stored as protein and must be broken down in the liver. From there the nitrogen is cleaved off in the urea cycle and excreted by the kidneys and the carbon skeletons are converted to glucose, glycogen, ketones, fatty acids or used as anaplerotic intermediate TCA cycle intermediate molecules.

Table 14.13   Metabolic Fates of Amino Acids in the TCA Cycle

Amino Acid	Intermediate Molecule	Metabolic Pathway	Energy Substrate Synthesised
Alanine			
Glycine	Pyruvate		
Serine			
Cysteine			
Aspartate	Oxaloacetate		
Asparagine			
Valine		Glucogenic Amino Acids	Glucose or Glycogen
Methionine	Succinyl CoA		
Aspartate	Fumarate		
Glutamine			
Glutamate			
Proline	Alpha- ketoglutarate		
Arginine			
Histidine			
Tyrosine	Acetoacetyl CoA		Ketones or Fatty Acids
	Fumarate		Glucose or Glycogen
Phenylalanine	Fumarate		Glucose or Glycogen
	Acetoacetyl CoA		Ketones or Fatty Acids
Tryptophan	Acetyl-CoA, Acetoacetyl CoA	Glucogenic & Ketogenic Amino Acids	Ketones or Fatty Acids
	Pyruvate		Glucose or Glycogen
Isoleucine	Acetyl-CoA		Ketones or Fatty Acids
	Succinyl CoA		Glucose or Glycogen
Threonine	Acetyl-CoA		Ketones or Fatty Acids
	Pyruvate		Glucose or Glycogen
Leucine	Acetyl-CoA, Acetoacetyl CoA	Ketogenic Amino Acids	Ketones & Fatty Acids
Lysine	Acetoacetyl CoA		

Note, threonine can be used to synthesise glucose by pyruvate or be utilised as a ketogenic amino acid for synthesis of ketones via acetyl-CoA. Some scientists report that threonine can also be converted to succinyl CoA and used to synthesise glucose via gluconeogenesis, but this third pathway for threonine is not included here.

The gluconeogenesis pathway converts pyruvate into glucose. Pyruvate can also be converted back to acetyl-Co A (using the enzyme pyruvate dehydrogenase PDH) to enter the TCA cycle. However, acetyl-CoA can-not be converted back to pyruvate as the step is irreversible and hence, fatty acids (unlike lactate, glycerol and amino acids) can-not be converted to glucose via gluconeogenesis.

In Figure 14.9 we saw how alanine is released from skeletal muscle and taken up by the liver for transamination back to pyruvate in the Alanine cycle. Pyruvate can then be used for gluconeogenesis. Alanine and to a lesser extent glutamine are the dominant amino acids contributing to gluconeogenesis.

Approximately 80% of nitrogen in the amine groups is lost in the urine as urea under normal (fed state) conditions. This falls during fasting but rises in acidotic conditions, for example too much meat intake.

There have been recent challenges to the consensus that alanine is the primary glucogenic amino acid with the liver holding a central position in gluconeogenesis.

However, there are reports demonstrating that glutamine and the kidneys have much more significance in glucose homeostasis than has been widely accepted hitherto and suggesting that because the main substrate for alanine synthesis is pyruvate formed during glycolysis, alanine should instead be considered as a substance that ensures glucose recycling between the liver and muscles via the "alanine cycle," rather than a substrate for gluconeogenesis. The roles of alanine and the liver on one side and glutamine and kidneys on the other in maintaining glucose homeostasis are not yet fully understood. [66]

It is important to examine the drawbacks and potential consequences of excessive protein breakdown to feed into gluconeogenic pathways. Breakdown of muscle protein yields amino acids causing a loss of electrolytes and around 0.75 litres of water. One kilogram of skeletal muscle tissue contains approximately 150 g of protein. Around half of this, or 75 g of glucose, can be synthesised via gluconeogenesis.

However, muscle also contains potassium, magnesium and phosphate and the kidneys must excrete this. Proteolysis of amino acids releases ammonia which has to be converted to urea requiring more energy. The urea also increases the osmotic load on the kidneys. Approximately 2 litres of water are required to excrete these substances in addition to the 0.75 litres from protein breakdown, a total of 2.75 litres and increased urination and potential dehydration results.

Starvation, long term fasting or deliberate carbohydrate deprivation increases plasma levels of fatty acids and acetyl-CoA to provide energy to the body and power the gluconeogenic process, increasing ketone bodies causes acidaemia. To counter severe acidosis, more ammonia is secreted into the renal tubules.

Excretion of ammonia avoids energy consumption due to urea synthesis, but significantly increases osmotic load of excreted nitrogen excreted, causing even greater urination. and significant depletion of blood volume may occur. High protein diets should be avoided.

14.22   Propionate

Another small contributing substrate for gluconeogenesis is provided by gut microbiota as part of the fermentation of non-digestible substrates (dietary fibres and endogenous intestinal mucus) from dietary carbohydrates.

It is the major source of gluconeogenesis in ruminant mammals. This fibre supports the growth of specialised species that produce short chain fatty acids (SCFAs). Recall this was introduced in Section 10.15 of Chapter 10 on carbohydrates.

The major SCFAs produced are acetate and butyrate, which is the main energy source for human colonocytes, where it is oxidized to ketone bodies and which also activates intestinal gluconeogenesis and also propionate, which is transferred to the liver where it regulates liver gluconeogenesis.

The liver extracts in the order of 90% of propionate into the hepatic portal vein from the circulation. The propionate is converted to propionyl-CoA and enters the TCA-cycle at the level of succinyl-CoA which enters the tricarboxylic acid (TCA) cycle and is converted to oxaloacetate, the precursor of gluconeogenesis.

This leads to an elevation in the levels of oxaloacetate, which is then converted to glucose.

Around 70% of the acetate is taken up by the liver, where it is used not only as an energy source but also to synthesise cholesterol. The kidneys, heart, adipose tissue, and muscle, metabolize the rest of the acetate.

Of course, the SCFAs propionate and butyrate (and acetate) are derived from carbohydrates and are thus strictly not non-carbohydrate sources of energy.

However, they can only be released by other organisms in the gut and qualify at least in that regard.

Therefore, accessible carbohydrates and the fibre provided are beneficial for the well-being and growth of microorganisms and consequently for the host in this symbiotic relationship.

14.23   Nutritional Status: From Feeding to Fasting

In Sections 14.24 to 14.26, we will aim to explore the changing nutritional states from the fed or absorptive state, through to the post absorptive fasting period then finally to early and worsening starvation. The various metabolic controls and adjustments will be reviewed, along with the body's protective mechanisms and processes to preserve the status quo and adapt to conditions of plenty and deprivation to protect the individual.

It will be seen that the metabolic processes discussed in other Chapters are fully integrated and regulated to achieve the goal of the cells organs tissues and the whole body.

14.24 Energy Provision in The Absorptive (Fed) State

After a standard mixed meal, carbohydrates are the preferred substrate and most of the fat eaten is stored in adipose tissue. After approximately four hours, some fat is mobilised from adipose via lipolysis. In the absorptive state, also referred to as the fed state, excess liver glucose is converted to:-

a) Glycogen (storage of glucose carbon)
b) Blood Glucose for Homeostasis
c) Pyruvate to Acetyl-CoA then…Fatty Acids (storage of glucose carbon)
d) Pyruvate to Acetyl-CoA then…Cholesterol
e) Pyruvate to Acetyl-CoA then…Triacylglycerol
f) Pyruvate to Acetyl-CoA then…Phospholipids
g) Pyruvate to Acetyl-CoA then…TCA Cycle to OXPHOS to ATP
h) The Pentose Phosphate Pathway (PPP) to Ribose 5 -Phosphate to Nucleotide Synthesis
i) The Pentose Phosphate Pathway (PPP) to Nicotinamide adenine Dinucleotide Phosphate

Glucose to fatty acids via pyruvate and acetyl CoA in c) above, only occurs if energy intake continues to exceed energy expenditure (overfeeding) more especially in obese individuals.

These fatty acids can be stored in the adipose tissue. Fatty acids, removed from chylomicrons and from the VLDL particles by lipoprotein lipase, can also be stored in the adipose tissue. These fatty acids provide a ready source of energy substrate for most body tissues. During the post absorptive state, and the longer fasting state, some exogenous (dietary) glucose from the intestine by passes the liver and circulates to other tissues.

Both carbohydrate (glucose) and fat (free fatty acids) are oxidized at rest and both produce acetyl-CoA to provide the basic energy required for basal metabolic processes in skeletal muscle and various body tissues, such as the heart and kidney cortex.

The proportions vary depending on conditions at the time but may for example be in the order of 62% fat and 38% carbohydrate overall oxidised during rest.

Changes in fuel substrates also occur at rest even without any change in activity, governed by the availability of each substrate and energy levels in each cell. Increasing blood glucose increases the uptake and oxidation of carbohydrate in muscle, while decreasing (β)-oxidation of fat and with no significant change in the metabolic rate. During exercise, the requirement for energy increases several times compared with the resting rate, and the oxidation of both fat and carbohydrate are triggered simultaneously.

Thereafter, during exercise, the interaction between oxidation of the 2 substrates varies and depends on both cellular and environment i.e., activity level. At low intensity, moderate and prolonged steady exercise, up to 40% of VO_2 max, fat will remain the predominant substrate oxidised although gradually declining against carbohydrate, in terms of proportion as exercise intensity increases. Weight for weight, total fat (β)-oxidation during exercise, reaches a peak at around 65% of VO$_2$ max, when proportional fat oxidation is at around 50% fat and 50% carbohydrate oxidised.

Thereafter, the contribution from fat decreases at higher power outputs as carbohydrate becomes the main substrate and increase in exercise intensity produces a gradual progressive shift from fat to carbohydrate as fat (β)-oxidation diminishes, until 95% of VO$_2$ max when the more rapidly mobilizable glucose becomes the sole energy source of substrate for skeletal muscles and will dominate the proportion of substrate oxidation, providing there are sufficient glycogen stores.

Although, as a proportion ratio, carbohydrate is the dominant substrate oxidised at a higher rate than fat at high intensity exercise, the body also oxidises more total substrate, even if a higher proportion of the energy required is provided from carbohydrates, thus the total amount i.e., mass of fat oxidised will also be higher at higher intensity exercise (Table 6.1).

14.25   Energy Provision in The Post Absorptive State

During the overnight fast, fat is the major source of energy. Note, Figure 14.10. The post absorptive state, also referred to as the fasted state, is from 18 to 48 hours after feeding.

The shift to gluconeogenesis during long fasting is signalled by glucagon hormone released from the pancreas alpha cells and the glucocorticoid hormones, for example, cortisol, released from the adrenal cortex.

Both glucagon and cortisol respond to low glucose in the blood. During this low substrate period, the liver effectively governs and spoon feeds the glucose dependent tissues of the rest of the body with glucose and ketone bodies.

The liver increases gluconeogenesis and gradual depletion of glycogen. Glycerol from lipolysis and also lactate from anaerobic metabolism provide gluconeogenic contributions and also, after 24 hours fasting, more protein (gluconeogenic) amino acids, principally alanine and glutamate, are hydrolysed in muscle cells for glucose synthesis in the liver.

Only leucine and lysine can-not contribute (Table 14.13) because these are solely used ketogenic.

Ketogenic amino acids do serve a purpose, however, as they allow the brain, heart and skeletal muscle to adapt if the starvation is prolonged. Early fasting is accompanied by a large daily loss of nitrogen from muscle breakdown and hepatic gluconeogenesis and nitrogen excreted in the urine.

14.26 Energy Provision in The Starvation State

We have seen how the first metabolic Starvation Priority of nutrition deprivation and starvation, or any conditions (including during sleep) where glucose is in short supply, is to provide sufficient glucose to the brain, RBCs, kidney medulla and other tissues that are absolutely and heavily dependent on glucose as fuel substrate. This is met by gluconeogenesis then ketogenesis. The second Starvation Priority is to preserve protein and the body accomplishes this by shifting the gluconeogenesis of substrate used from amino acids to glycerol, lactate, fatty acids and ketone bodies. Prolonged fasting (>24 hours after feeding) represents a major metabolic shift in order to spare body vital protein such as antibodies, enzymes and haemoglobin.

We will now review the processes and metabolic shifts and adaptations to prolonged starvation. Note Figure 14.10. During starvation, after 24 hours without feeding, the objectives of both Starvation Priority 1 and Starvation Priority 2 are achieved as follows.

---

1) An increase of lipolysis of triacylglycerol, triggered by glucagon
2) A rise in the level of (β)-oxidation fatty acids and acetyl-CoA, causing a build-up of pyruvate.
3) A rise in liver & kidney gluconeogenesis to provide carbons for glucose synthesis from glycerol

---

The glycerol is derived from the lipolysis of triacylglycerol, together with lactate, derived from anaerobic glycolysis and also amino acids, mostly alanine, initially derived from pyruvate via transamination of skeletal muscle, (>24 hrs.).

The next phase is after 24-48 hours. Blood cortisol levels rises and the degradation of skeletal muscle by proteolysis begins to provide amino acid contribution to gluconeogenesis.

There is a significant rise in ketone bodies in the liver, and after 3 days, the brain begins to adapt and uses obtains around 30% of its energy from ketones. This serves to preserve glucose, either from the remaining glycogen stores or synthesized from gluconeogenesis, for the RBCs and other dependent tissues. After 7 days, proteolysis of the amino acids is reduced. The level of gluconeogenesis falls and is fueled and maintained solely by glycerol and lactate. Continued high levels of (β)-oxidation of fatty acids and accumulated levels of acetyl-CoA in the liver causes a saturation of the TCA cycle, leading to the continued increased formation of ketone bodies.

The brain utilisation of ketones rises to a maximum of 80% of its energy needs. Also at this time, from around 7 days, many other tissues of the body, including the heart, are using ketones for energy. Hence ketogenesis reduces gluconeogenesis and therefore, preserves glucose and spares loss of skeletal muscle, helping to meet both fundamental survival priorities of the body.

After approximately 2 weeks, to preserve and shunt the ketones exclusively to the brain and RBCs, these tissues begin to rely solely on (β)-oxidation of fatty acids for energy.

Beyond this period, the brain continues to use ketones as the major energy source and by doing so, continues to ensure that any remaining glucose is preserved and available to the RBCs and other wholly glucose dependent issues for as long as possible. For further detail over time from 2 hours, we will now review chronological events referring to Figure 14.10.

Gluconeogenesis starts at low levels 2 hours after feeding then rises significantly after 12 hours then peaks at 24 hours when glycogen is depleted. This is maintained for around 1 week, then falls to a steady state to synthesise and maintain blood glucose for vital tissues and spare protein

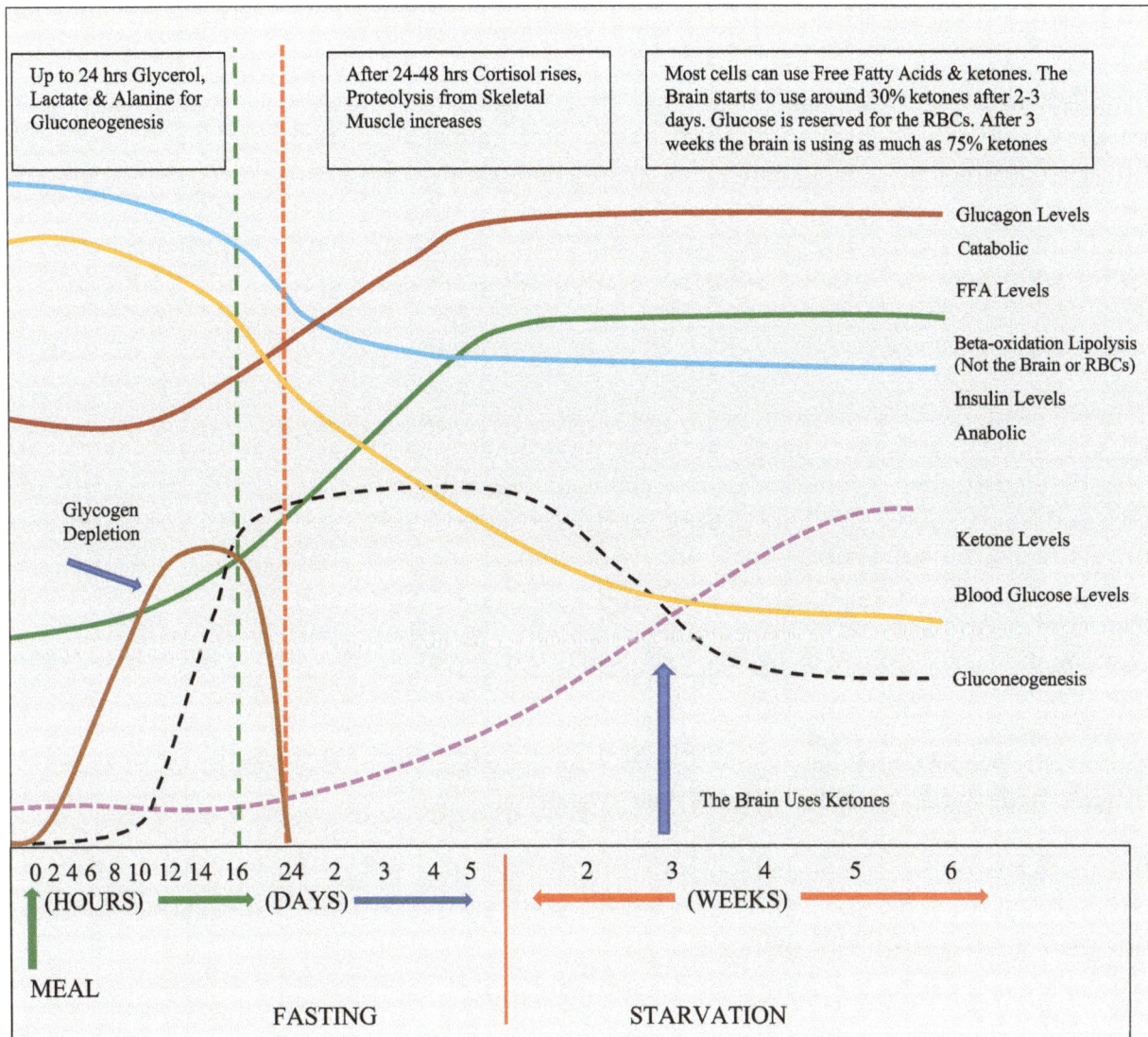

Figure 14.10 Metabolic Changes after feeding, fasting & starvation 6 week period.

Glucose levels fall sharply 4 hours after feeding, then fall steadily until around 10 days when gluconeogenesis reaches steady state to synthesise glucose. Amino acids used for gluconeogenesis are from transamination in the liver and not from structural or skeletal muscle. Fatty acids are mobilised via lipolysis around 12-14 hours after feeding. There is a gradual shift from gluconeogenesis to lipolysis. The fat stores become the major source of energy.

These continue to rise as an energy substrate via (β)-oxidation until around 3 weeks of starvation, then reaching a steady state in line with glucagon levels. Insulin levels fall gradually after feeding until around 10 days reaching a steady state. After 2 weeks of starvation, the brain principally uses ketones for energy to spare glucose for the RBCs. Ketones continue to rise until levels peak and stabilise at around 3-4 weeks. The fatty acids (FFA) are used for energy by the heart, liver and skeletal muscles. The brain may not be able to use FFA due to the blood brain barrier or for other reasons. However, fat breakdown also releases glycerol, which then becomes the major source of gluconeogenesis substrates for the brain.

Under these conditions, levels of glucagon rise and insulin fall. Both citrate and acetyl-CoA build up and glycolysis is effectively switched off or minimized to preserve glucose. TGAs are mobilised in adipose tissue and gluconeogenesis is stepped up in the liver. Fatty acids enter muscle cells freely but glucose entry is markedly reduced due to the drop in insulin. Consequently, at that stage, muscle relies entirely on fatty acids for fuel.

The brain and skeletal muscle also adapt to use ketones for energy. Eventually, the use of TCA cycle intermediates (pyruvate, succinyl CoA, oxoglutarate, malate fumarate etc) depletes the supply of oxalacetate. Low levels of oxaloacetate, plus a rapid rise in acetyl-CoA, due to fatty acid catabolism causes acetyl-CoA to build up and accumulate.

This builds up favour's formation of aceto-acetyl-CoA plus ketone bodies. During starvation, blood ketone levels increase sharply to around 5.5 mmol/L after 10 days, then peaking at 7 mmol/L after 30 days. Ketone bodies concentration rises in the blood. i.e., ketosis as the fuels is exported from the liver. The liver can-not use ketone bodies and they travel via the bloodstream to the muscle, brain and heart.

Here the ketone bodies are oxidised instead of glucose. As long as the ketone body high concentration is maintained by hepatic FFA beta (β)-oxidation, the need for glucose via gluconeogenesis is reduced.

This is why people who follow a low-carbohydrate or carbohydrate deprived diet, often see an increase in the production of ketones in their system.

Excess dietary protein and fat and low carbohydrate can lead to potential cellular and health issues. It is wiser to provide sufficient carbohydrates in order to fuel the TCA cycle and burn both carbohydrate and fat efficiently.

During prolonged starvation (ketosis), after 3 weeks, the brain will be using approximately 75% of its energy requirements from ketones, such as acetate and beta hydroxybutyrate and the remaining 25% from glucose synthesised in the liver from other non-carbohydrate precursors such as glycerol via gluconeogenesis.

Fatty acids are released by adipose tissues and converted to acetyl-CoA units by the liver. The liver then exports these as acetoacetate (a ketone body). High levels of acetoacetate in the blood signify an abundance of acetyl units. This leads to a decrease in the rate of lipolysis (fat breakdown and mobilisation) in the adipose tissue.

In addition to glucose sparing, ketogenesis also spares valuable, structural protein that has been derived from amino acid breakdown in gluconeogenesis, as survival for most animals depends on the ability to use muscle to move rapidly. Survival length of time depends on the quantity of fat stored and is normally around 90 days.

A highly obese person could theoretically survive for one year but prolonged unchecked ketosis can be a dangerous status for the body. Once fat reserves are gone, the body starts to use essential protein leading to muscle atrophy, loss of liver function, chest infection, impaired immunity, irreversible damage and ultimately, death would ensue.

The foregoing discussion illustrates the hormonal and metabolic control of glucose, fatty acids and ketone bodies and how the body works, using anabolic insulin and catabolic glucagon (amongst others) to preserve glucose via gluconeogenesis to provide the brain with alternative metabolites of ketone bodies via ketogenesis. These integrated metabolic survival mechanisms preserve glucose in the blood for the brain, CNS and any cells dependent on glucose for energy. Therefore, insulin resistance can be seen as a natural metabolic evolutionary process for the survival of animals during times of shortage.

Insulin resistance in modern western societies has now taken on the role of 'the bad man' due to over nutrition, diabetes and other health issues. Energy intake >Energy burnt is too high in most people and blood level free fatty acids (FFA) are raised. We have looked at metabolic changes from feeding to fasting then progressively to starvation.

The most pressing thing is that the opposite of this is relevant for many societies. The human body is simply not evolved for conditions that provide the opportunity for unlimited and continuous ad libitum over feeding.

If individuals are deliberately attempting to extend gluconeogenesis and increase metabolism and thermogenesis in order to lose weight or because, for some reason, they consider all carbohydrates are unhealthy, or they wish to gain muscle mass, then replacing carbohydrate with high quantities of dietary protein, such as meat and fat, may have health consequences.

Attempting to utilise excessive protein or fat for energy will produce by-products including toxic amounts of ammonia. Excessive quantities of methionine can undergo undesirable, erroneous methylation reactions and produce other toxic compounds.

Excess dietary protein can lead to potential cellular and health issues. It is wiser to provide sufficient carbohydrates in order to fuel the TCA cycle and burn both carbohydrate and fat efficiently and appropriately. Excess dietary fat has many other drawbacks to good health, especially for the cardiovascular system and diabetes. This is known to shorten lifespan.

14.27   Energy Substrates of Specific Body Tissues

The body uses both glucose and fat as a source of energy 24 hours per day. The ratio of energy used from these two macronutrients is dependent on cell energy, substrate availability and activity. The proportion of fat increases with moderate exercise then prolonged intensive exercise and starvation. The proportion of glucose increases immediately after feeding and also on short bursts of high intensity exercise.

No tissue is solely dependent on fatty acids for energy, though heart muscle (myocardium) and the hepatocytes in the liver and the kidney cortex all prefer fatty acids for most of their energy. Adipose tissue oxidises both fatty acids and glucose with more glucose oxidised during the absorptive stage and more fatty acids oxidised at the post absorptive stage.

Skeletal muscle cells use all three sources of glucose, fatty acids and amino acids and mostly fatty acids during rest and moderate intensity exercise. All three macronutrient energy substrates can be oxidised through the common pathway of the tricarboxylic (TCA) cycle to release energy. Because fatty acids within triacylglycerol's (TAGs) eaten in the diet are not used for energy, they are firstly taken up by the adipocytes and stored as TAGs.

Storing fat takes up very little energy and is completed with little effort and energy. The fatty acid profile of storage fat resembles the fatty acid profile of the fat in the diet and not that of de novo lipogenesis (DNL) from carbohydrate dietary sources. Fat eaten is fat stored in fat tissues.

As referred to Section 14.40, DNL conversion of glucose or even amino acids to fat is highly endergonic and the conversion of glucose into intracellular lipids via this process, is an inefficient and costly process requiring considerable metabolic energy and requires a number of additional proteins and enzymes that catalyse a number of complex reactions.

It is only significant in people with serious diseases such as NAFLD. In contrast, to store glucose and amino acids, they must first be converted to acetyl-CoA and then the acetyl-CoA units must be synthesised into fatty acids.

Both processes require more energy. If an energy imbalance (excess) continues, the adipocytes are enlarged. New adipocytes can be produced to accept the additional TAG, however, the number of fat cells increases most rapidly in late childhood and early puberty, whenever a positive energy balance exists, called Hyperplasia which is an enlargement of an organ or tissue. Later in life, whenever the fat cells become enlarged from overfeeding, the number of fat cells can increase. Obese people have more and larger adipocyte fat cells than slim people. If body fat is lost, the number of fat cells does not decrease, the cells just diminish in size. In dieting, both body fat and body muscle may be lost. The goal of losing body weight (body fat) over an extended period requires a negative energy balance.

The size and period of caloric deficit are the key drivers and influence adherence and long term sustainability. After a meal of approximately 90 g of glucose, approximately 50% of ingested glucose is converted to glycogen in the liver and skeletal muscle. Relative tissue absorption is shown in Table 14.14.

Table 14.14 Relative tissue absorption of glucose in the post absorptive fed state

Tissue	Glucose (g)	Glucose (%)
Skeletal Muscle for oxidised for energy	24 g	27%
Skeletal Muscle Stored as Glycogen	24 g	27%
The Brain	15 g	16%
The Kidneys (as lactate from the RBCs)	8 g	9%
Adipose Tissue (for conversion to TAG)	2 g	2%
Liver Stored as Glycogen	17 g	19%
Total	90 g	100%

Table 14.15 lists the amount of total net ATP yield from the principal energy substrates. The figures are calculated based on the full oxidation of substrates, including the TCA cycle ETC and OXPHOS where applicable, plus any additional ATP required to synthesise glucose in order to by-pass normally irreversible reactions of glycolysis or to remove nitrogen due to amino acid derived gluconeogenesis.

A comparison of ATP mass per 100 g of substrate is also calculated for glucose, fatty acids (Table 14.15) and the ketone bodies. Ketones produce similar amounts of ATP per mole as amino acids which is expected as these substrates are produced under nutrient shortage, substrate stress or starvation conditions (See Tables 15.2 and Section 15.10 to 15.12).

Note that ketone bodies produce more ATP per gram than glucose but the red blood cells or the retina of the eyes can-not use them for energy, nor can the liver, as it lacks the enzyme beta ketoacyl-CoA transferase. The brain still requires a minimum of 30% glucose to survive without damage. Note how fat produces in the order of twice the mass of ATP compared with other substrates.

Table 14.15 Estimated Metabolic Energy Yield from Various Substrates in Humans

Energy Substrate Fuel	Metabolic Pathway Process	Est. ATP (Per mole)	ATP (kg/100 g)
Fatty Acid (16-Palmitate)	(β)-oxidation + TCA/+ETC	106 ATP	20.96
Glucose	Glycolysis+TCA+ETC	30 ATP	8.34
Glycerol	Gluconeogenesis+TCA+ETC	27 ATP	
Alcohol	Oxidation Acetaldehyde/Acetate/AcetylCoA+TCA+ETC	25 ATP	
Lactate	Gluconeogenesis/Cori Cycle+TCA+ETC	24 ATP	
Pyruvate	Gluconeogenesis+TCA+ETC	24 ATP	
Ketone Beta-hydroxybutyrate	Ketogenesis/AcetylCoA+TCA+ETC	22 ATP	10.35
Ketone Aceto-Acetate	Ketogenesis/AcetylCoA+TCA+ETC	20 ATP	9.31
Amino Acids (Alanine)	Gluconeogenesis/Alanine Cycle/Urea Cycle+TCA+ETC	20 ATP	
Glucose (RBCs)	Glycolysis Anaerobic Pyruvate & Lactate only	2 ATP	

TCA Tricarboxylic Acid Cycle. TCA+ETC refers to Electron Transport Chain & Oxidative Phosphorylation and full oxidation to CO_2

Most cell types in the body possess mitochondria, an adequate supply of oxygen and access to all three main energy substrates; glucose, fatty acids and amino acids. However, certain specialised tissues rely on one particular, preferred sources of energy only in normal circumstances, unless other conditions dictate. Energy requirement of various key tissues and organs of the body are now reviewed.

14.28   Liver Energy Source

The liver requires a constant source of energy for biosynthetic functions and primarily uses fatty acid (β)-oxidation for energy. The liver derives up to 50% of its ATP released energy from (β)-oxidation of fatty acids. The energy generated may be used for gluconeogenesis, or the 75% of urea synthesised in the body in the urea cycle. During short-term fasting periods, the liver produces and releases glucose mainly through glycogenolysis. During prolonged fasting, glycogen is depleted, and hepatocytes synthesize glucose through gluconeogenesis using lactate, pyruvate, glycerol, and amino acids. When carbohydrates are plentiful and sufficient, the liver utilises glucose as a significant metabolic fuel and also converts a small amount of glucose into fatty acids. Hepatocytes also obtain fatty acids from the bloodstream, released from adipose tissue or absorbed from food digestion in the gastro intestinal tract. The urea cycle is important for the removal of ammonia (NH_3). The urea travels to the kidneys for excretion. Also, 25% of urea may be secreted from the blood into the intestinal lumen where it may be degraded by bacteria to yield the ammonium ion ($NH4^+$).

14.29   The Brain and CNS Energy Source

Glucose is the essential metabolic fuel for the brain. Normally, the brain uses glucose virtually exclusively for its energy needs. The brain can survive on ketones if necessary, during prolonged starvation but this is emergency status and not desirable. It is estimated that the adult human brain has 1 trillion cells, comprising 900 billion glial cells and 100 billion nerve cells. Hence, the brain has very high metabolism and the daily consumption of nerve cells is approximately 120g of glucose equivalent, which is an input of 420 kcal (1,760 kilojoules). This accounts for as much as 60% of glucose utilisation for the whole body and 20% of the total energy needs of the body in the resting state. The rate of blood flow to the brain is 750 ml/min/100g of tissue, compared with around 3 ml/min/100 g of tissue required for resting skeletal muscle and rising to 50ml/min/100g tissue during hard exercise. The brain and CNS normally therefore use only glucose to obtain energy depending on the supply of oxygen and glucose through blood.

The glucose transporter GLUT3 has special bidirectional characteristics that ensure satisfactory levels of glucose (euglycemia) amounts (2 to 4 mmol/L) reach the brain, regardless of saturation, during feeding or depletion, overnight, closely regulated; not too high and not too low.

If this were not the case, people would experience symptoms and fluctuations of hyperglycaemia and hypoglycaemia throughout the day. The brain also oxidises lactate where lactate acts as glucose sparing during strenuous exercise and lactate is present in large amounts. The brain also uses glycerol and the use of these other two substrates and ketones may help protect against glucose reactivity if the brain glucose levels become too low or too high. Unlike RBCs, the CNS cells are metabolically active and do have mitochondria. Thus, the CNS cells are able to fully oxidise glucose, generating greater amounts of ATP. Scientists now believe that it is not necessarily the blood-brain barrier that prevents the access of fatty acids and lipids to the cells of the CNS. There are several theories why the brain prefers glucose and not fatty acids for energy. One theory is that more damaging reactive oxygen species (ROS) are produced when metabolising fats. Another theory is that a substantial proportion of brain structure comprises fat and the preference of glucose would avoid potential pathways leading to degradation of brain tissue. Recall that the brain requires a minimum of 30% glucose to survive without damage.

Whatever the reason, from an evolutionary stand point, the brain and CNS cells normally rely solely on glucose as the fuel source. Under prolonged starvation or fasting, however, the brain and CNS can, if necessary, oxidise ketone bodies, provided by the liver as part of the continual metabolization of fatty acids and these can be converted to acetyl-CoA in the brain mitochondria and oxidised as fuel for ATP production through TCA and OXPHOS. Ketones become a major energy substrate for the brain after 2-3 weeks of starvation, principally to preserve glucose for RBCs and other tissues that rely solely on glucose (Figure 14.10).

14.30   Red Blood Cells (RBCs) Energy Source

There are approximately 25 trillion RBCs in the human body, or 84% of all the cells in the body at any one time. All of these cells are purely glycolytic and completely dependent on glucose for energy. They are, ultimately, totally dependent on glucose. The liver, brain, gluconeogenesis and ketogenesis, all have a key role to play in preserving glucose for the RBCs.
These cells are the most common cells that lack mitochondria (also lack nucleus and ribosomes) and although they are not highly metabolically active, they are very high in number and do rely in the continual uptake of glucose from the bloodstream for energy.

RBCs are unique compared with all other cells as they rely exclusively on anaerobic glycolysis for the generation of energy. They convert glucose to lactate anaerobically by lactic acid fermentation, producing a sufficient (2 molecules) of ATP for energy. Their energy needs are relatively low and anaerobic glycolysis of only 14-25 g/d of glucose is sufficient for their needs. They are the simplest cells and yet still highly complex transporting 1.1 litre of oxygen around the body with a combined total mass of 2.38 kg. [67] They are synthesised at a rate of around 2.4 million per second and monitored and regulated by the spleen. They have a life of approximately 3-4 months during which time, they will circulate 200,000 times in the body, travelling a total distance of 500 km, after which their components (97% haemoglobin dry weight) are degraded and recycled by splenic macrophage white blood cells. Their water content is 68% by mass. [67]

The RBCs depend on the anaerobic conversion of glucose by the Embden-Meyerhof pathway (EMP) for generation and storage of high-energy phosphates to provide energy.

Under normal circumstances 90% of glucose is catabolised anaerobically to pyruvate or lactate by the EMP, or glycolysis. The liver tidies up the lactate waste product via the Cori Cycle (Figure 14.8).

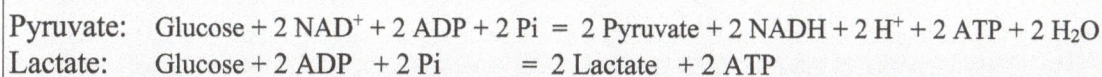

Pyruvate:	$Glucose + 2\,NAD^+ + 2\,ADP + 2\,Pi = 2\,Pyruvate + 2\,NADH + 2\,H^+ + 2\,ATP + 2\,H_2O$
Lactate:	$Glucose + 2\,ADP + 2\,Pi = 2\,Lactate + 2\,ATP$

Lactate generated by the RBCs is taken up by the liver via the Cori cycle for gluconeogenesis. RBCs use around 20 mg glucose/min and return this amount of lactate to the liver for synthesis of more glucose. RBCs also use the pentose phosphate pathway to generate the reducing equivalent NADPH and synthesis of 5 carbon sugars.

The PPP accounts for around 60% of NADPH production in the entire body. 2,3-bisphosphoglycerate (2,3-BPG) is one of the most important metabolites in the glycolysis pathway because it binds to oxygenated haemoglobin enabling the unloading of oxygen to tissues. [67]

14.31 Eye Retina and Cornea Energy Source

The retina probably has the highest oxygen rate of uptake rate in the entire body due to its high anabolic needs. The outer segment of the retina epithelial cells that contains the photosensitive pigment is also virtually devoid of mitochondria, as these organelles would interfere with light transfer and vision. Retina tissue primarily uses glucose, although there is evidence to show it is also supported by several alternative fuels such as lactate, malate, succinate and palmitate fat by (β)-oxidation TCA and OXPHOS and these may contribute up to 60% of energy when necessary. Normally, glucose is converted to pyruvate, which is then converted to lactate by lactic acid fermentation. Similarly, the cornea is also devoid of capillaries and mitochondria and also relies on glucose entry by diffusion for anaerobic metabolism & like the RBCs uses the pentose phosphate pathway to generate NADPH and synthesis of 5 carbon sugars. Using lactate yields 15 times less ATP compared with oxidising glucose.

14.32 Kidney Medulla Energy Source

The metabolic fate of glucose in the kidneys varies and is dependent on the type of the kidney tissue. The renal medulla has a low oxygen tension and low levels of oxidising enzymes and the oxygenated blood supply to the kidney medulla cells is insufficient for ATP synthesis, via OXPHOS. In contrast to the renal cortex, the renal medulla has little glucose phosphate capacity, but is rich in mitochondria and does have a high level of oxidative enzymes. These cells therefore rely on a continual uptake of glucose from the bloodstream for energy. Thus, the renal medulla is an obligate user of glucose for its energy source at very high rates. Much of this use of glucose is therefore, achieved anaerobically, via lactic acid fermentation with lactate as the end product rather than CO_2 and water and is the same process as that used by RBCs.

14.33 Kidney Cortex Energy Source

The cortex of the kidney requires a constant source of energy for the necessary, constant high energy demand process of reabsorption of glucose from the Glomerular Filtrate in the Proximal Convoluted Tubule. Consequently, the renal cortex does not use much glucose and the (β)-oxidation of free fatty acids (FFA) is the main dominant source of energy, supported by the ketone bodies, acetoacetate and 3-hydroxybutyrate and to a lesser extent, depending on conditions, the amino acid glutamine via gluconeogenesis.

14.34 Heart Cardiac Muscle Energy Source

For an average heart-beat of 100,000 times per day and all other energy needs, the heart uses an incredible 6 kg of ATP per day, or around 15 to 20 times its own weight each day. Given this staggering quantity of energy, the heart overall prefers fatty acids for energy at base rate and plays its own role in conserving glucose. The heart achieves its high energy demand by using a variety of fuels (fatty acids, lactate, glucose, ketones, pyruvate, and amino acids), primarily i.e., 95% by mitochondrial OXPHOS and 5% by glycolysis.

The use of different fuel substrates changes with concentration. Normal O_2 levels promote normal fatty acid oxidation but in ischemia, for example, O_2 depletion can block the ETC, leading to build up of AMP and AMP Kinase (AMPK) and acetyl-CoA. This inhibits PDH, uncoupling glycolysis to provide energy.

However, then lactate builds up causing proton free radicals and NADH accumulation, together with changes in heart enzymes and increased level of $[Ca^{2+}]$ and eventually intracellular acidosis.

Glucose normally provides around 10-40% of ATP production in cardiomyocytes. Between meals, cardiac heart muscle cells meet 90% of ATP energy demands by oxidizing fatty acids and to a lesser extent ketone bodies, acetoacetate and 3-hydroxybutyrate.

This proportion of non-glucose substrate for energy may fall to around 60%, depending on nutrient conditions and heart muscle contraction intensity, with 40% from glucose and during heavy loads, an increase of energy uptake from glucose via glycogen reserves but overall fatty acids are the major fuel consumed by cardiac muscle. Sertoli cells are essential for spermatogenesis and physical and nutritive support to developing germ cells.

Around 60-70% of ATP is used for the contractions of myocardial sarcolemma and around 40% ATP is used for ion pump by $[Ca^{2+}]$ ATPase and Na^+ and K^+. In the fasted state, acetyl-CoA accounts for 70% of the heart's fuel, including that from circulating free fatty acids. In the fed state, there is a much greater oxidation of glucose for fuel. Myocardial energy is stored as creatinine phosphate. Cellular TCA cycle and ETC OCPHOS full oxidation takes in the order of 10 minutes to kick in fully and hence the metabolic reason that a warm up is advised prior to strenuous exercise.

14.35 Skeletal Muscle Energy Source

Skeletal muscles also oxidise lipids and fatty acids are the main source of energy during rest and moderate exercise. As exercise intensity increases, glycolytic and glucose oxidation surpasses fatty acid oxidation and skeletal muscles also use considerable amounts of glucose from their own glycogen stores. Skeletal muscle normally oxidises fatty acids in preference to glucose and fatty acids inhibit glucose oxidation under prolonged exercise and starvation. This preference changes on more intensive activity.

Fast-twitch muscle fibers have very few mitochondria and require glucose. On highly intensive demand during the first few seconds (and up to a maximum of 6 seconds), muscle cells utilise Creatinine Phosphate via the phosphagen system, which is a small contributor overall but nevertheless very important, as the advantage in the phosphagen system lies in the extremely rapid rates at which it can regenerate ATP. A 100 metre sprinter will use creatinine for the first 5 or 6 seconds, then lactate for the remaining few seconds for this rapid burst of energy.

14.36   Adipose Tissue Energy Source

Adipose tissue mainly uses fatty acids through (β)-oxidation in the adipocyte mitochondria. It does utilise glucose for energy but only during the absorptive (post feeding) stage after a regular meal. Adipose tissue oxidises both fatty acids and glucose with more glucose oxidised during the absorptive stage and more fatty acids oxidised at the post absorptive stage.

14.37   The Ovaries and Testes Energy Source

In the ovaries, glucose from pyruvate and lactate are essential for oocyte development. Although glucose is usually the more preferable source of energy for cumulus oocyte complexes, fatty acid (β)-oxidation is also essential to oocyte growth and maturation. The amount of lipids present in the ooplasm is important and affects the sensitivity to fatty acid (β)-oxidation inhibition. The correct concentration of lipids in the ooplasm, as well as glucose, in the maturation environment is therefore necessary for proper oocyte developmental competence.

Fatty acid (β)-oxidation is also an important source of energy during oocyte maturation. The testes use both glucose and lactate for energy. Lactate produced by pyruvate generated from glycolysis is the main metabolic substrate for germ testes cells. Glycogen is also present in testicular cells and glycogen metabolism is a potential source of glucose to both testicular somatic (namely Sertoli and Leydig cells) and germ cells.

14.38   The Role of the Kidneys in Homeostasis

The kidney has many vital functions, probably second to the liver in terms of the number and importance of functions carried out by the main body organs. Special mention of these functions is included here as the kidney plays a key part in glucose homeostasis. In addition to filtration and removal of waste products from the body and balancing body fluids, the kidney is involved in the regulation of glucose homeostasis, via the release of glucose into the circulation i.e., from gluconeogenesis, the uptake of glucose from the circulation for its own energy needs and also the reabsorption of glucose from glomerular filtrate to conserve glucose. In a given day, the kidneys produce 15-55 g/d of glucose via gluconeogenesis (using glutamine) and they metabolise approximately 25-35 g/d of glucose. Therefore, in terms of glucose economy, renal reabsorption is the primary mechanism by which the kidneys influence glucose homeostasis. Normally, 180 litres of plasma are filtered by the kidneys each day. As the average plasma glucose concentration throughout a 24 hour period is around 5.5 mmol/L, this means that 180g of glucose is filtered by the kidneys each day. In healthy individuals virtually 100% of this filtrate is reabsorbed into the circulation and the urine is thus free from precious glucose.

Glucose filtered and reabsorbed	180 g/d
Glucose produced by gluconeogenesis	15 - 55 g/d
Glucose Metabolised for energy	25 - 35 g/d
Net Glucose made available for Homeostasis	160 - 210 g/d

There are at least 500,000 Glomeruli/Convoluted Tubules in each kidney. The kidneys require the same amount of energy as the heart (Table 14.1). Blood flow to the kidneys is approximately

200 litres/d. The normal range for 24 hour urine production is from 800 to 2,000 ml/d. The kidneys (and lungs) control the blood pH within a range from 7.35 to 7.45. The kidneys also produce vitamin D and the hormone erythropoietin (EPO) produced from the interstitial cells. EPO acts on RBCs to protect them against destruction and also stimulates stem cells in the bone marrow to increase the production of RBCs. The kidneys remove waste; ammonia HN4$^+$, uric acid, urea, creatinine, NaCl and excess fluid.

The kidneys also control blood pressure and blood volume via the Rennin-Angiotensin-Aldosterone System (RAAS). This is influenced by the hormones; anti-diuretic hormone (ADH), aldosterone and parathyroid hormone and the enzyme rennin and calcitonin (vitamin D). Rennin is produced in the kidney. Angiotensin is produced in the liver, Angiotensin-converting enzyme (ACE) is produced in the lungs and kidney. Aldosterone is produced in the zona glomerulosa of cortex of the suprarenal glands, ADH (or vasopressin) is produced in the pituitary gland and thyroid hormone is produced in the thyroid gland. The kidneys secrete Erythropoietin kidneys that stimulates RBC production due to reduced oxygen tension by acting on erythroid-committed stem cells in the bone marrow. Chronic kidney disease (CKD) is a significant risk factor for developing hypoglycaemia and this may occur even without diabetes. Hyperglycaemia and hypertension can also damage the small blood vessels in the kidneys and lead to diabetic nephropathy which can also be a complication of type 1 and type 2 diabetes and also poor lifestyle. This can ultimately lead to kidney failure which is life threatening.

14.39 Ketogenesis - Energy Back Up in Fasting & Starvation

We have reviewed the natural role of ketones in preserving glucose and protein and protecting the brain and ultimately the RBCs. We have also discussed the role of gluconeogenesis in maintaining blood glucose levels. Both natural processes are activated during stress situations, vital in preserving glucose and structural protein when real life conditions dictate. Ketone bodies are compounds that contain the ketone groups produced from fatty acids in the liver. The principal reason for the production of ketones during carbohydrate shortage is because the brain can oxidise ketone bodies for energy but is unable to oxidise fatty acids for these energy needs.

Table 14.16 A Summary of Ketogenesis

What is it?	The process through which ketone bodies are made from fatty acids
What is its Principle Role?	Prolong survival in times of nutrient shortage, guards the brain & heart
Why does it occur	To ensure the brain, muscles and heart have sufficient energy
When does it occur?	Low blood glucose, sleep, fasting and prolonged starvation
Where does it occur?	In the liver (within the mitochondrial matrix)
How is it stimulated?	High levels of acetyl-CoA & excessive fatty acid oxidation & glucagon
How is it inhibited?	A rise in insulin by reducing lipolysis and fatty acid levels
What is needed to action it	Thiolase, HMG-CoA synthase, HMG-CoA lyase, hydroxybutyrate DH
What are the start reactants?	Fatty acids, fatty acyl-CoA, acetyl-CoA,
What is the end product?	Acetoacetate, beta-hydroxybutyrate, acetone (ketone bodies)

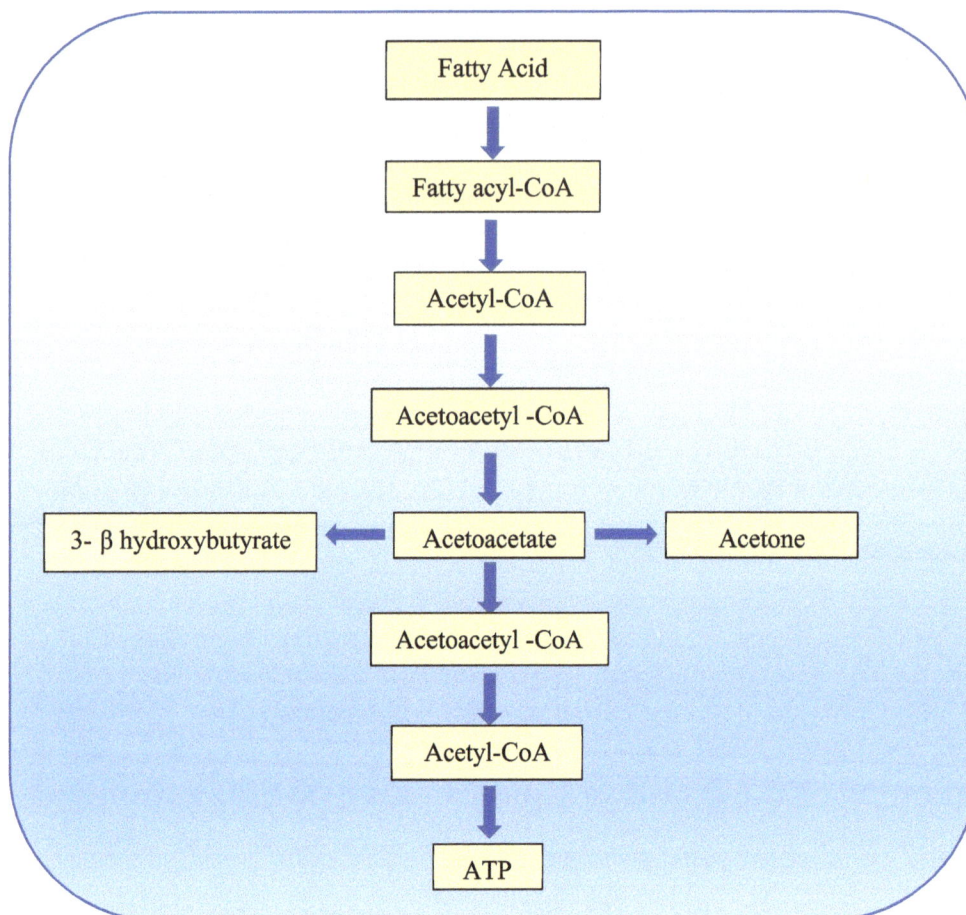

Figure 14.11
Overview of the pathway of Ketogenesis and Ketolysis from ketones to ATP

```
                    ┌──────────────┐
                    │  Fatty Acid  │
                    └──────┬───────┘
                           │
                           ▼
                    ┌──────────────┐
                    │ Fatty acyl-CoA│
                    └──────┬───────┘
                           │
                           ▼
                    ┌──────────────┐
                    │  Acetyl-CoA  │
                    └──────┬───────┘
                           │
                           ▼
                    ┌──────────────┐
                    │Acetoacetyl -CoA│
                    └──────┬───────┘
                           │
                           ▼
┌──────────────────┐ ┌──────────────┐ ┌──────────────┐
│3- β hydroxybutyrate│◄─│ Acetoacetate │─►│   Acetone    │
└──────────────────┘ └──────┬───────┘ └──────────────┘
                           │
                           ▼
                    ┌──────────────┐
                    │Acetoacetyl -CoA│
                    └──────┬───────┘
                           │
                           ▼
                    ┌──────────────┐
                    │  Acetyl-CoA  │
                    └──────┬───────┘
                           │
                           ▼
                    ┌──────────────┐
                    │     ATP      │
                    └──────────────┘
```

In this section we shall further explore both the natural and unnaturally deliberately enforced process of ketogenesis. Inadequate carbohydrate (the body's emergency status) leads to production of ketones such as beta-hydroxybutyrate. The body burns fats to preserve precious glucose and more so during emergencies such as starvation.

When sufficient carbohydrate stores are available, the main pathway used is glycogenolysis, i.e., the breakdown of glycogen from stores in muscle and liver. We have seen that gluconeogenesis, the production of glucose from non-carbohydrate sources such as lactate, is often used in addition, especially during exercise. In addition, the naturally evolved ketogenesis metabolic process, also reduces pressure on the body to promote gluconeogenesis and this shift gradually reduces protein degradation, via glucogenic precursor gluconeogenesis and spares the body's precious structural, hormonal and enzymatic protein.

If excessive acetyl-CoA levels are created from the (β)-oxidation of fatty acids and faster than the TCA cycle can use them, the mitochondrial TCA cycle becomes overloaded and ATP levels rise significantly.

This leads to a build-up of NADH and (and excess in the ratio of NADH to NAD$^+$) and electron levels in the ETC causing a slowing of the TCA cycle. In addition, increased gluconeogenesis uses up more oxaloacetate (OAA), further slowing the TCA cycle.

In this event (after 3 days of starvation or low carbohydrate intake) the excess acetyl-CoA is diverted from entering the TCA cycle. It is instead, shunted to the production of ketone bodies in the liver. (Figure 14.10). Ketogenesis (the second fasting state energy back up) is thus one of the body's natural biochemical mechanisms that has evolved to maintain glucose levels during conditions of stress and carbohydrate and energy scarcity.

It provides fuel substrates by breaking down fatty acids and ketogenic amino acids (for example, leucine) to produce ketones, providing a vital back up reserve of energy when required. Ketogenesis is entirely dependent on nutritional status.

When carbohydrates are scarce, for example, through fasting or genuine starvation or deliberate carbohydrate deprivation dieting, energy from carbohydrates is depleted and the body detects metabolic stress and gradually shifts into catabolic mode. Fatty acids start to become predominant in providing energy and the levels build up.

Therefore, the ketone bodies serve as a fuel source if glucose levels are too low in the body. Ketones are readily transported into tissues outside the liver, where they are converted into acetyl-CoA (acetyl-Coenzyme A) which then enter the TCA cycle (Krebs cycle) in the cells of these tissues and oxidized for energy.

The cells of those tissues preferring glucose, not able to utilise fats for energy, can then convert the ketone bodies back to acetyl-CoA to produce the required energy through TCA and OXPHOS. The liver is unable to use ketone bodies for energy as it lacks the enzymes required to catabolise them.

The ketones are released into the blood after glycogen stores in the liver have been depleted. Ketones are more readily oxidised than fatty acids, although ketogenesis is, in effect, an incomplete (β)-oxidation of fatty acids.

Whereas the complete oxidation of one molecule of palmitoyl-CoA produces 106 molecules of ATP energy (Section 15.11), its conversion to acetoacetate and beta hydroxybutyrate produces only 20 and 22.5 molecules of ATP respectively. In fact, ketones produce similar amounts of ATP per mole as oxidised amino acids, which is expected, as these substrates are produced under nutrient shortage, energy substrate stress or starvation conditions.

In emergency (ketogenesis) back up status, i.e., including the low carbohydrate, or so called, 'low carb' diets, the body converts non-carbohydrate foods into ketones for energy. Consider an overnight fast and a small degree of hepatic ketogenesis. This normally produces say 10 g/d of ketone bodies. However, following further fasting of 2 to 3 days, the liver produces much higher quantities of ketones, in the order of 100-150 g/d. The liver is capable of producing up to 185 g of ketones per day.

Over this period, the average oxidation rate of fat is approximately 3 g/kg fat of fat free mass/d (FFM) i.e., 3 g/kg/FFM/d. A large but not obese man, weighing 80 kg and 80% FFM (64 kg) and 20% fat mass (FM) (16 kg) oxidises a minimum of 192 g fat/d. via ketogenesis or fatty acid oxidation.

Around 60g of the free fatty acids (FFA) from lipid stores undergoes (β)-oxidation in the liver to yield an estimated 113 g/d of ketone bodies. 90% of these water soluble fuels undergo terminal oxidation by the brain and muscle and most of the other 10% is excreted in the urine.

Since the ketone bodies, beta-hydroxybutyrate and acetoacetate are excreted from the kidneys with equimolar ammonium NH_4^+, ketosis is an energetically inexpensive way to excrete nitrogen. Ketosis is the metabolic state characterised by detectable ketone levels in the blood or urine.

It is estimated that ketones may provide in the order of <5% of daily energy needs overall, termed normal nutritional ketosis. Much of this is probably within heart muscle. Generally, the body is continually synthesising a very small number of ketone bodies, converted from the break-down of fatty acids, to be used by the body for energy, for example, acetoacetate and 3-hydroxybutyrate provide energy for heart muscle and the kidney cortex and are normally found in circulation in relatively small quantities ranging between the baseline of 0.1 mmol/L in the postprandial (fed) state to 0.5 mmol/L.

In times of fasting, even overnight while sleeping, the level of ketone bodies in the blood increases and after a normal overnight fast, ketones may rise to around 1.0 mmol/L. The first 2 days of starvation or inadequate intake of carbohydrate, yields fatty acid breakdown to acetyl-CoA and the majority of protein degradation using glucogenic amino acids in gluconeogenesis.

When carbohydrate stores are significantly decreased or fatty acid levels increase, the ketogenic pathway is increased from around 24 hours and after 2 to 3 days, the number of ketone bodies rises up to around 1.8 mmol/L. Under conditions, when insulin is low, glucagon levels and lipolysis increase and the level of fatty acids and acetyl-CoA are high, ketone production becomes more predominant, up to 20% then 60% of the body's energy needs and even higher levels of total energy needs, after weeks of starvation. The brain can use up to 70-80% of ketone bodies for energy during prolonged starvation. Ketogenesis spares body structural protein and ultimately glucose for the RBCs and also reduces the need for gluconeogenesis and its role extends to that of protein sparing, extending healthy body life.

Thus, the liver produces large amounts of ketone bodies which are acidic. Severe acidosis (keto-acidosis) defined as >3 mmol/L, can impair tissue function, especially within the central nervous system. When oxaloacetate (OAA) is unavailable for condensation with acetyl-CoA, the acetyl-CoA is diverted to the formation of acetoacetate, D-3 hydroxybutyrate or acetone.

These 3 ketone bodies are found in high levels in untreated diabetics. Acetone can appear in the breath of someone who is a low carbohydrate/low sugar user, including some alcoholics. If starvation sets in, then after around 4 weeks, plasma levels of ketones will rise to a plateau of 6-8 mmol/L. As fasting progresses into starvation, gluconeogenesis is gradually reduced in the liver and shifted to a significant increase of gluconeogenesis via the amino acid glutamine in the kidneys.

Here, the increased glutamine nitrogen extraction generates ammonia $NH4^+$ which, together with increased production of bicarbonate, to offset the bicarbonate used in buffering ketoacid production, serves as the counterion to ketoacid salts produced by ketogenesis.

It is considered likely that the kidney's predominant role in acid base homeostasis may be even more important than the kidneys production of glucose from gluconeogenesis glutamine. Eventually, ammonia exceeds urea as the dominant excretion product.

Ketogenic diets tend to produce relatively low, mild levels of acidosis as long as insulin levels are normal. However, continued long term ketosis increases the risk of nephrolithiasis (kidney stones) and osteoporosis (reduction in bone mass and higher risk of fractures).

Also, as ketones are water soluble intermediates, this diet and state of ketosis is associated with dehydration risk, unless properly monitored. Individuals practicing this diet need to drink more water.

The current, popular 'low carb' diet, striving to invoke ketosis, can lead to a loss of weight very quickly as it reduces and depletes the body's stores of glycogen and water and often leads to lower total caloric intake (EI). However, prolonged deliberate ketogenesis and persistent ketosis can lead to other issues and be difficult to maintain over long periods of time.

In the long term, ketogenic diets suppress appetite and can be difficult to maintain for some, given the restriction of many healthy whole carbohydrates, such as grains, rice, wheat, oats, barley, vegetables, potatoes, sweet potatoes, peas bread, corn, apples, pears and bananas etc. It can therefore and often does lead to carbohydrate resistance.

This is effectively the inability to process any carbohydrates (including all the healthy carbohydrates) without invoking a high insulin response.

The diet encourages consumption of cheese, butter, eggs, cream, fish, meat, spinach, broccoli and can often lead to symptoms including; hunger, fatigue, low mood, irritability, constipation, headaches, sleep issues, anxiety increased stress and 'brain fog'.

Whatever the variety, all the various and 'keto' and extended fasting diets, all tend produce a whole range of metabolic stress responses throughout the body in addition to significant increased levels of circulating ketones (up to 3 mmol/L) and reactive oxygen species (ROS) in the mitochondria.

This tends to confound and add doubt to the so called metabolic benefits of long term ketosis, even assuming more serious ketoacidosis is prevented.

However, this diet is not recommended as a natural, evolutionary instinctive, healthy and sustainable long term way to lose and maintain weight loss.

Ketosis ('low carb') and high levels of ketone bodies in the blood can be dangerous for the body if prolonged, as it can disturb the delicately controlled acid base balance. In order to reach and stay in ketosis, the daily level of dietary carbohydrate must be kept under 50 grams of carbohydrates, which, assuming total dietary energy intake is sufficient for 2,000 kcal/d then equates to 10% dietary carbohydrate with 75% from dietary fat and 15% from dietary protein.

A diet so high in fat is unhealthy for so many reasons and the consensus of evidence is that this is highly likely to lead to many health issues in the short and long term. Ultimately, the evidence indicates that it is likely that, in terms of weight loss, keto diets produce similar results to other restrictive diets. The key is sustainability and practical application and adherence over long periods of time and healthy carbohydrate deprivation over a lifetime.

If people are comfortable on this kind of diet and feel healthy, have sufficient energy, are able to sustain this long term and the diet is producing no ill effects, then this 'keto' diet may work for some people. There is little evidence to show the very long term effects (>20 years) of continual high fat, ketogenic diets. It is up to the individual to consider the advice, information and risk of any extreme diet.

14.40   De Novo Lipogenesis - Synthesis of Fat from Glucose

De novo lipogenesis (DNL) is the process by which carbohydrate, or amino acid carbon precursors of acetyl-CoA, are synthesized into fatty acids in the liver. There is a great deal of misconception regarding the quantum of the total contribution of De novo lipogenesis (DNL) especially that derived from carbohydrate, turned into body fat. De novo lipogenesis is sometimes referred to as;

'De novo lipogenesis is the pathway of last resort' [21]

The relative impact of this process is the subject of one of the most commonly quoted nutrition myths, commonly referred to as *'carbohydrates turn into fat'*. De novo lipogenesis (i.e., DNL), mostly derived from carbohydrates (though not entirely exclusive to carbohydrates as it may also refer to protein) is indeed very limited in quantum.

Table 14.17   A Summary of De Novo Lipogenesis

What is it?	A process in which carbohydrates & protein can be converted to fat
What is its Principal Role?	To convert continuous, excessive levels carbs & protein to fat
Why does it occur	To conserve fat if fat is not readily available & to dispose of xs glucose
When does it occur?	During & after acute & sustained surplus of eating & xs energy balance
Where does it occur?	In the liver & adipose (xs fat stored in adipose & ectopic fat in organs)
How is it stimulated?	Excess glucose, high fructose, excess protein, increased insulin
How is it inhibited?	Dietary fat inhibits DNL in adipose tissue and low insulin levels
What is needed to action it	Acetyl-CoA carboxylase, malonyl-acetyl transferase
What are the start reactants?	Acetyl-CoA, malonyl-CoA, 7 ATP & 14 NADPH (per mole of palmitic)
What is the end product?	Fatty acids, VLDL and triacylgyclerides (TAG)

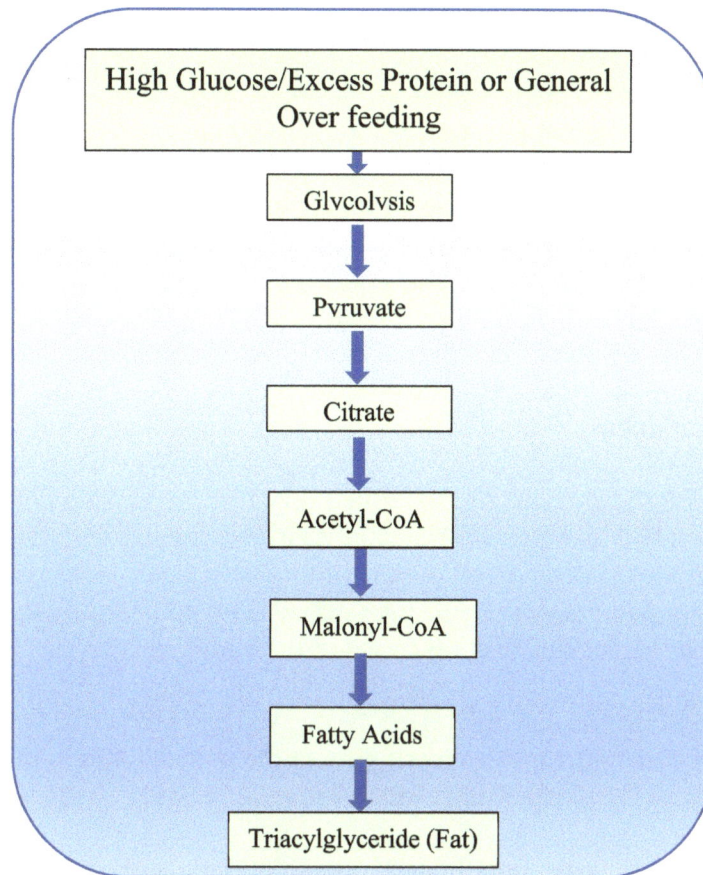

Figure 14.12 Overview of the pathway of the Pathway of De Novo Lipogenesis: fat from carbohydrate & protein

Contrary to common belief and confusion, in lean healthy individuals, DNL is a very minor contributor to whole-body lipid stores compared with the fat laid down and stored from dietary fat. In humans, the vast majority (>95%) of stored fat is derived from dietary fat.

The remainder (<5%) of stored fat is converted from protein, alcohol and carbohydrates via the process of de novo lipogenesis.

DNL occurs when there is repeated and constant caloric excess and the body has more than sufficient energy for its needs and particularly when the diet consists of simple, free sugars such as fructose, fruit juices, carbonated drinks, crackers, pastries, chocolate bars, sweets and sucrose (table sugar), fats, processed food and fast food, a description; aka the western diet.

Carbohydrate, or for that matter protein, does not become fat to any marked extent; however, carbohydrate does keep the body from burning fat when the body has sufficient or surplus reserves of energy (hence the term 'spare the fat' [14] or 'fat sparing carbohydrates').

The body prefers to burn the carbohydrates for day to day energy requirements and metabolism and 'spare the fat' for times of shortage. The bioenergetic cost of storing dietary fats as triacylglycerols is significantly lower than that of converting protein or carbohydrates into fat. There is a much greater bioenergetic cost to convert and store glucose and carbohydrate and protein to fat than the cost required to store fat in adipocyte fat cells.

For example, requiring approximately just 2 calories per 100 calories of ingested fat to store fat from the diet as opposed to costing 23 calories for every 100 calories of carbohydrate ingested.

The DNL average proportion of fat synthesis from carbohydrates normally ranges from 0.56 -1.32% of total energy intake (EI), a very small proportion of the total fat storage balance in humans consuming a typical diet.

In contrast, some animals are able to synthesise more fat from carbohydrates, for example, over fed pigs and cows. Honey bees make significant quantities of wax from nectar sugars which underpins much of their duties.

Higher fat diets diminish the amount of DNL. Even very low fat diets do not increase these low levels of DNL significantly. As noted above, even very lean humans carry sufficient fat (>10kg) for every day needs and for stored energy in case of food shortages. Non-essential saturated fatty acids, such as palmitate (palmitic acid), stearate (stearic acid) and oleic acid account for a little less than 50% of all the fat that can be synthesised in the body.

In humans, up to 20% of that fat can potentially be derived and converted from carbohydrate sugar sources via DNL, i.e., a net maximum of 10% in the body, although in reality the actual amount of fat stored from DNL is normally much lower i.e., <5%.

Most hepatic triacylglycerols arise from hepatic uptake of fatty acids from dietary fat (as plasma non-esterified fatty acids NEFA) and as triacylglycerol fatty acids in lipoprotein particles. Only in some pathological conditions does de novo lipogenesis make a more substantial contribution. For example, in those individuals with non-alcoholic fatty liver disease (NAFLD), fat accumulates in hepatocytes and is associated with liver damage.

NAFLD and high levels of obesity is closely associated with obesity and insulin resistance. In individuals with NAFLD, de novo lipogenesis may contribute around 25% of liver triacylglycerol with dietary fat the other 75%. Adipocyte de novo lipogenesis (i.e., DNL within the fat cells themselves) accounts for <2% of adipose triglyceride content, even less than hepatic DNL. Lipid accumulation in adipose (fat) tissue depends on the circulating fatty acid uptake. The relative level of human lipogenesis does increases considerably (3 fold) when individuals are continually and deliberately over fed (ad libitum) an extreme hypercaloric, high fructose, high free sugars, high-carbohydrate diet for a period of several days . The same metabolic conditions as for insulin resistance.

'The overwhelming evidence is compelling that a hypercaloric high fat, high protein, over nutrition, common in the western style diet, causes overloading of the liver and other organs, leading to the vast majority (>75%) of illnesses in these societies'.

Adipose tissue, rather than the liver, seems to be the major site for this de novo lipogenesis under this heavy nutritional load. However, the relatively large increase still makes up a very small proportion of fat mass storage compared with that stored from dietary fat. Studies [14] have shown that DNL synthesis from overloaded carbohydrate increases from 1.6 g/d (normal (isocaloric diet) to 3.7 g/d of fat (over feeding diet) or a total of just 14 kcal/d to 33 kcal/d. This means that the net change due to overfeeding 540 kcal of sugar in these studies was (3.7-1.6) or only 2.1 g/d or 19 kcal/d fat synthesised from DNL.

This is equivalent to 0.7% of total caloric intake (TEE). Given this rate of conversion, it would therefore take approximately 4 months to gain 1 pound of body fat by over feeding an additional (135g) 540 kcal/d of sugar and (50g) 450 kcal/d of dietary fat on top of the normal diet and clearly not significant by any stretch. Consuming excessive quantities of any foods, including both nutritionally healthy and complex food or unhealthy, refined carbohydrates and free sugars or fat or protein is, of course not recommended for health and nutrient quality reasons but this discussion is included here to illustrate the negligible impact of carbohydrate on overall body fat synthesis.

The excess or sufficient carbohydrate signals to the body that there is plenty of energy for body processes and survival and therefore less requirement to burn excess adipose tissue. The energy storage levels i.e., the energy stored in body fat thus remains. One of the biochemical reasons for this, is that in this fed, anabolic state, glucose raises insulin levels, which decreases fatty acid beta (β)-oxidation.

This increases the movement of fats to storage and upregulates malonyl-CoA. The malonyl-CoA effectively switches off fatty acid (β)-oxidation in the fed state. However, in fasting, times of stress, starvation or low carbohydrate diets, the opposite occurs, regulated by glucagon, where fatty acid (β)-oxidation increases. The vast majority of fat on the body is therefore due to dietary fat and overeating generally, thus, facilitating fat storage rather than fat burning.

Fat added to the body from dietary fat therefore ranges from 96.4% to 99.55% of total fat caloric intake, depending on the level of caloric excess. In contrast to the bioenergetic cost of storing fat from dietary fat, which costs only approximately 3% of the energy derived from the ingested fat, the conversion of dietary carbohydrate to fat in the liver via DNL is significantly energy expensive and endergonic, costing 25-30% of the energy due to loss of carbon atoms to CO_2.

The reason why such a large amount of energy is required for fat synthesis from glucose is because the process is a tortuous one requiring complex reactions, many enzymes and endergonic reactions. The reactions, starting with glucose, are as follows;

Glucose → acetyl-CoA → malonyl CoA → palmitic acid.
Glycerol-P + fatty acids → TAG (liver) → VLDL (liver) →VLDL (adipose tissue) → TAG hydrolysis and transport into adipose tissue → resynthesis in adipose

Therefore, in contrast to other species such as mice, de novo lipogenesis is, for the most part, a minor contributor to adipose tissue mass and whole-body energy balance. Repeated science research confirms that even substantial quantities of unhealthy, refined and processed carbohydrates contribute just a trivial amount to overall body fat. Recent studies have shown that endogenous de novo lipogenesis is an essential source of lipids for autophagy in adipocytes, particularly for membrane synthesis. In times of low energy, autophagy (cell maintenance and renewal) serves critical cytoprotective (protection of cells) and hormetic (an adaptive dose-response phenomenon to moderate stress, characterized by low-dose stimulation and high-dose inhibition) roles in virtually all cell types. DNL probably evolved to cater for a significant fall in fats consumed to a very low level, to ensure the maintenance of the required minimum levels of essential fats necessary for cellular lipid bi layer membrane and other function substances, such as cholesterol and hormones etc. That is preserving fat for these essential purposes but not for adipocyte (fat cell) adipose fat storage long term energy retention, as the ample body stores in the leanest of individuals (>10 kg) are there for just that purpose. To conclude, DNL is not significant for the vast majority of people. The following is an extract from 'Carbohydrate metabolism and de novo lipogenesis in human obesity' from the American Journal of Clinical Nutrition 1987; 45:78-85 [22].

'Fat balance is thus essentially equal to the discrepancy between fat oxidation (which is determined by the energy gap between carbohydrate-plus protein intake and total energy expenditure)' Evidence that dietary fat, rather than fat synthesis from carbohydrate, provides the precursors for adipose tissue triglycerides should help to alleviate the commonly held fear among weight conscious individuals that carbohydrate turns into fat. This erroneous belief may lead those individuals to avoid foods known to contain primarily carbohydrates in favour of foods known for their low carbohydrate content (i.e., high protein items). Such a pattern of food selection would be unfortunate, indeed, because it promotes the consumption of foods with a high fat content, which is much more likely to sustain a positive fat balance than dietary carbohydrate.

14.41 Nutrient Metabolic Processes

Table 14.18 lists the major nutritional energy metabolic processes that regulate and balance anabolic and catabolic homeostasis of the body in humans, from both exogenous (dietary) substrates and endogenous stored or tissue sources under normal, fed, fasting and starvation states.

Table 14.18 Summary of Nutrient Energy Metabolic Processes in Humans

Cellular Process	What is it & Function	Location	Under Which Conditions	Result
Glycolysis Section 15.3	The Anaerobic First step in glucose metabolism & conversion of glucose to Pyruvate, does not require oxygen	Within the cytosol of all cells including Red Blood Cells	Ongoing when cells require energy. Also stimulated by Insulin	2 Moles of pyruvate/mole Glucose & ATP, NADH & Water
Tricarboxylic Cycle (TCA) (Krebs cycle, Citric acid cycle) (Per mole of glucose) Section 15.5	The TCA cycle primary function is to create energy-carrying molecules that can transfer electrons to the electron transport chain from catabolism of energy rich substrates. Also provides metabolic framework for molecular anabolic synthesis of many vital molecules	Within the mitochondria of all cells using oxygen to produce energy	Ongoing when cells require energy. Also stimulated by Insulin and when O_2 is available	6 NADH 1 FADH$_2$ 2 GTP (2ATP) 4CO$_2$
Oxidative Phosphorylation Section 15.6	Oxidative phosphorylation is the final stage of aerobic cellular respiration. Comprises electron transport chain & ATP Synthase	Within the mitochondria of all cells using oxygen to produce energy	Ongoing when cells require energy. Also stimulated by Insulin and when O_2 is available	Most of the cell's energy; 30 Moles of ATP for glucose & 106 Moles of ATP for Fatty Acids

Lipolysis Section 14.6	Breakdown of Triacylglycerol fats to fatty acids for (β)-oxidation and glycerol	Within adipocytes of adipose tissue & other tissues for energy	Fasting, starvation, low glucose & low carbohydrate diets and moderate exercise & low glycogen when additional energy is required	Mobilisation of Fats for energy via beta oxidation to & within peripheral tissues
Glycogenesis Section 14.7	Formation of Glycogen from glucose molecules for energy storage	Within Liver Hepatocytes and Muscle Cells	Rest periods high blood glucose, post prandial well fed state, stimulated by insulin	Increase in Stored Glycogen
Glycogenolysis Section 14.9	Breakdown of glycogen to form glucose. In muscle cells provides glucose energy contraction for muscles. In hepatocytes provides glucose for all tissues in the body	Within Liver Hepatocytes and Muscle Cells	Fasting & intense or long endurance exercise. Stimulated by glucagon and epinephrine	Increase Blood Glucose, esp.' Brain & RBCs
Gluconeogenesis Section 14.16	Synthesis of glucose from endogenous (Not dietary) non-carbohydrate sources, glycerol, lactate, Amino Acids (Alanine & Glutamate), Propionate. Helps to maintain blood glucose levels. Preserves glucose for the body especially brain and ultimately RBCs	Within Liver Hepatocytes (75%) & Kidney Cortex (25%)	When glycogen stores are depleting, 8 hrs post fasting, or long starvation or when intake of carbohydrates is too low. It is an endergonic process and the energy for gluconeogenesis is provided by (β)-oxidation of fatty acids	Increased Blood Glucose esp.' Brain & RBCs
Ketogenesis Section 14.39	Break-down of fatty acids & 2 amino acids to form Ketone bodies for fuel. To preserve glucose for the body especially brain and ultimately RBCs	Within Liver Hepatocytes	Low levels normally, then increasing in Fasting, vigorous exercise, low oxaloacetate, low glucose, low glycogen, high fat diets & when intake of carbohydrates is too low & protein too low. Also, if insufficient Insulin	Rise In Ketone Bodies to supply energy to Brain heart & skeletal muscle
De Novo Lipogenesis Section 14.40	Process: Carbon precursors of acetyl-CoA synthesis to fatty acids. Mostly from carbs and amino acids' Minor contributor to whole-body fat stores, 1–3% of total fat balance in humans with typical diet	Within Liver Hepatocytes & Adipose Tissue	High fat & high carbohydrate diets, over nutrition & continued excessive feeding EI>TEE	Increase in Stored Fat 213approx. 1-3% of Fat Synthesis overall higher in obese people especially those with NAFLD

CHAPTER

From Light Energy to Chemical Energy

15

15.1 The Sun and Plants Provide Food and Energy

'The energy for life on Earth comes from the Sun as light energy and is driven by photons to electrons to protons to form chemical energy in food for plants and animals'

In photosynthesis, plants utilise photons as the source of light energy, released by the ultimate provider, the sun, plus carbon dioxide from the air, some of which, humans and other animals breathe out, as the source of carbon and also water from the by-product of animal respiration and all other sources of water on the earth..........all to generate food for energy and growth. The water is uniquely oxidised during the photolysis phase of photosynthesis and acts as the initial electron provider, donating electrons and protons to the coenzyme ($NADP^+$). The electrons reduce the $NADP^+$ to NADPH, which then provides the energy to reduce carbon dioxide, 'fixing' the carbon in the Calvin cycle, to produce food for energy in the form of glucose, plus other sugars, starch and cellulose. Oxygen atoms, released from this process, are combined to form molecular oxygen and released to the air as a by-product. This oxygen is used by both animals and plants, where it acts in reverse, as the final electron acceptor, in the process of cellular respiration. During respiration, the food synthesised by plants is broken down, yielding chemical energy, obtained from the oxidation of 2 electron donor coenzymes, NAD^+ and FAD, within the food. Water and carbon dioxide are issued as final by-products of respiration and can be absorbed by plants as the carbon cycle continues.

> Photolysis: $2 H_2O + 2 NADP^+ + 8$ photons (light) $= 2$ NADPH $+ 2 H^+ + O$
> Calvin Cycle: $3 CO_2 + 6$ NADPH $+ 9$ ATP $+ 5 H_2O = (G3P)^{23} + 6 NADP^+ + 9$ ADP $+ 8$ Pi

The ingenious plants also generate their own energy from cellular respiration, in a similar way to animals, thus plants do both. They synthesise food via photosynthesis (glucose $C_6H_{12}O_6$) and then break down this food, via respiration for energy and growth. Animals can only do one, i.e., use food obtained from the plants for energy via cellular respiration and growth. The plants are therefore the primary producers, i.e., the Autotrophs, that synthesise their own food. The animals, including humans, are the Heterotrophs i.e. (the consumers), unable to synthesise their own food within their own bodies. This cycle represents the continuous exchange of carbon (the carbon cycle), water and oxygen between living organisms and the environment on Earth.

> Photosynthesis: Light Photons + Chlorophyll $+12H_2O + 6CO_2 = C_6H_{12}O_6 + 6H_2O + 6O_2$
> Respiration: $C_6H_{12}O_6 + 6O_2 = 6CO_2 + 6H_2O$

Thus, in summary, light energy from photons produced by the sun is used by plants to generate food as chemical energy and stored and used in plants. This energy is available to animals in the form of chemical energy and used by animals for feeding and the sequential and gradual, controlled release of energy via cellular respiration. The metabolic pathways to store, manage and regulate this cellular energy were discussed in Chapter 14 and the biochemistry and mechanisms to release this energy will now be discussed.

15.2 Cellular Respiration

Cellular respiration is the process through which the nutrients synthesised by the autotroph plants are finally converted into useful energy, either within the plants themselves for their own energy requirements, or converted for useful energy within the animals that ingest the plants.

It is a set of metabolic reactions within the cells, to convert biochemical energy from the food carbon chemistry into the phosphorous chemistry of a chemical compound called Adenosine Triphosphate (ATP). ATP is the principal carrier of chemical energy in cells. It is not the source of energy, as the actual primary source of energy is within the reactions of breaking and resynthesis of carbon to carbon and carbon to hydrogen bonds in glucose and fatty acids. ATP captures chemical energy obtained from the breakdown of food molecules from cellular respiration and releases this energy to fuel all cellular processes and metabolism in the organism.

ATP is the principal carrier or coupling agent and immediate donor of free chemical energy in cells and stores and transfers energy in all living cells in the organism. The high energy phosphate bonds in ATP are broken by hydrolysis to yield chemical energy for all living things. The energy is derived from the net energy change in the reactions, where the reformation of the more stable carbon dioxide and water molecules gives back more energy than the energy required to oxidise the glucose or fat molecules. In a typical cell, an ATP molecule is used within 1 minute of its formation. Bond energy may be calculated as follows, example glucose -2816 kJ/mol (673.04 kcal/mol Table 15.2).

$$\Delta He_r = \sum (\Delta H°_B \text{ Bond Energy Reactants}) - \sum (\Delta H°_B \text{ Bond Energy Products})$$

Food molecules are converted into the universal molecule of acetyl-CoA, which then runs through a series of chemical reactions to yield ATP in the cells' mitochondria which are the power house organelles within most cells in the body. Fuel molecules are organic carbon compounds that store energy in their chemical bonds and are capable of being oxidised, that is losing electrons, principally, carbohydrates and fat. During cellular respiration, the carbon atoms derived from carbohydrates, fats and to a much lesser extent, protein, lose electrons and are oxidised. Glucose is the principal source and primary reactant of readily available energy and ATP and can be oxidised for energy aerobically, or if oxygen is low or not present, anaerobically, via glycolysis, the first step of cellular respiration, then converted into pyruvate, then into acetyl-CoA, before entering the TCA cycle and the next stage, OXPHOS if oxygen is present. The TCA cycle will stop if oxygen is not present for OXPHOS.

Triacyclglycerols (TAGs) i.e., fats, in the form of either free fatty acids or glycerol, are also regularly used to generate energy from ATP, providing oxygen is present, especially tissues such as heart muscle. In this process, fatty acids are also converted to acetyl-CoA, via (β)-oxidation to enter the TCA cycle and OXPHOS for energy. The glycerol backbone of TAG can be converted to glucose via gluconeogenesis for energy or converted to lactate for energy via glycolysis. The body will sometimes also use limited quantities of protein as a source of energy in low quantities day to day depending on cell energy conditions, via gluconeogenesis. This occurs to any significant extent only if there are excess amino acids in the diet and the body is in starvation state, or experiences carbohydrate deprivation. Normally, excess protein is converted to ketones, other amino acids lost in urine, or some is converted to fat and stored as fat. However, when it is necessary to use amino acids for energy, due to lack of both glucose and fats, the nitrogen is removed from the amino acids to be converted to urea in the liver then excreted leaving the carbon skeletons of the amino acids to be converted to acetyl-CoA then to glucose via gluconeogenesis and or to enter the TCA cycle as intermediates and OXPHOS for energy.

The function of the TCA cycle (also known as the Krebs or Citric Acid Cycle) is the harvesting of high energy electrons from carbon in fuels i.e., the food we eat. Acetyl-CoA is the principle metabolic intermediate molecule that is produced in the break-down of carbohydrates, via oxidative decarboxylation of pyruvate from glycolysis, or long chain fatty acids, via beta (β)-oxidation, or some amino acids from protein by oxidative degradation. The TCA cycle removes electrons from acetyl-CoA and uses these electrons to reduce coenzyme carriers NAD^+ and FAD to form NADH and $FADH_2$ (3 pairs or 6 electrons to NAD^+ and 1 pair or 2 electrons to FAD). Cellular respiration is the breakdown of energy from food fuel substrates and is essentially the transfer of 4 pairs of electrons to electron carrier coenzymes, FAD and NAD^+ (one pair to FAD and three pairs to NAD^+) These 4 pairs are transferred for each acetyl group oxidised. Then, in the Electron Transport Chain (ETC), a proton (H^+) gradient is generated, as electrons flow from the reduced $FADH_2$ and NADH. The proton gradient created is then used to synthesise ATP i.e., energy. To release the energy, the ATP becomes hydrolysed into ADP, or further to AMP, and free inorganic phosphate groups. The process of ATP hydrolysis to ADP is energetically favourable, yielding a Gibbs-free energy of -7.3 kcal/mol. The electrons are finally transferred to the final electron acceptor, oxygen, to generate water. There are several stages in respiration. The majority of ATP energy generated from cellular respiration takes place in the engine rooms of the cells, the mitochondria and consist of glycolysis, then pyruvate oxidation, the TCA cycle and OXPHOS and described as follows.

15.3   Glycolysis - The Breakdown of Glucose to Pyruvate

The first stage of cellular respiration of carbohydrate is glycolysis (Figure 15.1). This takes place in the cytoplasm or cytosol of the cell and not in the mitochondria. Glycolysis does not require oxygen. During glycolysis, one glucose molecule is split to 2 molecules of pyruvate generating 2 ATP molecules and 2 molecules of NADH.

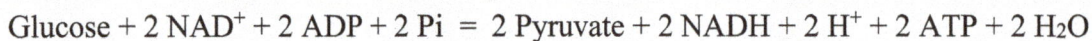

$$\text{Glucose} + 2\,NAD^+ + 2\,ADP + 2\,Pi \; = \; 2\,\text{Pyruvate} + 2\,NADH + 2\,H^+ + 2\,ATP + 2\,H_2O$$

The six-carbon sugar glucose molecule undergoes a series of chemical transformations. Ultimately, it is converted into two molecules of pyruvate, a three-carbon organic molecule. It is anaerobic as it evolved before sufficient oxygen was in the earth's atmosphere.

Figure 15.1 Glycolysis in animals. There are 10 stages as shown. In aerobic conditions Pyruvate enters the TCA cycle for full oxidation in the mitochondria with oxygen. In anaerobic conditions, the Pyruvate is converted to lactate, for example in Red Blood Cells. The number of carbon atoms per molecule is also shown as 6 or 3.

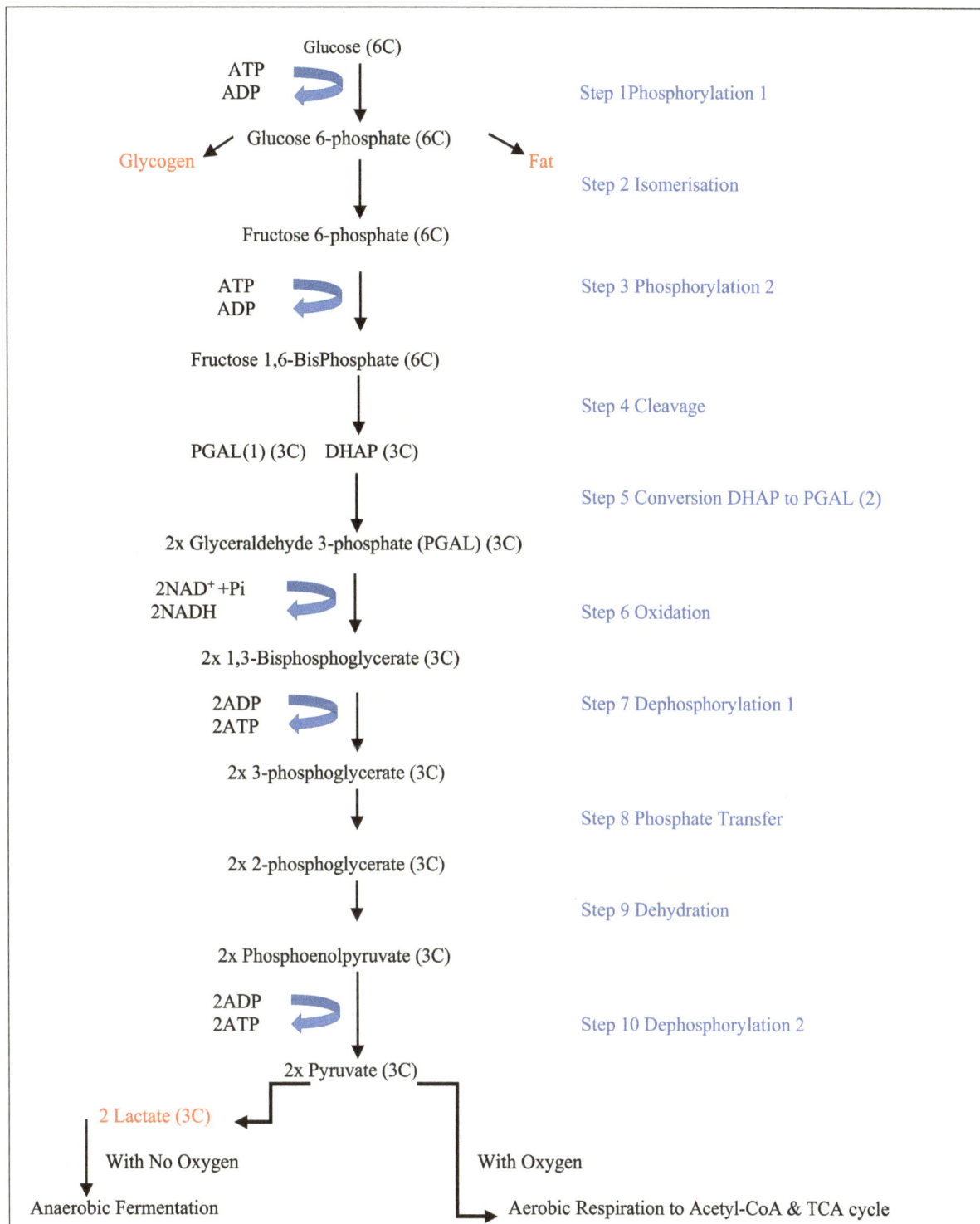

Under aerobic conditions, in cells that use aerobic respiration as the primary energy source, the pyruvate formed from the pathway can be used in the TCA cycle (citric acid cycle) then through to oxidative phosphorylation to be completely oxidised to CO_2. Under anaerobic conditions, pyruvate can be processed to form lactate in animals which is reversible, or ethanol in plants including yeast cells which is not reversible. In highly oxidative tissues, such as the heart, pyruvate production is essential for acetyl-CoA synthesis and L-malate synthesis. Each glucose molecule produces 2 pyruvate molecules. Pyruvate is an essential precursor to molecules such as lactate, alanine, and oxaloacetate and most pyruvate is provided by dietary carbohydrate. Glycolysis precedes lactic acid fermentation; the pyruvate made in the former process serves as the prerequisite for the lactate made in the latter process. Anaerobic lactate fermentation is the primary source of ATP in animal tissues with low metabolic requirements and few mitochondria. In RBCs, lactic acid fermentation is the only source of ATP, as the RBCs lack mitochondria and also, mature RBCs have relatively little demand for ATP compared with most other cells. Note the other routes from glucose 6-phosphate to glycogen storage or conversion to fat via DNL. The lens of the eye relies heavily on anaerobic glycolysis as it also has no mitochondria, as this would interfere with light transfer. Because glucose is such a precious fuel, metabolic products, such as pyruvate, lactate glycerol or amino acids can be salvaged to synthesise glucose via gluconeogenesis. The highly exergonic (release of energy), irreversible steps of glycolysis are thus by passed in gluconeogenesis as discussed in Section 14.16. The 2 processes of glycolysis and gluconeogenesis do not take place simultaneously in mammals.

15.4 Pyruvate Oxidation

We looked at the Cori cycle in the discussion of lactate in Section 14.19. The cycle allows glucose to be synthesised primarily from nonprotein precursors, therefore reducing proteolysis (muscle breakdown) from gluconeogenesis to a minimum in the early phase of food deprivation. When oxygen is present, there is plenty of NAD^+ and aerobic cells convert each of the 2 pyruvate molecules to 2 acetyl coenzyme A (CoA) for oxidation in the TCA acid cycle and aerobic respiration proceeds. Each of the two pyruvate molecule from glycolysis travels into the mitochondrial matrix. Here, pyruvate dehydrogenase (PDH) also known as pyruvate dehydrogenase complex (PDC) converts pyruvate into the two-carbon molecule bound to Coenzyme A, forming acetyl-CoA. PDC is one of the most important and largest and complex protein structures in the body, linking glycolysis with the TCA cycle. It converts pyruvate into acetyl-CoA irreversibly by a process called pyruvate decarboxylation. Acetyl-CoA can then be used in the TCA cycle, enabling cellular respiration and the release of significant energy for the cell. Alternatively, in the presence of abundant, accumulating acetyl-CoA, or high levels of ATP or, NADH, pyruvate can be acted on by a second enzyme, pyruvate carboxylase (PC), to replenish oxaloacetate (OAA) withdrawn from the TCA cycle for various pathways in an important anaplerotic reaction, portioning fuel towards gluconeogenesis. PC is involved in the election of pathways between glucose and fatty acid metabolism. In over feeding, or the well-fed state, or a high carbohydrate meal, insulin rises, PC is increased due to the high blood glucose and fatty acids are re-esterified and TAGs are stored in adipocytes. Carbon dioxide is released and NADH is generated. The main function of acetyl-CoA is to deliver the 2 carbon acetyl group, derived from pyruvate, to the next pathway step to condense with OAA in the TCA Cycle. Pyruvate is one of the major routes by which acetyl-CoA is generated.

$$\text{Pyruvate} + \text{NAD}^+ + \text{CoA} = \text{Acetyl-CoA} + \text{NADH} + \text{H}^+ + \text{CO}_2 \ \Delta \text{Go} = -8 \text{ kcal/mol}$$

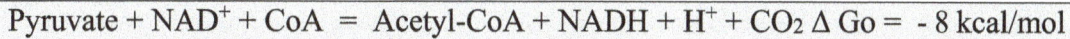

Because of the high exergonic nature of this reaction, this link between glycolysis and the TCA cycle is an irreversible reaction. The reverse reaction, i.e., the conversion of acetyl-CoA to pyruvate, does not occur and animals are unable to synthesise glucose from fatty acids. When oxygen is absent, NAD^+ levels can go down, and to prevent this from happening, lactate dehydrogenase uses NADH and pyruvate is converted to either lactate (animals) or ethanol (bacteria/yeast). In the Cori cycle (Figure 14.8), the skeletal muscle and RBCs consume glucose to produce lactate via glycolysis. Lactate is then transferred to the liver hepatocytes for gluconeogenesis acid and converted back to pyruvate to produce glucose, hence tidying up the liver lactate waste product and ensuring an appropriate level of lactate in the tissues. Anaerobic conversion of NADH to NAD^+ provides much less ATP energy to cells (2 ATP) compared with the ATP produced when oxygen is present (30 ATP). During hepatic gluconeogenesis 6 ATP is used up providing a net final consumption of 4 ATP in each cycle. This means that the Cori cycle can only be a stop gap process and can-not be a permanent process for energy production. Pyruvate is formed as the end product of glycolysis and then pyruvate dehydrogenase (PDH) converts pyruvate into the two-carbon molecule bound to Coenzyme A, forming the universal energy release intermediate molecule of acetyl-CoA.

When there is sufficient oxygen or when exercise intensity subsides, acetyl-CoA can then enter the TCA cycle facilitating cellular respiration and the release of significant energy for the cell from glycogen derived glucose or fatty acids released from lipolysis.

15.5 The Tricarboxylic Acid Cycle (Krebs/Citric Acid Cycle)

The Tricarboxylic acid cycle (TCA cycle), (Figure 15.2), is the pivotal metabolic pathway consisting of a cyclic series of chemical reactions that harness high-energy electrons from carbohydrate, fat and amino acid fuel sources. It is the body's primary pathway for oxidising fuels in the presence of oxygen. However, it may have evolved before earth's atmosphere became oxygenated and the original role of the TCA cycle was probably the biosynthesis of other molecules used throughout the body which it still does. Therefore, it is the key metabolic cycle for both the release of energy and synthesis of molecules. Enzymes catalyse the removal of electrons from energy substrates, such as pyruvate and acetyl CoA, and transfer these electrons to electron carriers NAD^+ and FAD, producing NADH plus H^+, and FADH_2 respectfully. The TCA cycle is the important evolutionary source of precursors to build many other molecules such as nucleotide bases, porphyrin and amino acids and as such, remains the key engine room for the whole body, coordinated by the liver and brain. For example, citrate can be used for fatty acid synthesis, while oxaloacetate can be used for nonessential amino acids (glucogenic amino acids or most of the nonessential amino acids, e.g., alanine, serine, cysteine, glycine). During low glucose conditions, these amino acids can produce glucose and α-ketoglutarate can be used for glutamic acid synthesis. Biologists have long discussed and still debate the chemical origins of life. The TCA cycle is pivotal in many of the theories developed, as this cycle supplies acetyl-CoA, pyruvate, oxaloacetate, succinate, and alpha-ketoglutarate, which are the five universal metabolic precursors for the five biosynthetic molecules, lipids, glucose, nucleic acids, amino acids, and cofactors.

The TCA cycle creates the pathway to generate energy from feeding. The majority (>90%) of the energy released from food is derived from oxidation via the TCA cycle and OXPHOS within the cell's mitochondria (Figure 15.2). The prime function of the TCA cycle is not to produce energy but to remove electrons from input molecules (acetyl-CoA) and transfer these electrons to electron carriers that deposit these electrons to the ETC used to power OXPHOS to produce cellular energy. It is the ETC that generates the most energy and not the TCA cycle itself. The reactions of the TCA cycle serve to oxidise the acetyl group in acetyl-CoA to CO_2 and generate NADH and $FADH_2$ and can be summarised as:

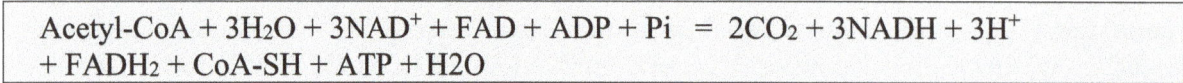

$$\text{Acetyl-CoA} + 3H_2O + 3NAD^+ + FAD + ADP + Pi = 2CO_2 + 3NADH + 3H^+ + FADH_2 + CoA\text{-}SH + ATP + H2O$$

It provides a common pathway for the final oxidation of all metabolic fuels, including directly as carbohydrates, fatty acids or amino acids, or indirectly, via pyruvate or acetyl-CoA for lactate, glycerol and ketone bodies. The final step of the TCA Cycle regenerates oxaloacetate (OAA). The OAA molecule is needed for the next revolution of the cycle. Two revolutions are required because glycolysis produces two pyruvate molecules when it splits glucose. (glycolysis and pyruvate. The TCA Cycle produces; 2 ATP, 8 NADH, 2 $FADH_2$, 6 CO_2 (2 from transformation of acetyl-CoA and 4 from the TCA Cycle). Figure 15.2.

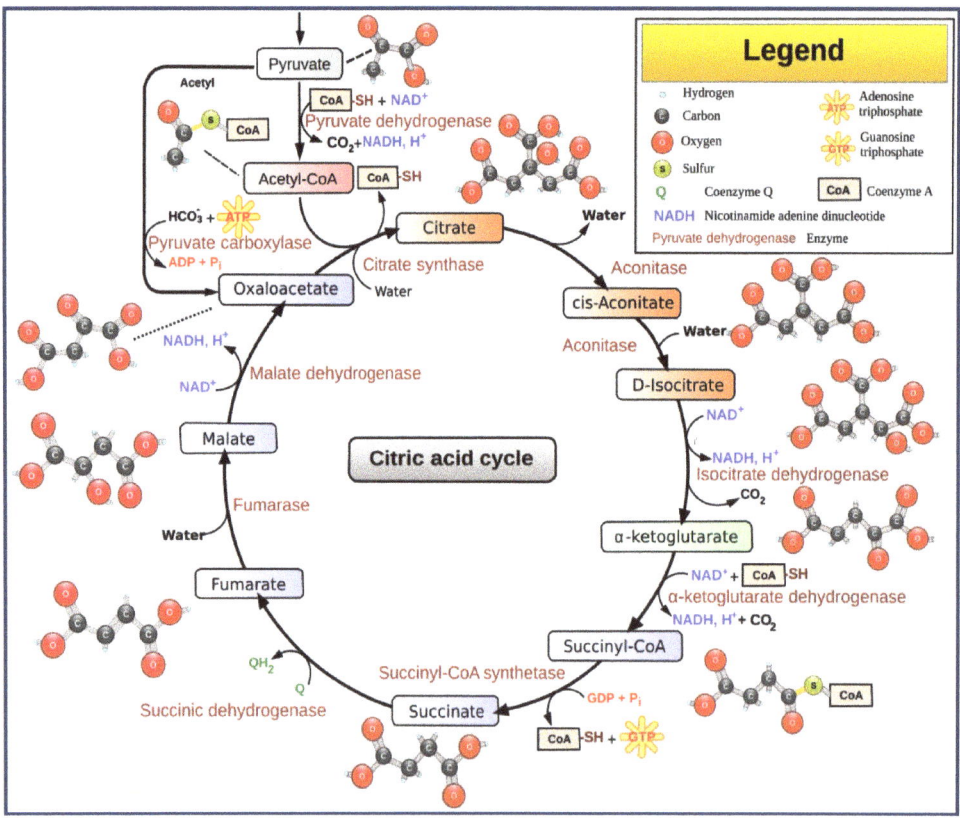

Figure 15.2 Citric acid (TCA cycle, Krebs Cycle, Tricarboxylic Cycle) Source Wikipedia WikiUserPedia, Wikimedia Commons. https://en.wikipedia.org/wiki/Author: Narayanese, Yassine Mrabet, Toto Baggins. Oct 2008 http://biocyc.org/META/NEW-IMAGE?type=PATHWAY&object=TCA
Wikipedia Creative Commons Attribution-Share Alike 3.0 Unreported

When the 2 carbon atom molecule of acetyl-CoA combines with the 4 carbon atom molecule of OAA, this produces the 6 carbon atom molecule citric acid. This important step is the reason why the TCA Cycle is also called the Citric Acid Cycle. After citric acid forms, there are a series of reactions that release energy. This energy is captured in molecules of ATP and the 2 electron carriers. The TCA Cycle has two types of energy-carrying electron carriers: NAD^+ and FAD. The transfer of electrons to FAD during the TCA Cycle produces a molecule of reduced $FADH_2$. CO_2 is also released as a by-product of these reactions. The rate-limiting step in the TCA cycle is the combination of oxaloacetate and acetyl-CoA producing citrate. It is for this reason that it is essential to have adequate quantities of pyruvate, oxaloacetate, carbohydrate and fatty acids in order to produce ATP and maintain cell viability. The acetyl-CoA, produced in pyruvate oxidation combines with the four-carbon molecule oxaloacetate. OAA goes through a cycle of reactions, ultimately regenerating the four-carbon starting molecule. OAA is fundamental to the energy generation process because sufficiency of the donor macronutrient, that is carbohydrate, provides the key to the most efficient oxidation i.e., burning of appropriate levels of both carbohydrate and fat, which is key to maintaining healthy weight. Free, or non-esterified free fatty acids, enter the TCA cycle as acetyl-CoA and are fully oxidised to carbon dioxide and water. This process of fatty acid (β)-oxidation can-not contribute carbon for the synthesis of glucose as the conversion of pyruvate to acetyl-CoA is irreversible (Section 14.12) in animals.

15.6 Oxidative Phosphorylation - The Generation of Most ATP

Following glycolysis and pyruvate oxidation, the next stage in energy synthesis, is oxidative phosphorylation (OXPHOS). OXPHOS is the final stage of aerobic cellular respiration and illustrated in Figure 15.3.

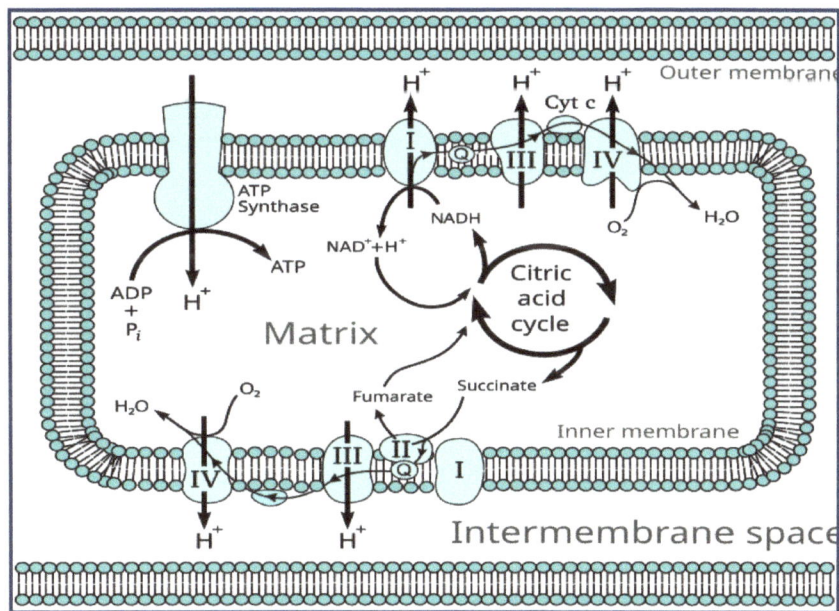

Figure 15.3 Oxidative Phosphorylation to release energy from ATP in the Mitochondria Author Fvasconcellos Sep 2007 Wikipedia Public Domain
Figure 15.3 Oxidative Phosphorylation to release energy from ATP in the Mitochondria Author Fvasconcellos Sep 2007 Wikipedia Public Domain

There are two distinct substages of OXPHOS; the electron transport chain (ETC) and Chemiosmosis.

In these final 2 stages, energy from NADH and $FADH_2$, is used to create a significant amount of ATP molecules, thereby releasing and generating the majority of energy from the food substrates.

This is where the vast majority of energy is released from fuel substrates and transferring electrons to the ultimate electron acceptor oxygen to form water.

The net ATP yield from the oxidation of one mole of glucose via glycolysis, the TCA cycle and OXPHOS is estimated to range from 30 to a maximum of 38 ATP. For this calculation we have used the more conservative 30 moles of ATP including 2.5 ATP per NADH and 1.5 ATP per $FADH_2$ and the glycerol 3 phosphate shuttle mechanism described in Section 15.10. We have also calculated 106 moles of ATP generated from the 16 carbon fatty acid rather than the often quoted 126 moles of ATP.

The delta-Gibbs energy of ATP hydrolysis equals -7.289 kcal/mol. Therefore 30 ATP provides 219 kcal or 916 kJ. This oxidation is around 32% efficient in terms of energy production. The remaining 68% is released as heat energy, largely to maintain core body temperature so it is not wasted as the body does not waste energy, or more correctly use energy, unless it serves another necessary purpose.

OXPHOS essentially allows the flow of electrons from NADH to $FADH_2$ and ultimately to the final electron acceptor, molecular oxygen that humans and other animals breathe in.

The NADH and $FADH_2$ produced in the previous stages of cellular respiration, deposit their electrons into the ETC, reverting to their oxidised forms NAD^+ and FAD. As electrons move down the chain, energy is released and used to pump H^+ protons from the mitochondrial matrix, forming a proton gradient.

Protons flow back into the matrix through the enzyme called ATP synthase, making ATP. At the end of the ETC, oxygen accepts electrons and takes up protons to form water, the opposite of its role as electron donor in photosynthesis.

$$4H^+ + 4e^- + O_2 = 2H_2O$$

Once energy is extracted, carbon dioxide (CO_2), derived from the carbon atoms in the food, are ultimately converted to CO_2 and exhaled. CO_2 and water are essentially unused by products of respiration. Therefore, the triad of final products of food break down is:-

1) ATP Chemical Energy and Thermal Heat Energy
2) Carbon Dioxide exhaled
3) Water

The ETC and Chemiosmosis are now discussed in further detail.

15.7   The Electron Transport Chain (ETC)

The electron transport chain (ETC) essentially generates an electrochemical gradient. In the ETC (Figures 15.3 and 15.4), high-energy electrons are released from NADH and FADH$_2$ from the TCA cycle, and move along the electron-transport chains in the inner membrane of the mitochondrion. The ETC is a series of molecules and protein complexes that transfer electrons from molecule to molecule by chemical reactions. These 2 electron carriers firstly become reduced by accepting electrons from the hydrogen atoms during the catabolic steps in the breakdown of glucose and fatty acids. The reduced forms of these coenzymes become NADH and FADH$_2$. NAD$^+$, in its oxidised form, accepts 2 electrons to become reduced NADH. FAD, in its oxidised form, accepts 2 hydrogen atoms, each with one electron, to become reduced FADH$_2$. These reduced coenzymes then donate and transport the electrons to the ETC.

The electrons from NADH are at a very high energy level and are transferred directly to electron carrier Complex I electron carriers thus becoming re-oxidised back to NAD$^+$.

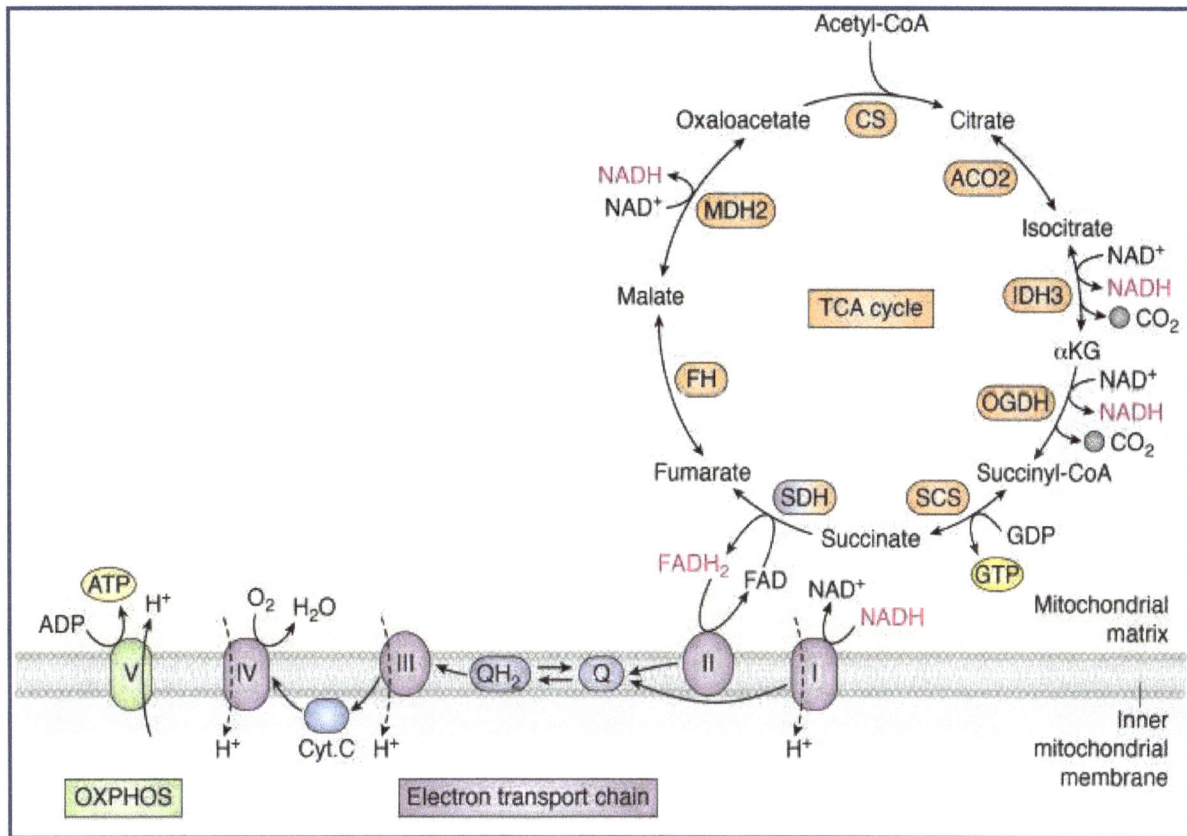

Figure 15.4 Overview of the tricarboxylic (TCA) cycle and electron transport chain (ETC) in the Mitochondria
Source: Regulation of the mammalian tricarboxylic acid cycle Paige K. Arnold and Lydia W.S. Finley Feb 2023 JBC National Library of Medicine kind permission open access article J Biol Chem. 2023 Feb;299(2):102838. doi: 10.1016/j.jbc.2022.102838. Epub 2022 Dec 26. PMID: 36581208; PMCID: PMC9871338.open access article under the CC BY license (http://creativecommons.org/licenses/by/4.0/).
see References [42].

The high energy electrons from the NADH are then used to pump protons from the matrix into the intermembrane space. The energy released in this process creates a proton gradient.

This ion transfer creates an electrochemical gradient that ultimately drives the ATP synthesis in Chemiosmosis. Enzymes catalyse the removal of 2 protons, (H^+) and 2 electrons ($2e^-$) from hydrogen atoms.

These electrons are transferred to two electron acceptor coenzymes; Nicotinamide Adenine Dinucleotide (NAD^+) and Flavin Adenine Dinucleotide (FAD). NAD^+ is derived from Vitamin B3 (niacin) and FAD is derived from Vitamin B2 (riboflavin).

The NAD^+ and FAD coenzymes are involved in reversible oxidation and reduction reactions.

The $FADH_2$ transferred electrons are in a lower energy state compared with those from NADH and can-not transfer electrons to Complex I but instead these electrons are supplied into the ETC via Complex II which does not pump protons across the membrane.

Because of this by pass system, each $FADH_2$ molecule causes fewer protons to be pumped and therefore $FADH_2$ contributes less to the proton gradient than NADH.

Beyond the first 2 electron carrier Complexes, electrons from both NADH and $FADH_2$ travel the same route.

Both Complex I and II pass their electrons to a mobile electron carrier called Ubiquinone Q which is then reduced to form QH2 and travels through the membrane delivering the electrons to Complex III. As electrons move through Complex III, more H^+ ions are pumped across the membrane, and the electrons are ultimately delivered to another 'mobile called cytochrome C (Cyt. C).

Cyt. C carries the electrons to Complex IV, where a final batch of H^+ ions is pumped across the membrane. (Figures 15.3 and 15.4)

Complex IV passes the electrons to O_2, which splits into two oxygen atoms and accepts protons from the matrix to form water H_2O. Four electrons are required to reduce each molecule of O_2, and two water molecules are formed in the process.

The 4 Complexes and cytochrome C are summarised as follows:

Complex I	NADH-Q oxidoreductase
Complex II	Succinate-Q reductase
Complex III	Q-cytochrome c oxidoreductase
Cytochrome Cyt	C Mobile Carrier
Complex IV	Cytochrome c oxidase

15.8 Chemiosmosis and ATP Synthase

Chemiosmosis is the pumping of hydrogen ions across the inner membrane. This action creates a greater concentration of these hydrogen ions in the intermembrane space compared with the concentration within the matrix, producing an electrochemical gradient that drives the synthesis of ATP via ATP synthase. This gradient then eventually causes the ions to flow back across the membrane into the matrix, where their concentration becomes lower. The flow of these ions occurs through a highly mobile protein Complex, known as the ATP synthase complex (referred to as Complex V). The ATP synthase (Figures 15.5 and 15.6) is a most remarkable complex enzyme protein that acts as a channel protein molecular motor and cellular turbine, facilitating the transfer of hydrogen ions across the membrane. The protein has a molecular weight of around 600,000 Daltons. Proton flow through ATP Synthase leads to the release of tightly bound ATP. The proton gradient is used to synthesise significant quantities of high energy adenosine triphosphate ATP via chemiosmosis. Protons are unable to diffuse through the phospholipid bilayer, so instead go through the enzyme ATP synthase.

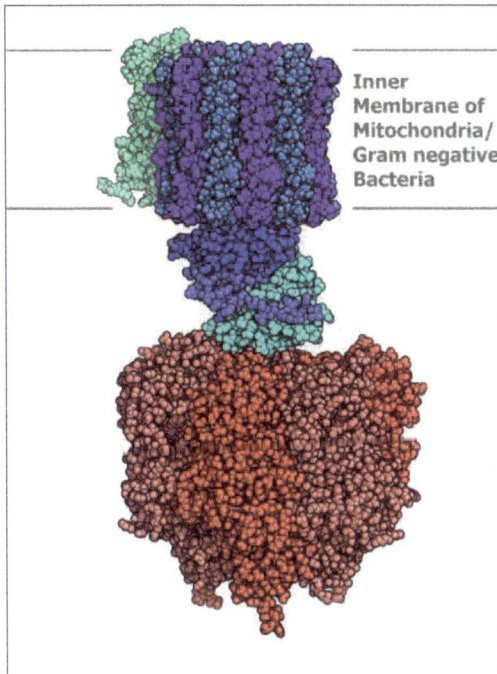

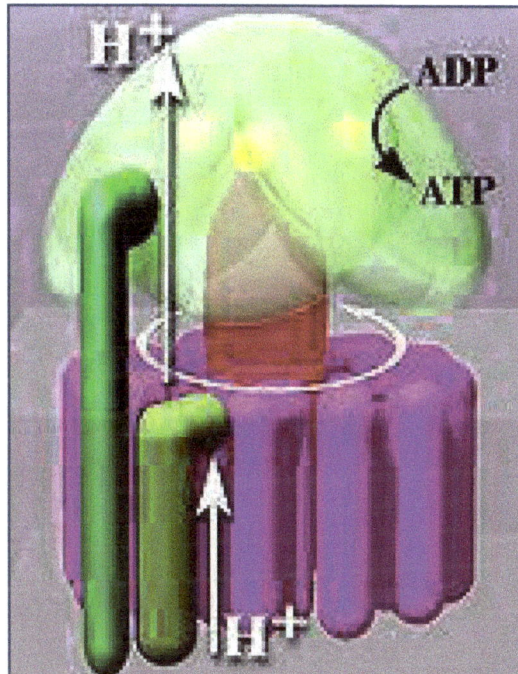

Figure 15.5 Molecular Model of ATP Synthase determined from X-ray crystallography Source: Wikipedia Commons Author Alex. X File licenced under the Creative Commons Attribution- Share Alike 3.0 Unreported Licence

Figure 15.6 Rotation Engine of ATP Synthase Source: Wikipedia Transferred from de.wikipedia to Commons Original uploader Asw-hamburg at German Wikipedia by Leyo using Commons Helper. File licenced under the Creative Commons Attribution- Share Alike 3.0 Unreported Licence

Proton movement provides potential energy, which rotates a section of ATP synthase, and causes phosphorylation of ADP (ADP + Pi →ATP). The overall reaction catalysed by ATP synthase is:

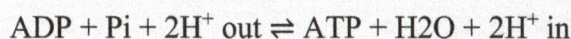

$$ADP + Pi + 2H^+ \text{ out} \rightleftharpoons ATP + H2O + 2H^+ \text{ in}$$

15.9   ATP Recycling

A sedentary lean male of 70 kg requires 8,400 kJ (2,000 kcal) for a day's activity (some males may require nearer 2,500 kcal/d but these figures will be used in this example). This requires 83 kg of ATP. However, humans only possess 250g of ATP at any given moment. This shortfall is compensated by recycling ADP to ATP. Each ATP molecule is recycled approximately 300 times per day. This recycling takes place primarily through OXPHOS. Anaerobic glycolysis; i.e., the 1st stage of glucose breakdown, occurs in the cytoplasm and results in a small amount of energy (<10% total energy released from cellular respiration). This is the sole source of energy for the RBCs. The brain and gastrointestinal tract (GI) derive most of their energy from aerobic glycolysis most derived from the tricarboxylic acid cycle (TCA) cycle. Most of the energy derived from carbohydrates is in the liver, muscle and adipose (fat) tissue.

We will now review the calculation of ATP derived from glucose, the fatty acid palmitate and 2 ketones, acetoacetate and beta hydroxybutyrate. Section 15.10 calculates the total ATP mole yield from the oxidation of 1 mole of the 6-carbon carbohydrate glucose.

15.10 ATP yield - Glucose Oxidation (per mole of glucose)

Glycolysis in cytoplasm Conversion of Glucose to Pyruvate

Phosphorylation of glucose	= - 1 ATP
Phosphorylation of fructose 6-phosphate	= - 1 ATP
Dephosphorylation of 2 molecules of 1,3-BPG	= + 2 ATP
Dephosphorylation of 2 molecules of phosphoenolpyruvate	= + 2 ATP

2 molecules of NADH are formed in the oxidation of 2 moles
of glyceraldehyde 3-phosphate

Conversion of pyruvate to acetyl CoA in mitochondria 2 moles NADH formed

TCA Cycle Inside Mitochondria

2 moles of adenosine triphosphate formed from 2 molecules succinyl CoA	= + 2 ATP

6 molecules of NADH are formed in the oxidation of 2 molecules of each of
isocitrate, alpha-ketoglutarate and malate
2 molecules of $FADH_2$ are formed in the oxidation of 2 molecules of succinate

Oxidative Phosphorylation Inside Mitochondria

2 molecules of NADH formed in glycolysis; each yield 1.5 moles of ATP (assuming transport of NADH by the glycerol 3-phosphate shuttle)	= + 3 ATP
2 molecules of NADH formed in the oxidative decarboxylation of pyruvate each yield 2.5 moles of ATP	= + 5 ATP
2 molecules of $FADH_2$ formed in the TCA cycle; each yield 1.5 moles of ATP	= + 3 ATP
6 molecules of NADH formed in the TCA cycle; each yield 2.5 moles of ATP	= +15 ATP

Net ATP Yield per Mole of Glucose   = **30 ATP**

The derived value of 30 ATP per molecule of glucose supercedes earlier and often quoted estimated values of 36 or even 38 net ATP. The stoichiometrics of proton pumping, ATP synthesis and metabolite transport should be regarded as estimates. Approximately 2 more molecules of ATP are formed per molecule of glucose oxidised when the malate-aspartate shuttle, rather than the glycerol 3-phosphate shuttle is used.

In contrast, even though glucose is more readily available, fats yield far higher overall yield of energy compared with glucose i.e., 106 moles of ATP a substantial 76 more molecules of ATP per molecule of fat compared with glucose or 2.5 times more ATP. By example, 15.10 below shows the ATP yield from one molecule of palmitate, a common 16 carbon fatty acid.

The principal reason why fats yield much more energy than carbohydrates is due to the high amount of hydrogen-carbon bonds in fats than carbohydrates. Therefore, fats are much more reduced than carbohydrates and carbohydrates are already more oxidised than fats. Fatty acids have a much higher ratio of hydrogen to oxygen atoms compared to glucose. Glucose is already partially oxidized (less reduced than fats); the carbon atoms already have oxygen atoms attached to them. Consequently, they have less energy to give in the reaction with oxygen to release energy compared with the more reduced fats. Fatty acids have a much higher ratio of hydrogen to oxygen atoms compared to glucose and hence produce more energy per mole.

In lipid (fat) (β)-oxidation both the carbon and hydrogen are oxidised but only the carbon is being oxidised in glucose. The 6 carbon atoms in glucose have oxidation states of -1, 0 or +1. For fats, however, the oxidation states are greater. For example, in palmitic acid, all but one of the 16 carbon atoms have oxidation states of -2 or -3. Hence, in addition to more than twice the number of carbon atoms, those carbon atoms in fatty acids have considerably more electrons available for reaction within their atomic orbitals. Energy is released when electrons move from an atom such as carbon, with low electronegativity, i.e., a low affinity for electrons, to one with a high affinity for electrons, like oxygen.

Significantly more total energy is released when fatty acids are oxidized compared with glucose, principally due to the higher number of electrons around the carbon atoms being transferred to oxygen and thus a lower number of electrons transferred when carbohydrates are oxidised. For example, the palmitate molecule contains just two oxygens per sixteen carbons, whereas glucose has six oxygen atoms per six carbons. It takes 6 moles of oxygen to oxidise 1 mole of glucose compared with 23 moles of oxygen to oxidise 1 mole of palmitate.

Carbohydrates (carbon and that is hydrated) have the approximate formula $(CH_2O)^n$ and the formulae for fats is $(CH_2)^n$ hence illustrating the relative sparsity of oxygen. Therefore, on a weight/density basis, a gram of fatty acids will yield much more energy than a gram of glucose. Section 15.11 calculates the total ATP mole yield from the oxidation of 1 mole of the 16-carbon palmitate fatty acid.

227

15.11 ATP yield - Fatty Acid Oxidation (per mole of fat)

Each Palmitate fatty acid compound has 16 carbons
Degradation of this requires 7 reaction cycles
In each (β)-oxidation cycle and acyl CoA is reduced by 2 carbon atoms
For each cycle,1 molecule each of $FADH_2$, NADH & acetyl CoA are formed
In the 7th cycle the C4 ketoacyl CoA is thiolysed to 2 moles of acetyl CoA
Approx. 2.5 moles ATP formed when resp' chain oxidises each NADH x 7 = +17.5 ATP
Approx. 1.5 moles ATP formed when resp' chain oxidises each $FADH_2$ x 7 = +10.5 ATP
Oxidation of acetyl CoA by the TCA cycle yields 10 moles ATP x 8 = +80.0 ATP

Total ATP produced = +108.0 ATP
But 2 ATP for palmitate initial activation energy = - 2.0 ATP
Net ATP yield from oxidation of 1 mole of palmitate fatty acid =   106 ATP

15.12 ATP yield from Ketone Bodies (per mole of ketone)

Two ketone bodies yield energy: acetoacetate and β-hydroxybutyrate.

Acetoacetate is converted into acetoacetyl-CoA by β-ketoacyl-CoA transferase
No energy is required as the CoA is transferred from succinyl CoA

Acetoacetyl-CoA is then converted into 2 acetyl-CoA by the thiolase reaction
No energy is required

The 2 acetyl-CoA enter TCA cycle generating 10 ATP per turn 10 x 2 = +20.0 ATP

β-hydroxybutyrate is converted into acetoacetate by β-hydroxybutyrate
dehydrogenase yielding 1 NADH and generates 2.5 ATP by OXPHOS = + 2.5 ATP
breakdown of acetoacetate yields 2 acetyl-CoA and no NADH = +20.0 ATP
Therefore, the yield from BHB is 2.5 more ATP than acetoacetate = +22.5 ATP

Net ATP yield from oxidation of 1 mole of acetoacetate = 20.0 ATP

Net ATP yield from oxidation of 1 mole of β-hydroxybutyrate = 22.5 ATP

15.13 Comparative Oxidation of Glucose and Fatty Acids

Table 15.1 shows the chemical properties of oxygen, hydrogen and carbon within food substrates and Table 15.2 shows the comparative yield of ATP and Oxygen required to oxidise glucose and the fatty acid palmitate to carbon dioxide and water. Table 15.3 shows the calculations.

Table 15.1 Chemical Properties of Oxygen, Hydrogen, Carbon within Food Substrates and ATP free energy

Factor		Units (SI)	Oxygen	Hydrogen	Carbon	ATP
(a)	Chemical formula (not gaseous)		O	H	C	$C_{10}H_{16}N_5O_{13}P_3$
(b)	Molar mass (molecular weight) [1]	g/mol	16.00	1.008	12.011	507.18
(c)	Molar mass of the gas	g/mol	32.00	2.016		
(d)	Mole per gram (mass/molar mass)	mol/g	0.0625	0.496	0.0833	
(e)	Free energy $\Delta G°$ hydrolysis of ATP	kcal/mol				-7.289
(f)	ATP $\Delta G°$ Intracellular conditions	kcal/mol				-13.623
(g)	Density of Oxygen	kg/m3				1.429

Table 15.2 Estimated Energy yield and Oxygen consumption for Glucose and Palmitate oxidation

Factor	Description	Units (SI)	Glucose	Palmitate
(h)	Chemical formula		$C_6H_{12}O_6$	$C_{16}H_{32}O_2$
(i)	Molar mass (molecular weight)	g/mol	180.156	256.43
(j)	Enthalpy of combustion of substrate	kcal/mol	677.21	2397.37
(k)	Substrate mole per gram	mol	0.0055507	0.0038996
(l)	ATP generated per mole oxidised [1]	mol/mol	30.0	106.0
(m)	Moles of ATP generated per gram oxidised	mol/g	0.1665	0.4133
(n)	Total ATP free energy per mole oxidised	kcal/mol	218.67	772.63
(o)	Energy content per gram of substrate	kcal/g	3.76	9.35
(p)	Moles of oxygen required to oxidise 1 mole	mol/mol	6	23
(q)	Mass of oxygen required to oxidise 1 gram	g/g	1.07	2.87
(r)	Volume of oxygen required to oxidise 1 mole	L/mol	134.4	515.2
(s)	Volume of oxygen required to oxidise 1 gram	L/g	0.75	2.00
(t)	Energy expenditure equivalent oxygen uptake	kcal/L	5.03	4.65
(u)	Oxygen uptake to generate each mol of ATP	L/mol	4.48	4.86
(v)	Mass of ATP produced per mass of substrate	g/g	84.45	209.65
(w)	Energy efficiency to generate ATP	%	32.29%	32.23%
(x)	Oxidation of glucose equation [2]		$C_6H_{12}O_6 + 6\,O_2 \rightarrow 6\,CO_2 + 6\,H_2O$ + energy	
(y)	Oxidation of palmitate equation [2]		$C_{16}H_{32}O_2 + 23\,O_2 \rightarrow 16\,CO_2 + 16\,H_2O$ + energy	

Notes: [1] From Section 15.10 & 15.11 [2] energy & heat

Table 15.3 Calculations for Table 15.2

Factor	To Calculate	Factor 1	Sign	Factor 2	Sign	Factor 3
(k)	Substrate mole per gram	1.	÷	(i)		
(m)	Moles of ATP mol generated per gram oxidised	(l)	÷	(i)		
(n)	Total ATP free energy per mole oxidised	(e)	x	(l)		
(o)	Energy content per gram of substrate	(j)	÷	(i)		
(p)	Moles of oxygen required to oxidise 1 mole	(x) or (y)				
(q)	Mass of oxygen required to oxidise 1 gram	(c)	x	(p)	÷	(i)
(r)	Volume oxygen required to oxidise 1 mole	(p)	x	(c)	÷	(g)
(s)	Volume of oxygen required to oxidise 1 gram	(q)	÷	(g)		
(t)	Energy expenditure equivalent oxygen uptake	(j)	÷	(r)		
(u)	Oxygen required to generate each mol of ATP	(r)	÷	(l)		
(v)	Mass of ATP produced per mass of substrate	(b)	x	(l)	x	(i)
(w)	Energy Efficiency to generate ATP	(n)	÷	(j)		

Compare the bioenergetic results and oxygen cost of oxidising these two substrates.

Oxidation equations for glucose and palmitate are as follows:

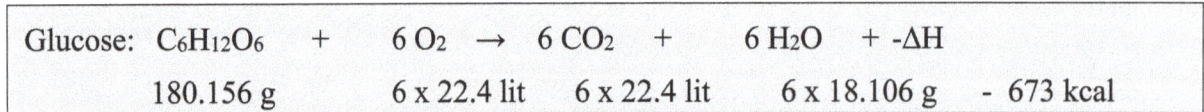

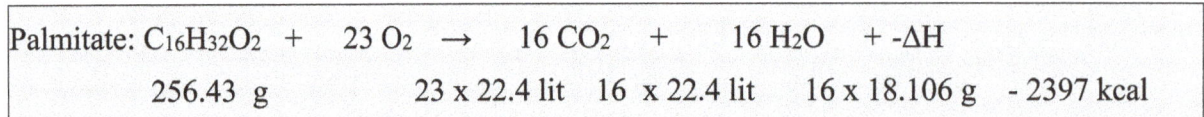

Glucose: $C_6H_{12}O_6$ + 6 O_2 → 6 CO_2 + 6 H_2O + -ΔH
180.156 g   6 x 22.4 lit   6 x 22.4 lit   6 x 18.106 g   - 673 kcal

Palmitate: $C_{16}H_{32}O_2$ + 23 O_2 → 16 CO_2 + 16 H_2O + -ΔH
256.43 g   23 x 22.4 lit   16 x 22.4 lit   16 x 18.106 g   - 2397 kcal

(ΔH° is the enthalpy change or heat produced. The negative sign means that heat is liberated).

Note the RQ for glucose 6 CO_2/6 O_2 is 1, whereas the RQ for palmitate is 16 CO_2/23 O_2 is 0.7

Note also that the value of the energy content per gram for glucose is calculated from the molar mass and enthalpy of combustion to yield 3.76. The widely accepted value of 4 kcal/g is based on the ground-breaking work of Atwater from over a century ago in 1910. They are averages of all carbohydrates and possibly included experimental errors. Therefore, the more accurate substrate specific value of 3.76 kcal/g for glucose is used here and the Enthalpy of combustion of substrate therefore, calculates to 677.21kcal/mol (2,833.45 kJ/mol) and not the often quoted value of 673.04 kcal/mol (2,816 kJ/mol). Sánchez-Peña et al., conducted very accurate and exhaustive studies derived from calorimetry in 2019. [68] What may we deduce from the information calculated in Table 15.2 regarding the comparison between the oxidation of glucose and the fatty acid palmitate?

(1) In terms of ATP and Energy Yield

Fat produces more ATP. Oxidation of palmitate fatty acid produces 3.5 times the amount of ATP (106 moles) compared than glucose (30 moles) and the heat of combustion of the fat (772 kcal/mol) is 3.5 times that of glucose (219 kcal/mol). Based on carbon atoms, palmitate fat produces one mole of ATP per 6.6 carbons, whereas one mole of glucose oxidises 5 carbon atoms per mole of ATP produced. Weight for weight, lipids have the highest energy density where palmitate produces 2.5 times the energy at 9.35 kcal/g compared with 3.76 kcal/g. Expressed per mol of ATP produced, 1 gram of glucose yields 0.166 mol of ATP and 1 gram of palmitate yields 0.413 mol of ATP. i.e., 2 and a half times as much ATP energy is generated by complete oxidation to carbon dioxide and water per gram of palmitate fatty acid than glucose. However, both are around 32% efficient. This ratio aligns closely with the ratio of 2.5 for substrate energy content per gram.

(2) In terms of Oxygen Consumption

Fat requires more oxygen to burn. Oxidation of palmitate fatty acid requires 3.8 times the total amount of oxygen at 23 moles/mole of palmitate compared to just 6 moles/mole glucose. Therefore, a total of 515 litres of O_2/mole is required to oxidise palmitate compared with 134 litres O_2/mol of glucose. On a weight for weight basis, palmitate has an oxygen demand 2.7 times the amount of oxygen compared with glucose, at 2.87g of O_2/g for palmitate, against 1.07 g of O_2/g for glucose. Similarly, on a volume per weight basis, palmitate requires 2.0 litres O_2/g palmitate compared with 0.75 litres of O_2/g for glucose.

When expressed as ATP energy produced per litre of oxygen consumed, there is an 8.2% higher ATP yield per litre of oxygen for the energy generated from glucose, at 5.03 kcal ATP/litre of O_2 against the lower yield of 4.65 kcal ATP/litre of O_2 from the oxidation of palmitate. Finally, the comparison of oxygen required per mole is 4.86 litres/mole O_2 required for the oxidation of palmitate compared with the lower oxygen required of 4.48 litres/mole O_2 for glucose and similar at a difference of 8.5%. Fat oxidation is 100% aerobic, whereas the glycolysis component of glucose oxidation is anaerobic. Of the total glucose energy expenditure of 5.03 kcal/O_2, this can be broken down to 4.88 kcal/L O_2 (93% of glucose oxidation for 28 mol ATP) as aerobic for the oxidation of pyruvate in TCA and OXPHOS and 0.15 kcal/O_2 (7% of glucose oxidation for 2 mol ATP) as anaerobic, during glycolysis. It should also be noted that both aerobic and anaerobic oxidation of fats and glucose entails inefficient component reactions, for example, ATP resynthesis and heat producing acid base reactions, where energy is lost to the environment. Hence, part of the 8% difference between fat and glucose oxidation yield and may well be due to glycolysis. Excluding the glycolysis component brings the substrates much closer in terms of oxygen requirement (4.88 kcal/mole for glucose and 4.65 kcal/mole for palmitate and the overall difference is probably <5%. The slower harvest lipolysis availability and then burning of fatty acids, necessitating approximately 7 rounds of beta oxidation before acetyl CoA is formed, compared to the less tortuous route of glucose oxidation is probably not therefore significant under most conditions at most times as oxygen availability is usually plentiful. However, this difference does become much more important during intensive exercise.

(3) In terms of Bioenergetics Efficiency

Calculated Gibbs enthalpy change in free energy $\Delta G°$ for the hydrolysis of one mole of ATP into ADP and inorganic phosphate Pi is −7.3 kcal/mol. This is true under standard conditions. Using the free energy $\Delta G°$ of −7.3 kcal/mol, calculates to and efficiency of 32% as follows:

Glucose	218.68/677.21	=	32.29%
Palmitate	772.63/2397.37	=	32.22%

Under standard conditions, glucose reacts with oxygen during respiration to release Gibbs free energy. This is an exergonic reaction and energy is coupled to controlled step by step cellular reactions and released during hydrolysis of ATP Standard conditions assume all reactants and products are in equal concentrations of 1 mole. However, intracellular conditions differ and for example, the concentrations of ATP are higher than ADP favouring the rate of hydrolysis. This indicates that the overall efficiency of oxidising glucose and fatty acids is virtually the same. This makes sense as the body relies on both glucose and fat for energy and if the pathway to oxidising fat was too onerous or too bioenergetically expensive, the body would be compromised during times of carbohydrate shortage or indeed to sustain tissues such as the heart overnight. For ATP hydrolysis in the cell, the theoretical $\Delta G°$ for the hydrolysis of one mole of ATP in intracellular conditions within a living cell mitochondrion is almost double the value of -7.3 kcal/mol at -13.62 kcal/mol at standard conditions.

Using this intracellular $\Delta G°$ for the hydrolysis of ATP yields efficiency of double at around 60% as follows:

Glucose	408/677.21	=	60.25%
Palmitate	1444/2397.37	=	60.23%

Most texts quote respiration oxidation efficiency levels ranging from 38% to 40% of converting chemical energy into external work. Most of these calculations are often built on the assumption that glucose produces 38 moles of ATP and fatty acids produce 129 mol of ATP.

It also depends on whether Gibbs enthalpy free energy $\Delta G°$ free enthalpy or intracellular free energy is used and most use the standard $\Delta G°$ of -7.3 kcal/mol.

Some texts even refer to 25% as the real metabolic cost efficiency. Theoretical efficiency and actual efficiency may differ widely because of the complexity of pathways within each cell and mitochondria. Within the cells there are a multitude of complex intermediate compounds and these metabolic pathways are used for other purposes.

For example, glucose oxidation connects with the processes that synthesis or break down an array of other biochemical compounds in cells. Some of these reactions do not run to fruition and there are waste products to deal with.

The five-carbon sugars are an integral part of nucleic acid structure made from intermediates in glycolysis. As we have seen, some non-essential amino acids are made from intermediates of both glycolysis and the TCA cycle.

Lipids, such as cholesterol and TAGs are also synthesised from intermediates in these pathways. Energy is also lost from the proton gradient, phosphate carriers and ATP/ADP exchangers within the inner membrane of the mitochondria. It would be more efficient if these reactions were carried out on the outer membrane to rely less on carriers.

This possibly developed as the original ancestral mitochondria (like their sister chloroplasts) were once free living prokaryote bacteria that carried out all their reactions within the cell and these prokaryotic mechanisms of symbiosis have remained. In the absence of definitive data, the author has assumed an efficiency of 32% and that the remaining delta energy at around 68% is generated heat for the body.

This assumption remains until corrections and evidence confirm the assessment of respiration efficiency should be higher and that the $\Delta G°$ of -13.62 kcal/mol or other proven value represents a more accurate estimate.

Where does the fat go when weight is lost. The answer is that it is reconverted to molecules of gas (carbon dioxide and oxygen) and water from whence it came. After all, recall how the plants built the food in the first place; from molecular CO_2, water and oxygen. The lungs are the principal pathway for the weight lost.

The same goes for daily fluctuations in weight and body content. The CO_2 oxidised and water and oxygen change are replaced each day with dietary energy intake. Any imbalance due to sustained weight loss from adipose tissue has been due to a net reduction in carbon atoms exhaled and excreted or egested.

To oxidise each molecule of food, the carbons in each substrate must combine with oxygen in the body to form carbon dioxide which is exhaled.

Likewise, the hydrogens in each substrate must combine with oxygen in the body to form water which may be utilised, excreted in urine, faeces, perspiration, breath, tears and can be replenished when drinking water. The oxygen is replenished with ventilation (breathing in).

The exhaled carbon, however, is replaced by consuming food or beverages containing energy. On average a person loses approximately 200 g of carbon each day by breathing out or exhaling.

Adding exercise may increase that to around 250g carbon/d, (exhaling). Full oxidation (i.e., mass lost) of 10 kg of human adipose tissue (triacylglycerol fat) would require 29 kg of inhaled oxygen and produces 28 kg of carbon dioxide and 11 kg of water.

Of the 29 kg of oxygen required, 19.6 kg, or 68% is incorporated and lost in exhaled carbon dioxide and 9.4 kg, or 32% is incorporated and lost within the water. Of the 10 kg of TAG fat lost, or 8.4 kg, or 84% of this fat is exhaled as CO_2 from the lungs and the remaining 1.6 kg, or 16% is lost as water through various means as described above.

As noted above, when expressed as moles of oxygen consumed, palmitate has an oxygen demand 2.7 times the amount of oxygen compared with glucose.

However, each mol of palmitate produces significantly more high-energy bonds (106) than a mol of glucose (30), and the caloric value of the stored high-energy phosphate bonds derived from the oxidation of a gram of palmitate at 9.35 kcal/g is 2.5 times greater than that derived from a gram of glucose at 3.76 kcal/g glucose, albeit at a greater relative cost in oxygen.

Hence, when oxygen is abundant and food is scarce, there is an advantage in utilizing fatty acids for fuel as opposed to using glucose.

The reverse, however, would occur when food is plentiful and oxygen is scarce. Variance of oxygen uptake between glucose and fatty acids is most probably dependent on the proportion of the types of bonds within each molecule and varies due to the variation in the enthalpy of molecular bond formation.

The larger fat molecule will have a more consistent proportion of the different types of bonds. Smaller glucose molecules will have a larger variance in the bond types as a proportion the total chemical energy and this will result in a greater variability in oxygen uptake.

CHAPTER

Weight Loss Studies & The Blue Zones $\quad$ 16

There are more than half a million rigorous scientific papers on obesity and more than a 100 new papers published every day. People consuming more plant-based diets have lower rates of overweight and obesity than those whose diets include or emphasise meat and refined foods. Plant-based food is lower in fat and cholesterol and promotes both weight loss and long-term healthy weight maintenance and improved satiety. Some examples are provided below.

16.1 The Adventist Health Study: Plant & Animal Diets

The largest study ever to compare the obesity rates of those eating plant-based diets was published in North America. [104] A study was undertaken to assess the prevalence of type 2 diabetes in people following different types of vegetarian diets compared with that in nonvegetarians, involving a population of 22,434 men and 38,469 women who participated in the adventist health study-2 conducted across North America over the period from 2002 to 2006. Meat eaters were the highest with an average body mass index (BMI) of 28.8 kg/m^2. Flexitarians (people who ate meat more on a weekly basis rather than daily) were lower, at a BMI of 27.3 kg/m^2, but were still overweight. With a BMI of 26.3, pesco-vegetarians (people who avoid all meat except fish) did better still. Even U.S. vegetarians tend to be marginally overweight, coming in at 25.7 kg/m^2. The only dietary group found to be of ideal weight were those eating strictly plant-based (the "vegans"), whose BMI averaged 23.6 kg/m^2. People who had once eaten vegetarian diets but then started to consume meat at least once a week were found in one study to experience a 231% increase in odds for weight gain. A dietary quality index was developed that simply reflects the percentage of calories people derive from nutrient-rich, unprocessed plant foods on a scale of 0 to 100. The higher the score, the more body fat may be lost over time and the lower the risk may be of abdominal obesity, high blood pressure, high cholesterol, and high triglycerides. The standard American diet was found to rate 11 out of 100.

According to U.S. Department of Agriculture estimates, 32 percent of US calories comes from animal foods, 57 percent from processed plant foods, and only 11 percent from whole grains, beans, fruits, vegetables, and nuts. On a scale of one to ten, the American diet would rate approximately one on the dietary quality index scale. Mean BMI was lowest in vegans (23.6 kg/m^2) and incrementally higher in lacto-ovo vegetarians (25.7 kg/m^2), pesco-vegetarians (26.3 kg/m^2), semi-vegetarians (27.3 kg/m^2), and nonvegetarians (28.8 kg/m^2). Prevalence of type 2 diabetes increased from 2.9% in vegans to 7.6% in nonvegetarians; the prevalence was intermediate in participants consuming lacto-ovo (3.2%), pesco (4.8%), or semi-vegetarian (6.1%) diets. After adjustment for age, sex, ethnicity, education, income, physical activity, television watching, sleep habits, alcohol use, and BMI, vegans (OR 0.51 [95% CI 0.40–0.66]), lacto-ovo vegetarians (0.54 [0.49–0.60]), pesco-vegetarians (0.70 [0.61–0.80]), and semi-vegetarians (0.76 [0.65–0.90]) had a lower risk of type 2 diabetes than nonvegetarian. The study concluded that the 5-unit BMI difference between vegans and nonvegetarians indicated a substantial potential of vegetarianism to protect against obesity.

Increased conformity to vegetarian diets protected against risk of type 2 diabetes after lifestyle characteristics and BMI were taken into account. Pesco- and semi-vegetarian diets afforded intermediate protection.[77]

16.2 The Ad Libitum Traditional Hawaii Diet

Researchers in Hawaii tested a pre-western diet via feeding 20 Native Hawaiians to ad libitum satiety for 21 days on the Waianae Diet Program comprising high carbohydrate (78%) and low fat (7%) and moderate protein (15%). Average intake fell from 2594 kcal/d to 1569 kcal/d with an average weight loss of 7.8 kg and an added benefit of a significant reduction in blood pressure 11.5 mm Hg systolic and 8.9 mm Hg diastolic (see further detail on Hawaii below).[105]

16.3 The BROAD Study

2017 BROAD study found that a Whole Food Plant based diet resulted in similar weight loss at 12 months compared to other diets such as low-carbohydrate and low-fat diet, which led to significant BMI and cholesterol improvements and improvements to other risk factors. was a randomised controlled trial using a low-fat WFPB diet in a community setting for obesity, ischaemic heart disease or diabetes. The intervention group was not required to restrict calories or undertake physical activity and was compared to a control group that continued their usual care. The intervention led to a significant and sustained reduction in weight and BMI at all time points when compared to the control group and out until the 12-month follow-up. This is notable, as it was a community intervention such that what people ate was not controlled in any way other than through upfront dietary education for the intervention group. The authors concluded that this study 'achieved greater weight loss at 6 and 12 months than any other trial that does not limit energy intake or mandate regular exercise'.[106]

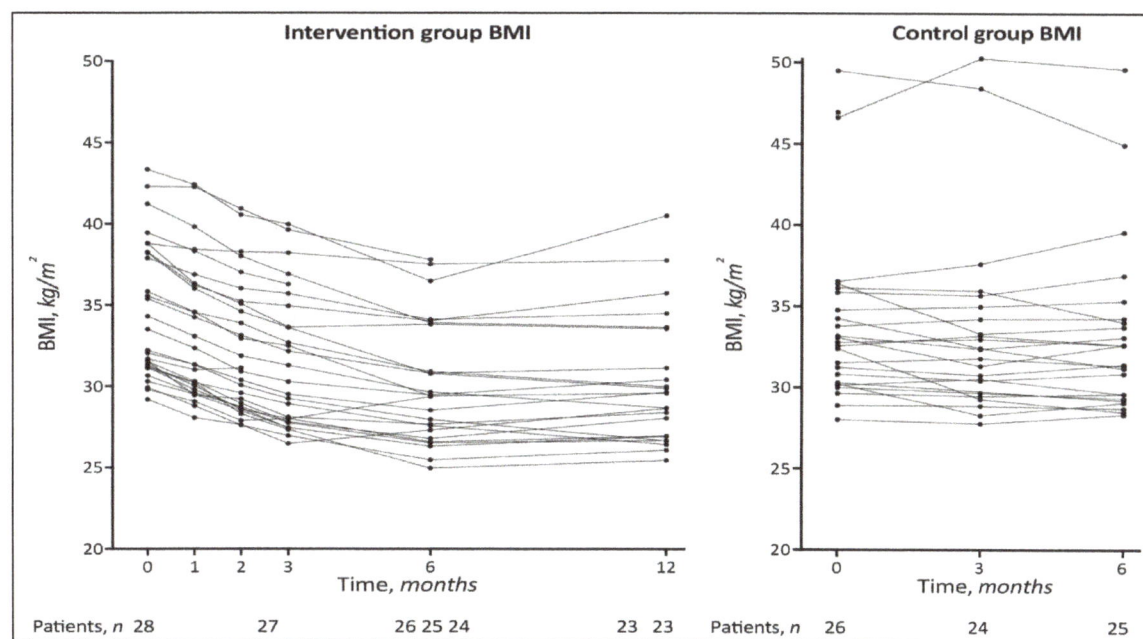

Figure 16.1 The BROAD Study reduction and sustained change in BMI and follow up over 12 months

16.4 Plant Based Dietary Patterns and Obesity Risk

This study provides data from 9 prospective cohort studies that have used the Plant Based Diet Index (PDI) to assess the impact of plant-based diets on weight management. The results confirm that adherence to a plant-based diet is associated with lower body weight, with a greater impact when a healthy plant-based diet is followed. However, the Unhealthful Plant Based Diet Index (uPDI) showed that a diet composed of unhealthy plant foods had no advantage for body weigh All studies were conducted in Western populations except for one study in South Korea. [107]

16.5 The PCRM Randomised Study of WFPB

The Physicians Committee for Responsible medicine (PCRM) conducted a study of a WFPB diet versus the National Cholesterol Education Program recommended diet (control group). 64 overweight women were included. There was no calorie restriction, food was not provided, and exercise was not required. After 14 weeks, the mean weight loss in the WFPB diet group was 5.8 kg, which was significantly greater than in the control group (weight loss 3.8 kg).

At 1 year and 2 years follow up, the WFPB diet group had maintained the weight loss, whereas the control group had regained the weight. In addition, the WFPB diets group showed improvements in glucose tolerance and insulin sensitivity. [108]

16.6 The Diet Fits Study

To counter the myths around weight loss performance requiring low carbohydrate intake and also discussed in Section 1.11, this 12-month study involving 609 participants conducted and reviewed from 2013 to 2016 concluded that there was no significant difference in weight change between a healthy low-fat diet (-5.3 kg) vs a healthy low-carbohydrate diet (-6.0 kg), and neither genotype pattern nor baseline insulin secretion was associated with the dietary effects on weight loss. [109]

A prospective cohort in Europe of 11,554 participants with normal weight (BMI<25 kg/m^2) at baseline found that greater adherence to a plant-based diet (not necessarily entirely plant-based) significantly reduced the risk of developing overweight or obesity by 18% after a median Vegetarian Nutrition Research Brief: Plant-Based Diets and Obesity 3 follow up of 10.3 years. The dietary pattern that emphasized less healthy plant foods (i.e., saturated fats, and refined grains) showed a 7% nonsignificant increase in risk (Source Vegetarian Nutrition Research Brief: Plant-Based Diets and Obesity). [110]

16.7 The EPIC-Oxford Study

Body mass index (BMI) was compared in four diet groups (meat-eaters, fish-eaters, vegetarians and vegans) in the Oxford cohort of the European Prospective Investigation into Cancer and Nutrition (EPIC-Oxford) and to investigate lifestyle and dietary factors associated with any observed differences. The study, conducted in 2003, involved a total of 37,875 participating healthy men and women aged 20-97 years.

Age-adjusted mean BMI was significantly different between the four diet groups, being highest in the meat-eaters (24.41 kg/m^2 in men, 23.52 kg/m^2 in women) and lowest in the vegans (22.49 kg/m^2 in men, 21.98 kg/m^2 in women). Fish-eaters and vegetarians had similar, intermediate mean BMI.

Differences in lifestyle factors including smoking, physical activity and education level accounted for less than 5% of the difference in mean age-adjusted BMI between meat-eaters and vegans, whereas differences in macronutrient intake accounted for about half of the difference. High protein (as percent energy) and low fibre intakes were the dietary factors most strongly and consistently associated with increasing BMI both between and within the diet groups. The study authors concluded that Fish-eaters, vegetarians and especially vegans had lower BMI than meat-eaters. Differences in macronutrient intakes accounted for about 50% of the difference in mean BMI between vegans and meat-eaters. High protein and low fibre intakes were the factors most strongly associated with increasing BMI.[111]

16.8 The Blue Zones Geographic Areas

The Blue Zone areas, or more correctly the Longevity Blue Zones, are regions around the world where people tend to have a lower BMI and a high proportion of long life spans well beyond the age of 80. The concept of the Blue Zones was developed in 2000 by Michael Poulain, Professor Emeritus of the Universite Catholique de Louvain in Belgium, then supported by Dan Buettner (who follows a 98% plant based diet) who was responsible for studying these high longevity areas, also from 2000.

A 2016 study showed that people residing in these areas reach the age of 100 at a rate that is ten times than in the USA. There are 5 such areas across the world discussed; Nicoya Costa Rica, Sardinia, Italy, Ikaria, Greece, Okinawa, Japan, Loma Linda, California USA and more recently a sixth, comprising several communities in Florida that have also achieved Blue Zone Status. These areas are defines as the having the highest percentage of centenarians of the whole world where people reach the age of 100 at 10 times greater than in the USA.

The phenomena of increased longevity is generally considered to be due the combination of a diet which is rich in locally sourced wholesome wholegrain foods, predominantly from plants and an absence of processed foods, saturated fat and very little meat and dairy.

All this combined with low levels of stress moderate levels physical activity, low levels of alcohol and enhanced social circles all contributing to lowered incidence of diseases common across the west including diabetes and heart disease, where diet and lifestyle is markedly different as we have explored in this book. The populations in these areas have tended to shun more modern culture, instead adhering to a more traditional and obviously healthier diet, food production and community and family based lifestyle the antithesis of which has plagued more modern cultures throughout the rest of the world. The Blue Zone lifestyles all have the common characteristics of predominantly plant based diets, low levels of meat, fewer excess calories, daily moderate exercise, sufficient and quality sleep, building good family ties and relationships and finding a purpose for life and a respect for the Earth.

The diet tends to be high in vegetables, heathy whole grains, potatoes, sweet potatoes, fruit, legumes, nuts, olive oil, some oily fish and if meat is eaten it is eaten sparingly, for example 3 to 4 ounce portions as an occasional side Critics of the claims that these areas are healthier site that there is little or no scientific evidence to support the claims that regarding the precise reasons for the increased longevity. However, the author firmly believes that there is sufficient observational and epidemiological evidence to suggest that the combination of a healthier plant based diet, lack of meat and dairy and overall healthier lifestyle improves metabolic processes, regulation of energy substrates and much healthier levels of antioxidants and anti-inflammatory nutrients from plants ,when compared against diseases and longevity across the rest of the world, where obesity and poor health is spiralling and worsening.

Given the knowledge available currently regarding the impact of excess fat, excess protein, the predominance of meat and dairy, free sugars, low fibre low vegetable and fruit intake processed food, lack of exercise, it is clear that this is not the diet and lifestyle to follow to achieve a healthier longer life. A prime comparator example of this compares the life expectancy of American people on the Standard American Diet (SAD) consuming, on average 35% fat, 16% meat, 47% carbohydrates, of which 14% only is wholegrain. The longevity of people residing in the USA has fallen to its lowest level since the turn of the Century. The various Blue Zone Areas all share the healthy predominantly plant-based diet as recommended by the author and the first Blue Zone areas researched are now discussed in further detail. [99, 100, 101]

16.9 Sardinia, Italy

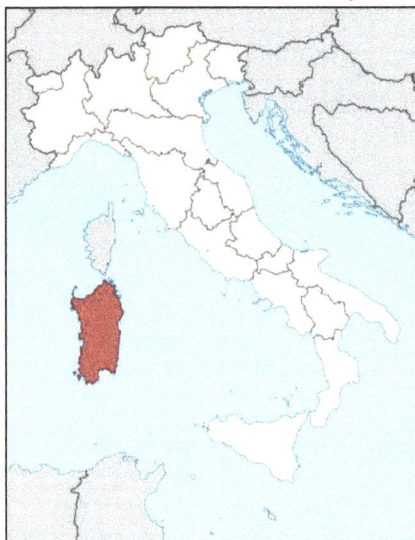

The percentage of male centenarians is much higher in Sardinia compared to the other Blue Zone Areas. Sardinia is the location of the world's longest men. Because of the relative isolation of the island there is a preponderance to bake foods that were originally created to cope with famines, for example whole grain acorn bread made with acorns and clay and ancient abbamele honey, pane carasau sourdough flat bread with a very long shelf life. These so called famine foods are correlated with the high longevity and as well as highly nutritious can be made with low levels of stress at an easy pastoral type pace of life, emphasising quality of food rather than quantity. The diet consists primarily of garden vegetables and fruits when in season and these are dried and pickled to eat as 'famine foods' when out of season.

Location of Sardinia in Italy. Wikipedia Commons, GNU Free Documentation Licence Author TUBS, https://creativecommons.org/licenses/by-sa/3.0/ includes elements that have been taken or adapted from this map: Italy location map.svg (by NordNordWest).

A wide variety of wild plant are also eaten containing a wide range of antioxidants. The Sardinian diet consists of very little animal protein, (c5%) of the diet, eaten sparingly on Sundays only or on occasions as a treat or as part of a special celebration.
The Sardinians do however, consume significant dairy products from goat and sheep milk products. A significant 47% of calories in the Sardinian diet comprises wholegrains (compare

14% on the SAD). The Sardinians love to eat a daily handful of nuts such as pistachios, walnuts and almonds high in Omega 3 fats.

Only 3% of calories are derived from refined sugars compared with around 24% on the SAD. Consumption of vegetables is around 12% of DI in Sardinia and this is generally higher than the rest of Europe at around 1% to 6% of DI. Young men drink wine moderately which is associated with reduction in stress.

The moderate levels of natural exercise, for example, shepherding sheep and walking an average 5 miles each day, is certainly considered to contribute to longevity and once again demonstrates that gymnasium membership is not necessarily the answer to a healthier life.

16.10 Okinawa Japan

Possibly the most researched and discussed Blue Zone area, Okinawa Island is the largest of the Ryukyu Islands and is located off the coast of mainland Japan between the East China and Philippine seas.

Japan location map with Author modified by Dr. Blofeld Source http://www.maps-for-free.com/GNU General Public Licence Free Software Foundation. This file is licensed under the Creative Commons Attribution 3.0 Unported license.

It is the home of the world's longest lived women has a remarkable and longest disability free longevity record across the globe with more centenarians than anywhere else.

For original Okinawan elders, the average BMI has ranged from 18 to 22 throughout their lifespan and again all tend to exercise moderately practising yoga and gardening regularly and a preponderance of hobbies and good levels of social networking.

In Okinawa, a 60 year old male has better than a four-fold chance of reaching the age of 90 than a person of the same age in the USA. Heart disease is 6 to 12 times lower than in the USA.

Expenditure on public health is 20% lower than in the USA. It is also one of the poorest regions of Japan. Although the longevity is similar with the rest of mainland Japan, the quality of health demonstrated by the Island inhabitants is far higher than their mainland compatriots, with a significantly lower risk of heart disease, stroke, cancer, osteoporosis and Alzheimer's disease. The incredible benefits are considered to be due to diet and lifestyle. The traditional diet practised for generations was 90% plant based 1% fish, 1% dairy and eggs, low in calories, low in fat, high in carbohydrates including purple and orange sweet potatoes, legumes and very rich in vegetables, where the volcanic soil bestows high levels of polyphenols and vitamin C, fibre and potassium.

Okinawa Isalnd, Okinawa Prefecture, Japan, from Space
File in the public domain in the United States because it was solely created by NASA
https://eol.jsc.nasa.gov/SearchPhotos/photo.pl?mission=ISS042&roll=E&frame=165588
Modifications: cropped and color adjusted. Modifications made by Kugel~commonswiki.

By contrast there can be excessive levels of sodium in the diet (even up to 3.2 g/d) compared with the recommended maximum of 1.5 g/d and perhaps the high intake of potassium serves to offset this. Turmeric and mug wort are also important in the kitchen. Okinawans eat a staggering 40% fewer calories than consumed in the average western world diet. The diet also includes small amounts of noodles, rice fish and pork. Average consumption of carbohydrates is 400g/d or a minimum of 70% of total Dietary Intake (DI) and the proportion of carbohydrates in the elder population is around 10:1 carbohydrate to protein ratio. Consumption of vegetables is a very healthy 300 g/day. Okinawan women tend to have a high intake of natural oestrogens in their diet mainly from soy which contains phytoestrogen flavonoids. It is known that soy isoflavones slow menopause bone loss. Elderly Okinawans generally have very healthy arteries showing little sign of atherosclerosis and cholesterol and very low levels of coronary heart disease compared to less healthy populations across the rest of the world. The Okinawans practice something called the *moai*, a form of long lasting social meeting and gathering framework, for example known to last 0ver 90 years, that improves social networks. Uniquely, the Okinawans recite the *Hara hachi bu* prior to each meal which reminds them to eat to just 80% of fullness. These elderly folk are very positive thinkers and highly respected across Japanese society. It has also been shown that the Okinawans possess a specific gene influencing mitochondria and metabolism. However, although the healthier gene is bound to have an impact (probably around 20% influencing), the most widely accepted reason for the longevity is considered to be due to the diet. All this may change and alas, be a matter for history and record books. Sadly, in more recent years, since 2010, fast food has reached the shores and the modernisation of food production has led to an increase in the consumption of protein and fat. The typical original diet consisted of 85% carbohydrate, 9% protein and 6% fat (2% saturated fat). Now the diet for many, especially younger Okinawans has changed to only 57% carbohydrates, 15% protein and a massive increase to 28% fat, consisting of 7% saturated fat.

It is sad to report that the western type food consumption and behaviours has even spread to Okinawa and many of the younger population are now starting to adopt relatively unhealthy eating and lifestyle patterns which is leading to significant increases in average BMI. This is, in itself, a clear message and evidence how food influences weight.

16.11 Ikaria Greece

Ikaria (see Island 2 below) is one of the thousands of Greek Islands situated in the Aegean Sea with a population of approximately 8,000. It was the last of the original Blue Zone has one of the highest life expectancy rates in the whole of Greece with a female to male ratio very close to 1. People here live an average 8 years longer than people in the USA and experience half the rate of heart disease and dementia is virtually non-existent. A Mediterranean diet is followed as well as high levels of physical activity.

Location Map of Greece. Own work (Original text: own work, using United States National Imagery and Mapping Agency data) Author Lencer. This file is licensed under the Creative Commons Attribution-Share Alike 3.0 Unported license. Author Lencer.

The diet comprises whole grains, homegrown vegetables, olives, high levels of potatoes and a little fish, honey and cheese and goats milk. Also, herb tea, a little red wine and olive oil. The lifestyle is very low stress and relaxed with afternoon naps.

Exercise is mainly walking between villages and gardening, weeding, digging, sowing and harvesting. Grandparents play an important role in the upbringing of the grandchildren and the whole household. Very few people live alone. Studies are continuing to explore all the factors influencing longevity.

Municipalities of Samos prefecture, Greece; 1: Dimos Samou, 2: Dimos Ikarias, 3: Dimos Korseon GNU Author Pitichinaccio; Free Documentation Licence Creative Commons Attribution Share-Alike

16.12 Nicoya Peninsula Costa Rica

Nicoya Peninsula in Costa Rica. This image is in the public domain because it is a screenshot from NASA's globe software World Wind using a public domain layer, such as Blue Marble, MODIS, Landsat, SRTM, USGS or GLOBE. File usage on Commons

The Nicoya Peninsula is situated in the North-Western region of Costa Rica. Inhabitants share a 20% lower mortality and a greater average height and lower body mass index (BMI) and lower disability and lower levels of suicide compared with the rest of Costs Rica. Health expenditure is 15% less than that in the USA.

The inhabitants are more than twice as likely to reach the age of 90 than Americans. The diet is predominantly plant based and adherence to traditional foods such as antioxidant rich fruit, wholegrains, high fibre, beans, rice and a little beef, chicken and pork and little or no processed food. The traditional Nicoya diet includes black beans and white rice, homemade corn tortillas, a variety of squash, yams, papaya, bananas, and pejibayes. Meat, fish, and poultry is limited to around 5% of the usual diet.

Also around 5 ounces of fruit is eaten daily. The Nicoya people also drink significantly less milk (an average 0.5 glass per day compared to 0.7 glass by other Costa Ricans). Other notable factors are that the drinking water has high levels of calcium and magnesium and in common with other Blue Zone areas, a relatively stress free lifestyle, no doubt helped in part by the free healthcare introduced by the government. Faith and family is highly regarded in the Peninsula.

16.13 Loma Linda California USA

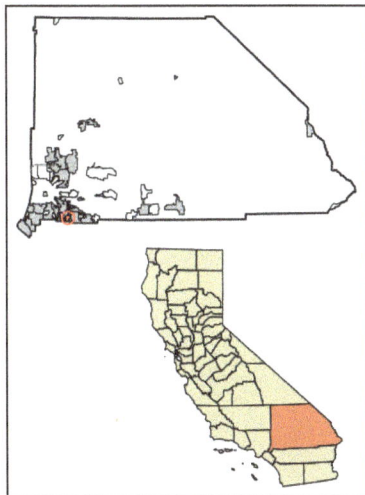

Loma Linda is a city in San Bernadino County located in California, USA. The majority of the population of Loma Linda follow the Seventh Day Adventists faith. People in this area and community live an average 10 years longer than the average American. Most of the inhabitants are vegan or vegetarian and primarily plant-based, adopting their diet directly from Genesis 1:29 of the Bible. Their staple foods are whole grains, beans, green vegetables and nuts. The community generally tend to eat specific quantities at certain times each day. Large breakfast and lower quanta of foods as the day develops with light dinners in the evening, no doubt contributing their low BMI. They abstain from eating pork or shellfish and not smoking or drinking any alcohol and no caffeinated beverages such as tea or coffee.

Location of Loma Linda in San Bernardino County, California. Source English Wikipedia Creative Commons https://creativecommons.org/licenses/by-sa/4.0/deed.en Author DeomocraticLuntz

They have complete rest on Sundays. All this done on a very restricted budget. The diet is rich in antioxidants, sufficient mono unsaturated fats, polyunsaturated fats vitamin A and E and essential minerals. There is very little, if any processed food in the diet. The people have low levels of cholesterol, low cardiovascular disease and low blood pressure, They have regular walks on most days. It is interesting to note that Seventh Day Adventists all around the world who follow this diet are thriving suggesting that the diet plays the key role in their very good health.

16.14 Naples Florida USA

The 8 Southwest Florida communities of Naples, Immokalee, Ave Maria, Bonita Springs, Estero, Golden Gate, East Naples and Marco Island have recently achieved Blue Zone Community certification. The initiative commenced in 2015 where local leaders, 800 organisations and volunteers engaged approximately 275,000 local residents to eat healthier, develop social circles and move more.

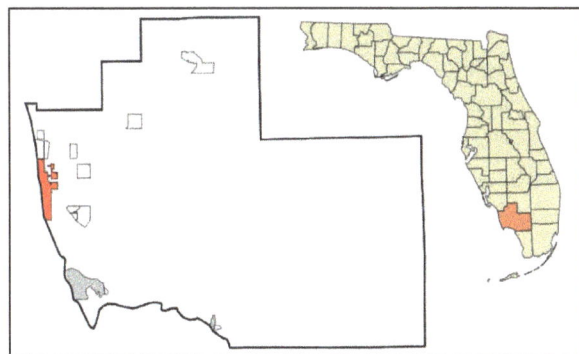

The results and improvements have led to a reduction in hospital emergency visits and health care costs plus an increase in overall well-being compared with the rest of Florida and the USA. The number of residents who report they are thriving in daily life has risen from 27% to 79% of the population, a 20% increase in exercise, the rescue, resupply and waste prevention of 310,000 pounds of food a reduction of $190 million in avoided medical care and lost productivity costs.

Location of Naples in Florida. Incorporated and unincorporated areas in Collier County, Florida, highlighting Naples in red. Source English Wikipedia Commons. https://en.wikipedia.org/wiki/GNU_Free_Documentation_License. Author Arkyan own work, based on public domain information. Based on similar map concepts by Ixnayonthetimmay)

All this achieved during the pandemic and two major hurricanes. How the adopted new lifestyle will unfold over decades and whether or not true record long healthier longevity remains to be seen and proven as in the other original Blue Zone areas but the benefits so far are certainly encouraging as a blue print for other to monitor and adopt.

16.15 Hawaii Populations & Changing Diets

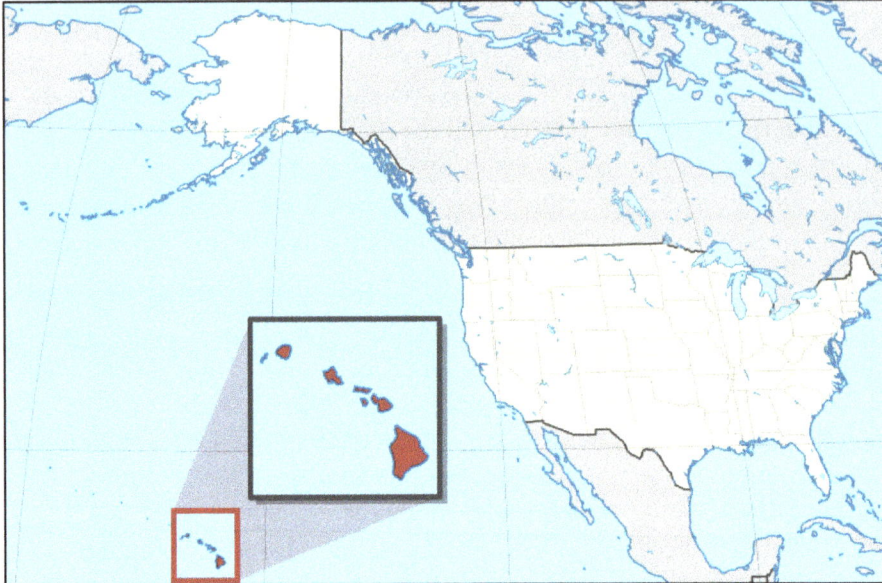

Location of Main Islands of the state of Hawaii in the United States Wikipedia Commons, GNU Free Documentation Licence Author TUBS, https://creativecommons.org/licenses/by-sa/3.0/

The traditional diet of Hawaiian culture is healthy and often quoted as one of the healthiest around in the world, predominantly centred around vegetables, high starch and fibre, low saturated fat and low cholesterol, including bread fruit, roots, yams, shellfish, seaweed and some pork, at 70-80% carbohydrate, 12% protein and 18% fat. Hawaiians also traditionally exercised regularly and consistently.

Unlike western society today, fatty (high caloric) foods were reserved for high ranking elite members of the community with the associated weight gain, whereas those commoners lower in the social orders retained their more healthy food choices and hence, slimness. However, as immigrants arrived on Hawaiian shores from around 1945 on, the foreigners also imported their own food choices. This has gradually embedded in food choices throughout Hawaii, to such extent that the current diet consists of many elements of the American diet and high quantities of fast food, at 45% carbohydrate, 15% protein and 40% fat are now available to the entire population.

Traditional foods were replaced with processed foods, leading to dietary shifts that have contributed to obesity. Before Europeans arrived, native Hawaiians had just one form of starch in their diet which is taro. For many thousands of years the diet was limited to the food which was available on their island chain and this was low in free sugars and fat. After European contact, processed food, rice and bread was integrated into the diets of people who for thousands of years never had these foods. This switch has led to an obesity epidemic where 18% of the total population are obese which is seemingly low compared with the USA and the UK for example, but this level is very high compared with obesity levels traditionally before this spread of unhealthy choices, especially amongst the younger generation. Statistics comparing obesity levels in Hawaii with other states of the US and the rest of the world are skewed due to the presence of so many inhabitants of Asian ancestry residing in Hawaii. Latest data in 2023 shows that the total Hawaii population is 25-30% obese, with most of the rest of the US at 30-40%. Just 2 states Colorado (24.9%) and District of Columbia (23.5%) are in the lowest range of 20-25%. West Virginia (41.2%) and Arkansas (40.0%) are the highest.

Native Hawaiian people only comprise 6% of the total population. However around 40% of native Hawaiians are obese or overweight and yet the Hawaii's Chinese population are less than 7% overweight. Japanese migrants to Hawaii have become four times more likely to be obese than people in Japan. Pacific Island countries are amongst the highest in overweight countries across the world. The general population inhabiting small islands and Polynesian populations, with relatively low abundance of food in times gone by and with a hot climate across the equatorial region, probably evolved to be sustained on very little food and to store fat easily supporting the thrifty gene hypothesis discussed earlier. Since the influx of migrants and the access to cheaper, unhealthy processed foods incumbent islanders including Hawaiians tended to eat much larger portions more often than their bodies were used to gained weight gain. There are many other reasons compounding this including poor socioeconomic opportunity and homelessness.

16.16 The North Karelia Project Finland

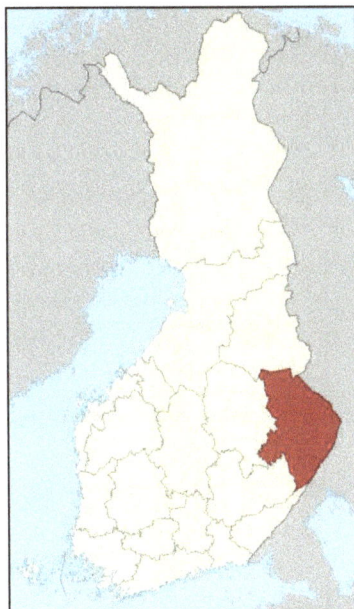

North Karelia is a region in east Finland bordering the regions of Kainuu, North Savo, South Savo, South Karelia and the Russian Republic of Karelia. It is the easternmost region of Finland and shares a 300 km border with Russia. The North Karelia Project commenced 52 years ago and originally ran for 5 years from 1972 to 1977 and was instigated due to the extremely high mortality rates as a result of cardiovascular disease amongst men. The project dramatically reduced annual cardiovascular disease mortality and improved general public health through lifestyle changes, making it a model for successful population-based lifestyle intervention.[102, 103] International studies confirmed that Finnish men had the highest mortality from heart disease in the world during the 1960s. At the time, approximately 700 per 100,000 men aged from 35-64 died of cardiovascular disease in North Karelia. The factors contributing to these diseases included high serum cholesterol, high consumption of dietary fat, hypertension and smoking.

North Karelia on a Map of Finland Source Wikipedia Creative Commons Author Fenn-O-maniC, https://creativecommons.org/licenses/by-sa/3.0/

The diet consisted of high intake of salt and animal fat with the bare minimum of vegetables and fruit. The project, led by Dr Pekka Puska, MD, PhD, MPolSc, a professor, physician, and former member of the Finnish Parliament who served as the director and principal investigator of the project and the North Karelia Project ran for 25 years through socio-behavioural framework community based interventions and national legislation, targeted lifestyle changes to reduce cardiovascular disease risk factors.

This included work in health centres, schools , the private sector and media. Over the 5 year period more than 1,000 newspaper articles were published on nutrition and other risk factors.

There were many recommendations on the improvement of diet including lowering saturated fats and reduction in salt intake.

Reduction in smoking was also targeted. The significant success of the project after 5 years interventions led to the expansion of the project to cover the whole of Finland. There were significant reductions in hypertension, serum cholesterol levels, smoking prevalence and cardiovascular disease mortality.

Nationally, policy measures were introduced in the food industry including tighter tobacco legislation. The project across Finland formally ended after 25 years in 1997. Over the 40 year period from 1972 to 2012, the use of butter on bread had diminished by 80%. Cholesterol levels dropped by more than 20% in both men and women in North Karelia over the same 40 year period.

Consumption of saturated fat decreased from 20% to 12% i 2007 but rose a little to 14% by 2012. The decline in cholesterol was evidenced by the change in diet rather than due to increase prescriptions for statins.

In North Karelia, sodium levels fell from 13 grams to 9.5 grams and from 10 grams to 7.4 grams for women. As a result mean systolic blood pressure fell from 149 to 134 mmHg and 153 to 127mm Hg in men and women respectively from 1972 to 2012.

In the first 10 years of the extended project, smoking was cut from 51% to 36% of the population in North Karelia. Nationally, smoking fell from 60% in the 1960's to 16% by 2016. Regarding cardiovascular disease, the level at 2011 was 100 per 100,000 a seven fold reduction. From 1969 to 2011, CVD mortality reduced by greater than 80% amongst men aged 35 to 64. in both North Karelia and Finland.

Overall life expectancy rose by around 10 years due to the project. Numerous experts around the world have visited Nort Karelia to learn how to implement improved health initiatives across communities and the work has significantly contribute to the WHO work on Global Strategy on the Prevention and Control of Noncommunicable Diseases.

Several project leaders have continued to serve as experts to assist similar projects in many other countries.

16.17 Food Choices and Care for Our Environment

This journey began with the principal reasons for obesity and over-nutrition. We then looked at healthier nutrition and how to achieve weight loss goals and lifestyle improvements, reviewing the various macro and micronutrients and the nutritional value of the food we choose to eat.

We explored some of the major metabolic pathways and the regulation of homeostasis governing weight loss and weight gain and finally, examining the release of cellular energy from the various food substrates.

There is still much to understand in many areas. Ultimately, it is an appreciation of how nature and the body works, together with compelling and robust, evidenced based science and ongoing research, that should always guide and inform us towards our choices and goals, most importantly, using our own intuition and common sense.

We should of course not forget that our personal choices of the food we eat not, only impacts on our own weight and health but has a very considerable impact on the whole environment and this book has been written primarily for people that wish to understand how best to lose and sustain weight loss, how the body works and to help people further understand how the nutritional quality and the quantity of food impacts their own lives.

However, it would be remiss of the author, who is most passionate about the environment, all animals and plants and the future of the planet we all share, not to comment on this much bigger picture.

Indeed, this task should quite rightly be left to all the very fine environmentalists, authors and experts who have written extensively on this vast and most important topic these past 60 years.

Therefore, a very small sample of some of the pressing issues and our responsibilities to change is provided here.

Human consumption of meat is the greatest threat to wildlife. Humanity has wiped out 60% of mammals, birds, fish and reptiles over the past 55 years since 1970.

Of the land that humanity uses, 75% is devoted to pasture for livestock. From this land humans derive only 7% of total calories and only 15% of protein including dairy.

Excluding dairy only cows, red meat provided by cattle, sheep and goats provides just 2% of human calories and 7% of protein. 49% of protein eaten by humans is provided by plants.

No less than 68% of wildlife population has been destroyed by human expansion and the need for food especially meat. 50% of plants are gone and 50% of fish are gone.

The estimated average per capita protein intake in the United States is 90-100 g/d. Of that 70% to 85% is derived from animal protein. Replacing just 3% of this with plant based protein improves healthy ageing outcomes by 38%.

People's diets not only affect their own health and the incredible growing pressure on our overloaded health services but also very significantly impact the health of the whole planet.

Diet change is imperative if humans are to sustain a habitable planet.

Cattle farming is the greatest driver of biodiversity loss. 80% of agricultural land is devoted to livestock and feed.

Agriculture, largely due to land cleared for grazing livestock for meat production, causes >90% of deforestation. 84% of South American deforestation is for cattle grazing and feed crops for those cattle.

For a sense of the enormity of the change accompanying human population growth to 8.2 billion in 2025, livestock biomass is now 7 times more than the level 10,000 years ago.

Soon there will be 9 billion humans to feed and 90 billion livestock to feed. Just for meat on a plate.

Of all the mammals across the entire world, farmed livestock and pets represent some 62%, with humans at 32%, totalling 96% and the entirety of all the other wild mammals now stands at just 4% and falling.

Water, needed to grow animal protein is 6 times that required to grow protein from plants and 20 times the volume of water is required to grow beef calories, compared to grain or potatoes.

The earth's oceans seas, lakes and rivers have not escaped this onslaught, apart from pollution, acidification and eutrophication, due to livestock waste, over fishing and bottom trawling has now eliminated 90% of human consumption fish stocks in the oceans.

Palm oil supplies 40% of the world's vegetable oil on just 6% of the land to produce all other vegetable oil.

It is used in animal feed, is also a major driver of deforestation and habitat destruction for many endangered species and churns out millions of tonnes of greenhouse gases into the atmosphere.

Palm oil is now in 50% of all packaged foods and products in supermarkets, including make-up, deodorant, toothpaste, lipstick, pizza and doughnuts.

Many scientists believe that the '6th mass extinction' is already here and that to avert further global environmental damage, humans need to change eating habits; substantially away from animal grown foods, including meat and dairy products, changing to plant-based sources and to maximise the use of the RSPO global standard, to ensure that palm oil production has the least environmental impact.

By improving our nutrition and eating more plant and less animal products, we can not only maintain a healthier weight but also improve our health, reduce the pressure on our health services, reduce animal suffering and finally begin to start significantly healing our fragile planet.

References

1. Dietary Guidelines for Americans, 2020-2025 (DGA), December 2020

2. Phrase used by Professor Marc Hellerstein, professor of endocrinology, metabolism and nutrition at the San Francisco General Hospital, University of California, San Francisco & the Dr. Robert C. and Veronica Atkins Chair in Human Nutrition at Berkeley

3. Quantification of the effect of energy imbalance on body weight Hall KD, Sacks G, Chandramohan D, et al.

4. Aetiology of type 2 diabetes in people with a 'normal' body mass index: testing the personal fat threshold hypothesis; Professor Roy Taylor Clinical Science (London) 2023 Aug137(16: 1333-1346

5. mTOR-Mammalian Target of Rapamycin Reduces Lifespan & coordinates eukaryote cell growth & metabolism & nutrient intake, inhibited by stress to ensure that cells grow only in favourable conditions. It is a nutrient sensing pathway. Lowering leucine & methionine slows mTOR

6. pKRAS-Kirsten Rat Sarcoma Viral Oncogene Homolog-A gene that makes a protein that is involved in cell growth, cell maturation and cell death

7. IGF-1-Insulin like Growth Factor

8. Sirtuins-Family of signalling proteins involved in metabolic regulation that are anti-inflammatory & delay cellular senescence & extend lifespan, shut off by ingestion of high amounts of animal protein0

9. What We Eat in America. US Dept of Agriculture (USDA) Food Research Group (FRSG) Agricultural Research Service (ARS) Beltsville Human Nutrition Research Center 2012

10. Values derived from Metabolizable Energy System

11. Time-Restricted Eating: Benefits, Mechanisms, and Challenges in Translation Prashant Regmi, Leonie K. Heilbronn

12. Professor David Swain Target HR for the development of CV fitness

13. Miller et al Indiana University, formula for calculation of VO2 Max vs MHR

14. Phrase used by Professor Marc Hellerstein, professor of endocrinology, metabolism and nutrition at the San Francisco General Hospital, University of California, San Francisco & the Dr. Robert C. and Veronica Atkins Chair in Human Nutrition at Berkeley

15. UK Dietary Reference Values A Guide Department of Health 1991 ISBN 0 11 321396 4 set by The committee on Medical Aspects of Food and Nutrition Policy (COMA), 1991

16. Saturated fats and health SACN Scientific Advisory Committee on Nutrition, July 2019

17. NICE Guideline Cardiovascular disease: risk assessment and reduction, including lipid modification National Institute for Health Care Excellence, December 2023

18. Source: nutritionix database

19. Ancient DNA reveals insights into starch digestion Oct 2024 Science Journal University of Buffalo and Kackson Laboratory (JAX) Professor Omer Gokcumen, Professor Charles Lee, Robert Alvine, Peter Pajic, Kendra Scheer

20. Hall KD. Energy compensation and metabolic adaptation: "The Biggest Loser" study reinterpreted. Obesity (Silver Spring). 2022 Jan;30(1):11-13. doi: 10.1002/oby.23308. Epub 2021 Nov 23. PMID: 34816627.

21. Phrase used by Professor Marc Hellerstein, professor of endocrinology, metabolism and nutrition at the San Francisco General Hospital, University of California, San Francisco & the Dr. Robert C. and Veronica Atkins Chair in Human Nutrition at Berkeley

22. Carbohydrate metabolism and de novo lipogenesis in human obesity Kevin J Acheson, PhD, Yves Schutz, PhD, Thierry Bessard, MD, JP Flatt, PhD and eric Jequier, MD Am J Clin Nutrition 1987; 45:78-85 1987 American Society for Clinical Nutrition

23. G3P =Glyceraldehyde-3-phosphate is the three carbon sugar precursor of the 6 carbon sugar glucose

24. Calorie for Calorie, Dietary Fat Restriction Results in More Body Fat Loss than Carbohydrate Restriction in People with Obesity. Kevin D. Hall, Thomas Bemis, Robert Brychta, Mary Walter, Peter J. Walter, Laura Yannai Cell Metabolism, Hall et al., 2015, Cell Metabolism 22, 427–436 September 1, 2015 a2015 Elsevier Inc.

25. Effect of a plant-based, low-fat diet versus an animal-based, ketogenic diet on ad libitum energy intake; Randomized Controlled Trial Nat Med 2021 Feb;27(2):344-353. doi: 10.1038/s41591-020-01209-1. Epub 2021 Jan 21 Kevin D Hall , Juen Guo , Amber B Courville James Boring , Robert Brychta, Kong Y Chen , Valerie Darcey , Ciaran G Forde , Ahmed M Gharib , Isabelle Gallagher , Rebecca Howard , Paule V Joseph , Lauren Milley , Ronald Ouwerkerk , Klaudia Raisinger , Irene Rozga , Alex Schick , Michael Stagliano , Stephan Torres Mary Walter , Peter Walter , Shanna Yang , Stephanie T Chung. This study was registered on ClinicalTrials.gov as NCT03878108

26. Energy expenditure and body composition changes after an isocaloric ketogenic diet in overweight and obese men, Am J Clin Nutr. 2016 Aug; 104(2): 324-333; Kevin D Hall, Kong Y Chen, Juen Guo, Yan Y Lam, Rudolph L Leibel, Laurel ES Mayer, Marc L Reitman, Michael Rosenbaum, Steven R Smith, B Timothy Walsh, and Eric Ravussin

27. Target HR for the development of CV fitness' - Medicine & Science in Sports & Exercise Swain et al (1994) ', 26(1), 112-116

28. Predicting max HR' Miller et al (1993) - '- Medicine & Science in Sports & Exercise, 25(9), 1077-1081

29. The effect of sex, age and race on estimating percentage body fat from body mass index Jackson, A., Stanforth, P., Gagnon, J. et al.: The Heritage Family Study. Int J Obes 26, 789

30. Time-Restricted Eating: Benefits, Mechanisms, and Challenges in Translation; Prashant Regmi, Leonie K. Heilbronn iScience Vol 23, issue 6 26 June 2020, 101161

31. Dietary Energy Partition: The Central Role of Glucose; Xavier Remesar and Marià Alemany; Department of Biochemistry and Molecular Biomedicine Faculty of Biology, University Barcelona, 08028 Barcelona, Spain; IBUB Institute of Biomedicine, University of Barcelona: CIBER Obesity and Nutrition, Institute of Health Carlos III. Int J Mol Sci. 2020 Oct 21(20): 7729 National Library of Medicine PubMed Central

32. Enzyme Structure MIT News Feb. 20 2018: Massachusetts Institute of Technology MIT research 3.3-Å resolution cryo-EM structure of human ribonucleotide reductase with substrate and allosteric regulators bound. Edward Brignole, former Scripps Research Institute postdoc Kuang-Lei Tsai, JoAnne Stubbe, the Novartis Professor of Chemistry Emerita at MIT, and Francisco Asturias, associate professor of biochemistry at the University of Colorado School of Medicine

33. Monounsaturated Fatty Acids and Risk of Cardiovascular Disease: Synopsis of the Evidence Available from Systematic Reviews and Meta-Analyses Lukas Schwingshackl and Georg Hoffmann. Published: 11 December 2012. Nutrients 2012 ,4 1989-2007; doi 10. 3390/nu4121989. Department of Nutritional Sciences, Faculty of Life Sciences, University of Vienna, Althanstrasse 14, 1090 Vienna, Austria; E-Mail: lukas.schwingshackl@univie.ac.at

34. Effects of dietary fat on post prandial oxidation and on Carbohydrate and Fat balances Kevin J Acheson, PhD, Yves Schutz, PhD, Thierry Bessard, MD, JP Flatt, PhD and eric Jequier, MD. JP Flatt, PhD, Eric Ravussin, Kevin J Acheson and eric Jequier, MD, Institute of Physiology University of Lausanne, 1011, Switzerland and Department of Biochemistry, University of Massachusetts Medical School, Worcester Massachusetts, American Society for Clinical Investigation inc. 0021-9738/85/09/1019/06, Volume 76 September 1985, 1019-1024

35. Cellular mechanisms of insulin resistance Gerald I. Shulman; J Clin Invest. 2000;106(2):171-176

36. De novo lipogenesis in humans: metabolic and regulatory aspects M K Hellerstein 1999 Apr:53 Suppl 1:S53-65. doi: 10.1038/sj.ejcn.1600744. Department of Nutritional Sciences, University of California at Berkeley, 94270-3104, USA

37. Fat and carbohydrate overfeeding in humans: different effects on energy storage. Tracy J Horton, Holly Drougas, Amy Brachey, George W Reed, John C Peters and James O Hill. Am J Clin Nutr 1995; 62:19-29 1996 American Society for Clinical Nutrition

38. Carbohydrate metabolism and de novo lipogenesis in human obesity Kevin J Acheson, PhD, Yves Schutz, PhD, Thierry Bessard, MD, JP Flatt, PhD and eric Jequier, MD Am J Clin Nutrition 1987; 45:78-85 1987 American Society for Clinical Nutrition

39. Glycogen storage capacity and de novo lipogenesis during massive carbohydrate overfeeding in man. Kevin J Acheson, PhD, Yves Schutz, PhD, Thierry Bessard, Krishna Amantharaman, PhD; MD, JP Flatt, PhD and eric Jequier, MD. Am J Clin Nutr 1988; 48:240-7 1988 American Society for Clinical Nutrition

40. Dietary fat and the regulation of energy intake in human subjects
Lauren Lissner. PhD; David A Levitsky, PhD; Barbara J Strupp, PhD; Heidi J Kalkwarf, MNS; and daphne A Roe, MD. Am J Clin Nutr 1987; 46:886-92 1987 American Society for Clinical Nutrition

41. The metabolism of "surplus" amino acids. Published online by Cambridge University Press: 01 August 2012, Professor David A. Bender. Division of Biosciences and Division of Medical Education, University College London, Robert l. Jungas, Mitchell L. Halperin, and John T. Brosnan, Physiological reviews Vol. '72, No. 2, April 1992 Printed in U.S.A. Department of Physiology, University of Connecticut Health Center, Farmington, Connecticut; Department of Medicine, St. Michael's Hospital, University of Toronto, Toronto, Ontario; and Department of Biochemistry, Memorial University, St. John k, Newfoundland, Canada

42. Regulation and function of the mammalian tricarboxylic acid cycle. Paige K. Arnold1,2 and Lydia W.S. Finley1,National Library of Medicine National Center for Biochemical Information J Biol Chem 2023; 299(2): 102838

43. Fad Diets: Facts and Fiction 2022; 9: 960922. Published online 2022 Jul 5. doi: Aaiza Tahreem, Allah Rakha, corresponding author , Roshina Rabail, Aqsa Nazir, Claudia Terezia Socol, Cristina Maria Maerescu, corresponding author and Rana Muhammad Aadil corresponding author

45. Tracking the carbons supplying gluconeogenesis Journal of Biological Chemistry Volume 295, Issue 42 October 2020 14419-14429, Ankit M. Shah, Fredric E. Wondisford American Diabetes Association Volume 62, Issue 5 May 2013

46. Dietary Proteins Contribute Little to Glucose Production, Even Under Optimal Gluconeogenic Conditions in Healthy Humans; Claire Fromentin; Daniel Tomé; Françoise Nau; Laurent Flet; Catherine Luengo; Dalila Azzout-Marniche; Pascal Sanders; Gilles Fromentin; Claire Gaudichon National Library of Medicine Diabetes Obesity Metabolism 2004 Jul; 6(4): 239-48 Department of Endocrinology, Metabolism and Pathobiochemistry, University of Tubingen, Germany

47. Intramyocellular Lipids and Insulin Resistance Jürgen Machann 1 , Hans Häring, Fritz Schick, Michael Stumvoll National Library of Medicine Diabetes Obesity Metabolism 2004 Jul; 6(4): 239-48 Department of Endocrinology, Metabolism and Pathobiochemistry, University of Tubingen, Germany: 2004

48. Dietary recommendations for prevention of atherosclerosis Gabriele Riccardi , Annalisa Giosuè, Ilaria Calabrese, Olga Vaccaro. Oxford University Press Oxford Academic Cardiovascular Research: 2021

49. Effects of Plant-Based Diets on Weight Status National Library of Medicine PMC PubMed Central: A Systematic Review 2020

50. Effectiveness of plant-based diets in promoting well-being in the management of type 2 diabetes: a systematic review BMJ Journals 2018

51. Plant-Based Dietary Patterns and Incidence of Type 2 Diabetes in US Men and Women: Results from Three Prospective Cohort Studies PLOS Medicine: 2016

52. Vegetarian diets and incidence of diabetes in the Adventist Health Study-2 National Library of Medicine PMC PubMed Central: 2014

53. Vegan Diet Advice Might Benefit Liver Enzymes in Non-alcoholic Fatty Liver Disease: an Open Observational Pilot Study National Library of Medicine PMC PubMed Central: 2021

54. A Look at Plant-Based Diets National Library of Medicine PMC PubMed Central 2021

55. Plant-Based Diets and Incident CKD and Kidney Function National Library of Medicine PMC PubMed Central: 2019

56. Protein Sources and Risk for Incident Chronic Kidney Disease National Library of Medicine PMC PubMed Central: Dietary: Results From the Atherosclerosis Risk in Communities (ARIC) Study 2016

57. Critical Reviews in Food Science and Nutrition Oxford University Nuffield Department of Population

58. Saturated Fat is More Metabolically Harmful for the Human Liver than Unsaturated Fat or Simple Sugars American Diabetes Association: 2017

59. The Randle cycle revisited: a new head for an old hat. Louis Hue and Heinrich Taegtmeyer 01 SEP 2009https://doi.org/10.1152/ajpendo.00093.2009

60. Dietary Protein and Amino Acids in Vegetarian Diets—A Review: François Mariotti, Christopher D Gardner. PMCID: PMC6893534 PMID: 31690027
61. Eating our way to extinction Movie 2020

62. The Multifaceted Pyruvate Metabolism: Role of the Mitochondrial Pyruvate Carrier
by Joséphine Zangari, Francesco Petrelli, Benoît Maillot and Jean-Claude Martinou
Department of Cell Biology, Faculty of Sciences, University of Geneva, 30 Quai Ernest Ansermet, 1211 Geneva 4, Switzerland Biomolecules 2020, 10(7) 1068

63. Dietary protein intake in midlife in relation to healthy aging - results from the prospective Nurses' Health Study cohort. Andres V Ardisson Korat, M Kyla Shea, Paul F Jacques, Paola Sebastiani, Molin Wang, A Heather Eliassen, Walter C Willett, Qi Sun. American Journal of CLINICAL NUTRITION 119 (2024) 272-282PMID: 38309825 PMCID: PMC10884611 DOI: 10.1016/j.ajcnut.2023.11.010

64. Mechanisms decreasing glucose oxidation in diabetes and starvation: Role of lipid fuels and hormones Philip J. Randle, Alan L. Kerbey, Joseph Espinal November 1988 Nuffield Department of Clinical Biochemistry, University of Oxford, John Radcliffe Hospital, Oxford, United Kingdom

65. Regulatory interactions between lipids and carbohydrates: the glucose fatty acid cycle after 35 years. P Randle, Dec 1998 Nuffield Department of Clinical Biochemistry, University of Oxford, Radcliffe Infirmary, Oxford, U.K. PMID 10095997

66. Origin and Roles of Alanine and Glutamine in Gluconeogenesis in the Liver, Kidneys, and Small Intestine under Physiological and Pathological Conditions by Milan Holeček ORCID, June 2024 Department of Physiology, Faculty of Medicine, Charles University, 500 03 Hradec Kralove, Czech Republic

67. Erythrocyte metabolism Panagiotis N. Chatzinikolaou, Nikos V. Margaritelis, Vassilis Paschalis, Anastasios A. Theodorou, Ioannis S. Vrabas, Antonios Kyparos. Acta Physiologica Wiley On Line. First published: 25 January 2024

68. Heats of combustion representative of the carbohydrate mass contained in fruits, vegetables, or cereals Ana-Guadalupe Martínez-Navarro, Eulogio Orozco-Guareño, María-Judith Sánchez-Peña, Edgar-José López-Naranjo, Priscilla Muñiz-Mendoza, Luis-Javier González-Ortiz Pub Med Central Food Science & Nutrition August 14 2019

69. Organ and Tissue Contribution to Metabolic Rate In Elia, M. (1992). In: Kinney, J.M. and Tucker, H.N., Eds., Energy Metabolism: Tissue Determinants and Cellular Corollaries, Raven Press, New York, 61-79.

70. Evaluation of specific metabolic rates of major organs and tissues: comparison between men and women. Wang Z, Ying Z, Bosy-Westphal A, Zhang J, Heller M, Later W, Heymsfield SB, Müller MJ. Am J Hum Biol. 2011 May-Jun;23(3):333-8. doi: 10.1002/ajhb.21137. Epub 2010 Dec 22. PMID: 21484913; PMCID: PMC3139779.

71. Resting Energy Expenditure: From Cellular to Whole-Body Level, a Mechanistic Historical Perspective 11211 Steven B. Heymsfield , Brooke Smith , Jared Dahle , Samantha Kennedy , Nicole Fearnbach , 344 Diana M. Thomas , Anja Bosy-Westphal , and Manfred J. Müller

72. Metabolic Slowing and Reduced Oxidative Damage with Sustained Caloric Restriction Support the Rate of Living and Oxidative Damage Theories of Aging Redman LM, Smith SR, Burton JH, Martin CK, Il'yasova D, Ravussin E.. Cell Metab. 2018 Apr 3;27(4):805-815.e4. doi: 10.1016/j.cmet.2018.02.019. Epub 2018 Mar 22. PMID: 29576535; PMCID: PMC5886711

73. Changes in energy expenditure resulting from altered body weight.Leibel RL, Rosenbaum M, Hirsch J. N Engl J Med. 1995 Mar 9;332(10):621-8. doi: 10.1056/NEJM199503093321001. Erratum in: N Engl J Med 1995 Aug 10;333(6):399. PMID: 7632212.

74. Metabolic slowing with massive weight loss despite preservation of fat-free mass Johannsen DL, Knuth ND, Huizenga R, Rood JC, Ravussin E, Hall KD.. J Clin Endocrinol Metab. 2012 Jul;97(7):2489-96. doi: 10.1210/jc.2012-1444. Epub 2012 Apr 24. Erratum in: J Clin Endocrinol Metab. 2016 May;101(5):2266. doi:

75. A Plant-Based High-Carbohydrate, Low-Fat Diet in Overweight Individuals in a 16-Week Randomized Clinical Trial: The Role of Carbohydrates. Nutrients. Kahleova H, Dort S, Holubkov R, Barnard ND. 2018 Sep 14;10(9):1302. doi: 10.3390/nu10091302

76. The effects of a low-fat, plant-based dietary intervention on body weight, metabolism, and insulin sensitivity Barnard N.D., Scialli A.R., Turner-McGrievy G., Lanou A.J., Glass J.. Am. J. Med. 2005;118:991–997. doi: 10.1016

77. Effects of 7 days on an ad libitum low-fat vegan diet McDougall J., Thomas L.E., McDougall C., Moloney G., Saul B., Finnell J.S., Richardson K., Petersen K.M.: The McDougall Program cohort. Nutr. J. 2014;13:99. doi: 10.1186/1475-2891-13-99

78. The role of dietary fat in body fatness: Evidence from a preliminary meta-analysis of ad libitum low-fat dietary intervention studies. Astrup A., Ryan L., Grunwald G.K., Storgaard M., Saris W., Melanson E., Hill J.O.. Br. J. Nutr. 2000;83:S25–S32. doi: 10.1017/S0007114500000921

79. A randomised controlled trial using a whole food plant-based diet in the community for obesity, ischaemic heart disease or diabetes Wright N., Wilson L., Smith M., Duncan B., McHugh P. The BROAD study:. Nutr. Diabetes. 2017;7:e256. doi: 10.1038/nutd.2017.3

80. Randomised comparison of diets for maintaining obese subjects' weight after major weight loss: Ad lib, low fat, high carbohydrate diet v fixed energy intake Toubro S., Astrup A.. BMJ. 1997;314:29–34. doi: 10.1136/bmj.314.7073.29

81. Clinical Evidence and Mechanisms of High-Protein Diet-Induced Weight Loss Moon J, Koh G.. J Obes Metab Syndr. 2020 Sep 30;29(3):166-173. doi: 10.7570/jomes20028. PMID: 32699189; PMCID: PMC7539343

82. Effects of energy-restricted high-protein, low-fat compared with standard-protein, low-fat diets: a meta-analysis of randomized controlled trials. Wycherley TP, Moran LJ, Clifton PM, Noakes M, Brinkworth GD.. Am J Clin Nutr. 2012;96:1281–98. doi: 10.3945/ajcn.112.044321.

83. Effects of higher- versus lower-protein diets on health outcomes: a systematic review and meta-analysis. Santesso N, Akl EA, Bianchi M, Mente A, Mustafa R, Heels Ansdell D, et al. Eur J Clin Nutr. 2012;66:780–8. doi: 10.1038/ejcn.2012.37

84. Randomized trial on protein vs carbohydrate in ad libitum fat reduced diet for the treatment of obesity. Skov AR, Toubro S, Rønn B, Holm L, Astrup A. Int J Obes Relat Metab Disord. 1999;23:528–36. doi: 10.1038/sj.ijo.0800867.

85. The Effects of High-Protein Diets on Kidney Health and Longevity. Ko GJ, Rhee CM, Kalantar-Zadeh K, Joshi S.. J Am Soc Nephrol. 2020 Aug;31(8):1667-1679. doi: 10.1681/ASN.2020010028

86. Long-term effects of high-protein diets on renal function. Kamper AL, Strandgaard S: Annu Rev Nutr 37: 347–369, 2017

87. Effect of a high-protein diet on kidney function in healthy adults: Results from the OmniHeart trial. Juraschek SP, Appel LJ, Anderson CA, Miller ER 3rd: Am J Kidney Dis 61: 547–554, 2013.

88. Protein intake and kidney function in the middle-age population: contrast between cross-sectional and longitudinal data. Cirillo M, Lombardi C, Chiricone D, De Santo NG, Zanchetti A, Bilancio G:. Nephrol Dial Transplant 29: 1733–1740, 2014

89. The ketogenic diet for obesity and diabetes-enthusiasm outpaces evidence. Joshi S, Ostfeld RJ, McMacken M: [published online ahead of print July 15, 2019]. JAMA Intern Med doi:10.1001/jamainternmed.2019.2633

90. Dietary fat intake does affect obesity George A Bray, Barry M Pokin American Journal of Clinical Nutrition Dec 1998, Version of record March 2023

91. Elevated Fat Intake Increases Body Weight and the Risk of Overweight and Obesity among Chinese Adults: Wang L, Wang H, Zhang B, Popkin BM, Du S. 1991-2015 Trends. Nutrients. 2020 Oct 26;12(11):3272. doi: 10.3390/nu12113272. PMID: 33114561; PMCID: PMC7694029

92. Effects of Macronutrient Distribution on Weight and Related Cardiometabolic Profile in Healthy Non-Obese Chinese: A 6-month, Randomized Controlled-Feeding Trial. Wan Y., Wang F., Yuan J., Li J., Jiang D., Zhang J., Huang T., Zheng J., Mann J., Li D. EBio-Medicine. 2017;22:200–207. doi: 10.1016/j.ebiom.2017.06.017

93. The role of low-fat diets in body weight control: A meta-analysis of ad libitum dietary intervention studies. Astrup A., Grunwald G., Melanson E., Saris W., Hill J. Int. J. Obes. 2000;24:1545. doi: 10.1038/sj.ijo.0801453.

94. Effects of low-carbohydrate vs low-fat diets on weight loss and cardiovascular risk factors: A meta-analysis of randomized controlled trials. Nordmann A.J., Nordmann A., Briel M., Keller U., Yancy W.S., Brehm B.J., Bucher H.C. Arch. Intern. Med. 2006;166:285–293. doi: 10.1001/archinte.166.3.285

95. A Randomized Trial of a Low-Carbohydrate Diet for Obesity
Authors: Gary D. Foster, Ph.D., Holly R. Wyatt, M.D., James O. Hill, Ph.D., Brian G. McGuckin, Ed.M., Carrie Brill, B.S., B. Selma Mohammed, M.D., Ph.D., Philippe O. Szapary, M.D., Daniel J. Rader, M.D., Joel S. Edman, D.Sc., and Samuel Klein, M.D.Author Info & Affiliations
Published May 22, 2003, N Engl J Med 2003;348:2082-2090. DOI: 10.1056/NEJMoa022207. VOL. 348 NO. 21
96. A Low-Carbohydrate as Compared with a Low-Fat Diet in Severe Obesity
Authors: Frederick F. Samaha, M.D., Nayyar Iqbal, M.D., Prakash Seshadri, M.D., Kathryn L. Chicano, C.R.N.P., Denise A. Daily, R.D., Joyce McGrory, C.R.N.P., Terrence Williams, B.S., Monica Williams, B.S., Edward J. Gracely, Ph.D., and Linda Stern, M.D.Author Info & Affiliations Published May 22, 2003 N Engl J Med 2003;348:2074-2081 DOI: 10.1056/NEJMoa022637 VOL. 348 NO. 21

97. Effects of a low-carbohydrate diet on weight loss and cardiovascular risk factor in overweight adolescents. Sondike SB, Copperman N, Jacobson MS. J Pediatr. 2003 Mar;142(3):253-8. doi: 10.1067/mpd.2003.4. PMID: 12640371.

98. A Randomized Trial Comparing a Very Low Carbohydrate Diet and a Calorie-Restricted Low Fat Diet on Body Weight and Cardiovascular Risk Factors in Healthy Women. Bonnie J. Brehm, Randy J. Seeley, Stephen R. Daniels, David A. D'Alessio, The Journal of Clinical Endocrinology & Metabolism, Volume 88, Issue 4, 1 April 2003, Pages 1617–1623

99. Poulain, M., Herm, A. (2024). Exceptional longevity in Okinawa: Demographic trends since 1975, Journal of Internal Medicine, 295(4), pp. 1_13. doi.org/10.1111/joim.13764

100. Buettner, D. (2017). 'Blue Zones' Where the World's Healthiest People Live. National Geographic, 6 Apr. 2017, http://www.nationalgeographic.com/books/features/5-blue-zones-where-the-worlds-healthiest-people-live/.

101.Poulain, M., Herm, A. & Pes, G.M. (2013). The blue zones : Areas of exceptional longevity around the world. Vienna Yearbook of Population Research 11, 87–108. https://www.jstor.org/stable/43050798

102. Puska P, Vartiainen E, Nissinen A, Laatikainen T, Jousilahti P. Background, principles, implementation, and general experiences of the North Karelia Project. Glob Heart. 2016;11:173-178.

103. Puska P, Jaini P. The North Karelia Project: Prevention of Cardiovascular Disease in Finland Through Population-Based Lifestyle Interventions. Am J Lifestyle Med. 2020 Mar 19;14(5):495-499. doi: 10.1177/1559827620910981

104.Tonstad S, Butler T, Yan R, Fraser GE. Type of vegetarian diet, body weight, and prevalence of type 2 diabetes. Diabetes Care. 2009 May;32(5):791-6. doi: 10.2337/dc08-1886. Epub 2009 Apr 7. PMID: 19351712; PMCID: PMC2671114.

105. Shintani TT, Hughes CK, Beckham S, O'Connor HK. Obesity and cardiovascular risk intervention through the ad libitum feeding of traditional Hawaiian diet. Am J Clin Nutr. 1991 Jun;53(6 Suppl):1647S-1651S. doi: 10.1093/ajcn/53.6.1647S. PMID: 2031501.

106. Wright N, Wilson L, Smith M, Duncan B, McHugh P. The BROAD study: A randomised controlled trial using a whole food plant-based diet in the community for obesity, ischaemic heart disease or diabetes. Nutr Diabetes. 2017 Mar 20;7(3):e256. doi: 10.1038/nutd.2017.3. PMID: 28319109; PMCID: PMC5380896.

107. Jarvis SE, Nguyen M, Malik VS. Association between adherence to plant-based dietary patterns and obesity risk: a systematic review of prospective cohort studies. Appl Physiol Nutr Metab. 2022 Dec 1;47(12):1115-1133. doi: 10.1139/apnm-2022-0059. Epub 2022 Aug 19. PMID: 35985038.

108. Barnard ND, Scialli AR, Turner-McGrievy G, Lanou AJ, Glass J. The effects of a low-fat, plant-based dietary intervention on body weight, metabolism, and insulin sensitivity. Am J Med. 2005;118(9):991-997. doi:10.1016/j.amjmed.2005.03.039

109. Gardner CD, Trepanowski JF, Del Gobbo LC, et al. Effect of Low-Fat vs Low-Carbohydrate Diet on 12-Month Weight Loss in Overweight Adults and the Association With Genotype Pattern or Insulin Secretion: The DIETFITS Randomized Clinical Trial. JAMA. 2018;319(7):667–679. doi:10.1001/jama.2018.0245

110. Gómez-Donoso, Clara, et al. "Healthful and Unhealthful Provegetarian Food Patterns and the Incidence of Overweight/Obesity in the Seguimiento Universidad De Navarra (SUN) Cohort (OR33-05-19)." Current Developments in Nutrition, vol. 3, no. Suppl 1, 2019, p. nzz039.OR33-05-19-, https://doi.org/10.1093/cdn/nzz039.OR33-05-19..

111. Spencer EA, Appleby PN, Davey GK, Key TJ. Diet and body mass index in 38000 EPIC-Oxford meat-eaters, fish-eaters, vegetarians and vegans. Int J Obes Relat Metab Disord. 2003 Jun;27(6):728-34. doi: 10.1038/sj.ijo.0802300. PMID: 12833118.

Healthy Nutrition Advice 🍎 HNA

Appendix I: 7 Day Pre-Change Diet Record

Name:

Start & End Date:

Don't worry too much about the results or make any changes to what you usually eat & drink. Just be totally honest with yourself & include everything as this will really help the review process

Monday	Time	What I ate & drank at each meal or snack & how much I ate & drank [1]	Location [2]	Prepared [3]

Tuesday	Time	What I ate & drank at each meal or snack & how much I ate & drank [1]	Location [2]	Prepared [3]

[1]**What I ate/drank: Record any individual Meal or Snack. e.g., fried chicken & chips, Water, Coffee milk & 2 sugars, 2 pints lager, Double cheese burger & large fries & Large coke**
[2]**Location: Where Consumed H = Home, C = Café, W = Work, R = Restaurant, P = Pub, F = Family/Friends House, G = Garden/Visitors Centre, E = Event (Wedding/Christening/Show etc)**
[3]**Prepared: H = Cooked/Prepared at home/ friends from Whole ingredients, S = Ready Packaged meal Supermarket/Shop, W = Work, M = Meal Out, T = Takeaway/delivered/ collected**

Wednesday	Time	What I ate & drank at each meal or snack & how much I ate & drank [1]	Location [2]	Prepared [3]

Thursday	Time	What I ate & drank at each meal or snack & how much I ate & drank [1]	Location [2]	Prepared [3]

Friday	Time	What I ate & drank at each meal or snack & how much I ate & drank [1]	Location [2]	Prepared [3]

[1]**What I ate/drank: Record any individual Meal or Snack. e.g., fried chicken & chips, Water, Coffee milk & 2 sugars, 2 pints lager, Double cheese burger & large fries & Large coke**
[2]**Location: Where Consumed H = Home, C = Café, W = Work, R = Restaurant, P = Pub, F = Family/Friends House, G = Garden/Visitors Centre, E = Event (Wedding/Christening/Show etc)**
[3]**Prepared: H = Cooked/Prepared at home/ friends from Whole ingredients, S = Ready Packaged meal Supermarket/Shop, W = Work, M = Meal Out, T = Takeaway/delivered/ collected**

Saturday	Time	What I ate & drank at each meal or snack & how much I ate & drank [1]	Location [2]	Prepared [3]

Sunday	Time	What I ate & drank at each meal or snack & how much I ate & drank [1]	Location [2]	Prepared [3]

[1]**What I ate/drank: Record any individual Meal or Snack. e.g., fried chicken & chips, Water, Coffee milk & 2 sugars, 2 pints lager, Double cheese burger & large fries & Large coke**
[2]**Location: Where Consumed H = Home, C = Café, W = Work, R = Restaurant, P = Pub, F = Family/Friends House, G = Garden/Visitors Centre, E = Event (Wedding/Christening/Show etc)**
[3]**Prepared: H = Cooked/Prepared at home/ friends from Whole ingredients, S = Ready Packaged meal Supermarket/Shop, W = Work, M = Meal Out, T = Takeaway/delivered/ collected**

Healthy Nutrition Advice 🍎 HNA

Appendix II: 7 Day Post-Change Diet Record

Name:

Start & End Date:

Don't worry too much about the results or make any changes to what you usually eat & drink. Just be totally honest with yourself & include everything as this will really help the review process

Monday	Time	What I ate & drank at each meal or snack & how much I ate & drank [1]	Location [2]	Prepared [3]

Tuesday	Time	What I ate & drank at each meal or snack & how much I ate & drank [1]	Location [2]	Prepared [3]

[1] **What I ate/drank**: Record any individual Meal or Snack. e.g., fried chicken & chips, Water, Coffee milk & 2 sugars, 2 pints lager, Double cheese burger & large fries & Large coke

[2] **Location**: Where Consumed H = Home, C = Café, W = Work, R = Restaurant, P = Pub, F = Family/Friends House, G = Garden/Visitors Centre, E = Event (Wedding/Christening/Show etc)

[3] **Prepared**: H = Cooked/Prepared at home/ friends from Whole ingredients, S = Ready Packaged meal Supermarket/Shop, W = Work, M = Meal Out, T = Takeaway/delivered/ collected

Wednesday	Time	What I ate & drank at each meal or snack & how much I ate & drank [1]	Location [2]	Prepared [3]

Thursday	Time	What I ate & drank at each meal or snack & how much I ate & drank [1]	Location [2]	Prepared [3]

Friday	Time	What I ate & drank at each meal or snack & how much I ate & drank [1]	Location [2]	Prepared [3]

[1] What I ate/drank: Record any individual Meal or Snack. e.g., fried chicken & chips, Water, Coffee milk & 2 sugars, 2 pints lager, Double cheese burger & large fries & Large coke

[2] Location: Where Consumed H = Home, C = Café, W = Work, R = Restaurant, P = Pub, F = Family/Friends House, G = Garden/Visitors Centre, E = Event (Wedding/Christening/Show etc)

[3] Prepared: H = Cooked/Prepared at home/ friends from Whole ingredients, S = Ready Packaged meal Supermarket/Shop, W = Work, M = Meal Out, T = Takeaway/delivered/ collected

Saturday	Time	What I ate & drank at each meal or snack & how much I ate & drank [1]	Location [2]	Prepared [3]

Sunday	Time	What I ate & drank at each meal or snack & how much I ate & drank [1]	Location [2]	Prepared [3]

[1]**What I ate/drank: Record any individual Meal or Snack. e.g., fried chicken & chips, Water, Coffee milk & 2 sugars, 2 pints lager, Double cheese burger & large fries & Large coke**
[2]**Location: Where Consumed H = Home, C = Café, W = Work, R = Restaurant, P = Pub, F = Family/Friends House, G = Garden/Visitors Centre, E = Event (Wedding/Christening/Show etc)**
[3]**Prepared: H = Cooked/Prepared at home/ friends from Whole ingredients, S = Ready Packaged meal Supermarket/Shop, W = Work, M = Meal Out, T = Takeaway/delivered/ collected**

Appendix III

Weight Loss Record:

Name: [   ]

Date: [--/--/----] **Baseline Weight:** [kg] **Weight Goal:** [kg]

Date	Weight (kg)	Change (+/- kg)	Comment
	kg	kg	
	kg	kg	
	kg	kg	
	kg	kg	
	kg	kg	
	kg	kg	
	kg	kg	
	kg	kg	
	kg	kg	
	kg	kg	
	kg	kg	
	kg	kg	
	kg	kg	
	kg	kg	
	kg	kg	
	kg	kg	
	kg	kg	
	kg	kg	
	kg	kg	
	kg	kg	
	kg	kg	
	kg	kg	
	kg	kg	
	kg	kg	
	kg	kg	
	kg	kg	
	kg	kg	
	kg	kg	
	kg	kg	
	kg	kg	
	kg	kg	
	kg	kg	

Index

A

acetate, 92, 129, 189, 194
acetyl-CoA, 17, 18, 92, 95, 129, 149, 155, 157, 161, 162, 163, 164, 165, 168, 169, 174, 175, 177, 178, 183, 186, 187, 188, 190, 192, 194, 196, 199, 201, 204, 206, 208, 211, 213, 215, 216, 218, 219, 220, 221, 228
acidification, 248
Ad Libitum Hawaii Diet, 12, 235
adenosine triphosphate, 154, 225, 226
adipocytes, 111, 117, 132, 140, 141, 142, 147, 148, 183, 196, 211, 213, 218
adipose tissue, 17, 18, 19, 22, 23, 25, 40, 44, 45, 52, 58, 59, 92, 111, 112, 113, 114, 133, 134, 135, 136, 137, 138, 139, 140, 142, 143, 144, 145, 146, 148, 149, 151, 152, 153, 159, 161, 162, 167, 171, 180, 182, 183, 189, 190, 194, 198, 211, 212, 213, 232, 233
adrenalin, 160
adventist health study, 234
Adventist Health Study, 12, 234, 252
agricultural land, 247
alanine, 95, 159, 175, 178, 179, 180, 181, 183, 184, 185, 187, 188, 191, 192, 218, 219
Alanine cycle, 175, 181, 184, 188
alcohol, 13, 17, 19, 22, 26, 29, 37, 46, 48, 52, 67, 72, 74, 77, 81, 85, 109, 128, 129, 209
amino acids, 18, 19, 21, 32, 83, 87, 94, 95, 96, 98, 99, 100, 101, 103, 104, 105, 117, 137, 141, 145, 146, 155, 156, 157, 160, 163, 164, 168, 171, 173, 175, 179, 180, 183, 184, 185, 186, 187, 188, 191, 192, 196, 197, 198, 201, 205, 206, 213, 216, 218, 219, 220, 232, 252
AMP Kinase, 201
amylin, 33, 144
anabolism, 164
anaerobic glycolysis, 86, 181, 192, 199, 218
anaplerotic, 95, 163, 174, 187, 218
animal feed, 248
animal products, 81, 119, 248
animal suffering, 82, 248
antioxidants, 78, 83, 105
atherosclerosis, 19, 88, 252
Atkins, 16, 249, 250

ATP, 12, 15, 87, 95, 114, 145, 149, 150, 152, 153, 155, 157, 158, 159, 161, 162, 163, 168, 169, 170, 171, 173, 174, 175, 177, 181, 182, 183, 184, 185, 187, 190, 197, 198, 199, 200, 201, 204, 205, 212, 214, 215, 216, 218, 219, 220, 221, 222, 224, 225, 226, 227, 228, 229, 230, 231, 232
ATP synthase, 222, 225
ATP yield, 12, 150, 175, 197, 222, 226, 227, 228, 231
avocado oil, 73, 117, 119, 120

B

basal metabolic rate, 37, 57
β-hydroxybutyrate, 228
BCAA, 105, 106
beta hydroxybutyrate, 194, 205, 226
beta oxidation, 213, 231
bile, 162
bioenergetic cost, cost of burning fat, cost of energy, cost of storing fat, 19, 38, 43, 59, 98, 161, 175, 185, 209, 211, 230
efficiency, 13
bioenergetically , unfavourable, 33, 99, 173, 175, 178, 185, 231
blood glucose level, 18, 25, 44, 52, 87, 99, 101, 111, 138, 141, 147, 150, 154, 155, 160, 171, 172, 178, 185, 203, 213
Blue Zones, 12, 234, 237, 256
Blue Zones Areas, 256
BMI, 13, 22
BMR, 7, 37, 38, 39, 40, 51, 57, 131, 133, 134, 135, 136, 144, 156
body mass index, 249, 250
brain, 15, 17, 18, 20, 21, 32, 45, 52, 65, 73, 86, 88, 91, 94, 110, 111, 112, 113, 116, 117, 122, 127, 128, 132, 134, 135, 137, 138, 141, 142, 146, 153, 157, 158, 159, 160, 164, 168, 169, 172, 177, 178, 180, 183, 185, 191, 192, 193, 194, 195, 197, 198, 199, 203, 206, 207, 213, 219, 226

glycogen depletion, 40
glycogen stores, 18, 19, 52, 58, 59, 95, 99, 151, 153, 159, 167, 177, 178, 191, 192, 201, 205, 213
glycogenesis, 92, 137, 142, 151, 152, 167, 180, 181
glycogenolysis, 139, 141, 151, 153, 154, 157, 158, 160, 177, 178, 179, 180, 181, 198, 204
glycolysis
anaerobic, 87, 92, 94, 138, 149, 155, 157, 163, 167, 168, 169, 170, 173, 174, 175, 177, 181, 183, 188, 194, 197, 199, 200, 201, 202, 215, 216, 218, 219, 220, 221, 222, 226, 231, 232
ATP, 87, 92, 94, 138, 149, 155, 157, 163, 167, 168, 169, 170, 173, 174, 175, 177, 181, 183, 188, 194, 197, 199, 200, 201, 202, 215, 216, 218, 219, 220, 221, 222, 226, 231, 232
pyruvate, 87, 92, 94, 138, 149, 155, 157, 163, 167, 168, 169, 170, 173, 174, 175, 177, 181, 183, 188, 194, 197, 199, 200, 201, 202, 215, 216, 218, 219, 220, 221, 222, 226, 231, 232
growth hormone, 33, 83, 141, 143, 149, 167, 173
gut microbiome, 15, 91, 92

H

Hawaii Diets, 235
healthy snacks, 69, 70
heart, 15, 17, 18, 22, 37, 52, 58, 59, 60, 61, 62, 83, 88, 94, 111, 114, 115, 116, 122, 127, 128, 135, 136, 138, 153, 157, 159, 160, 161, 162, 168, 169, 178, 182, 189, 190, 191, 192, 194, 196, 200, 201, 202, 206, 213, 215, 218, 231
heart disease, 115, 116
homeostasis, 102, 144, 153, 157, 169, 178, 188, 202, 207, 212, 247
hormones, 26, 33, 57, 64, 87, 92, 96, 99, 101, 102, 109, 112, 113, 122, 131, 137, 139, 140, 145, 146, 149, 154, 160, 168, 171, 177, 191, 203, 212, 253
hydration, 74, 127
hypercaloric excess, 22
hyperglycaemia, 152, 160, 198
hypertriglyceridemia, 139
hypoglycaemia, 45, 88, 154, 160, 167, 172, 178, 198, 203

hypothalamus, 141, 142, 168

I

IGF-1, 83, 125, 143, 249
Ikaria Greece, 241
insulin, 5, 19, 22, 26, 32, 33, 34, 35, 44, 83, 89, 92, 95, 99, 101, 113, 114, 138, 139, 140, 141, 142, 143, 144, 146, 148, 150, 160, 161, 162, 165, 167, 168, 169, 170, 171, 173, 177, 178, 187, 194, 195, 206, 207, 210, 211, 213, 218, 251
insulin levels, 35, 89, 95, 138, 143, 148, 177, 187, 207, 211
insulin like growth factor, 83
insulin resistance, 19, 22, 26, 33, 44, 83, 113, 139, 140, 141, 167, 171, 177, 178, 195, 210, 251
iso-caloric, 146

K

keto, 5, 16, 21, 88, 94, 155, 167, 172, 206, 207
ketogenesis, 18, 32, 34, 87, 168, 175, 177, 180, 192, 195, 199, 204, 205, 207
ketogenic, 34, 35, 83, 162, 186, 187, 191, 205, 206, 207, 208, 250
ketone bodies, 92, 95, 138, 140, 156, 164, 168, 186, 188, 189, 191, 192, 194, 195, 197, 199, 200, 201, 203, 205, 206, 207, 220, 228
ketones, 16, 18, 21, 31, 32, 138, 155, 164, 168, 169, 177, 183, 185, 186, 187, 192, 193, 194, 198, 201, 203, 205, 206, 207, 216, 226
ketosis, 16, 18, 32, 169, 172, 194, 195, 206, 207
kidney, 17, 19, 21, 59, 73, 74, 79, 86, 88, 90, 94, 99, 110, 111, 116, 122, 123, 124, 126, 127, 135, 138, 149, 153, 157, 159, 160, 161, 162, 178, 179, 182, 184, 190, 192, 196, 200, 202, 203, 206, 207
kidney cortex, 17, 59, 94, 111, 157, 159, 160, 161, 162, 190, 196, 206
kidney medulla, 21, 86, 138, 157, 159, 161, 182, 192, 200
kidney stones, 19, 127, 207
Kidneys, 11, 133, 134, 178, 197, 202, 253
aldosterone, angiotensin, 11, 133, 134, 178, 197, 202, 253
renin, 11, 133, 134, 178, 197, 202, 253
Krebs cycle, 205, 212

HEALTHY NUTRITION ADVICE ᴴᴺᴬ http://healthynutritionadvice.net

www.ingramcontent.com/pod-product-compliance
Lightning Source LLC
Chambersburg PA
CBHW042338030426
42335CB00030B/3385